Y0-DMX-433

CRC SERIES IN AGRICULTURE

Editor-in-Chief

Angus A. Hanson, Ph.D.
Vice President-Research
W-L Research, Inc.
Highland, Maryland

HANDBOOK OF SOILS AND CLIMATE IN AGRICULTURE

Editor
Victor J. Kilmer (Deceased)
Chief
Soils and Fertilizer Research Branch
National Fertilizer Development Center
Tennessee Valley Authority
Muscle Shoals, Alabama

HANDBOOK OF PLANT SCIENCE IN AGRICULTURE

Editor
B. R. Christie, Ph.D.
Professor
Department of Crop Science
Ontario Agricultural College
University of Guelph
Guelph, Ontario, Canada

HANDBOOK OF PEST MANAGEMENT IN AGRICULTURE

Editor
David Pimentel, Ph.D.
Professor
Department of Entomology
New York College of Agricultural
and Life Sciences
Cornell University
Ithaca, New York

HANDBOOK OF ENGINEERING IN AGRICULTURE

Editor
R. H. Brown, Ph.D.
Chairman
Division of Agricultural Engineering
Agricultural Engineering Center
University of Georgia
Athens, Georgia

HANDBOOK OF TRANSPORTATION AND MARKETING IN AGRICULTURE

Editor
Essex E. Finney, Jr., Ph.D.
Assistant Center Director
Agricultural Research Center
U.S. Department of Agriculture
Beltsville, Maryland

HANDBOOK OF PROCESSING AND UTILIZATION IN AGRICULTURE

Editor
Ivan A. Wolff, Ph.D. (Retired)
Director
Eastern Regional Research Center
Science and Education Administration
U.S. Department of Agriculture
Philadelphia, Pennsylvania

CRC Handbook of Plant Science in Agriculture

Volume I

Editor

B. R. Christie

Professor
Department of Crop Science
Ontario Agricultural College
University of Guelph
Guelph, Ontario
Canada

CRC Series in Agriculture

A. A. Hanson, Editor-in-Chief
Vice President-Research
W-L Research, Inc.
Highland, Maryland

CRC Press, Inc.
Boca Raton, Florida

Library of Congress Cataloging-in-Publication Data

Handbook of plant science in agriculture.

(CRC series in agriculture)
Bibliography: p.
Includes index.
1. Crops.   2. Agriculture.   I. Christie, B. R.
(Bertram R.), 1933-   . II. Series.
SB91.H36 1987   631   86-12937
ISBN 0-8493-3821-2

This book represents information obtained from authentic and highly regarded sources. Reprinted material is quoted with permission, and sources are indicated. A wide variety of references are listed. Every reasonable effort has been made to give reliable data and information, but the author and the publisher cannot assume responsibility for the validity of all materials or for the consequences of their use.

All rights reserved. This book, or any parts thereof, may not be reproduced in any form without written consent from the publisher.

Direct all inquiries to CRC Press, Inc., 2000 Corporate Blvd., N.W., Boca Raton, Florida, 33431.

© 1987 by CRC Press, Inc.

International Standard Book Number 0-8493-3821-2 (set)
International Standard Book Number 0-8493-3822-0 (v. 1)
International Standard Book Number 0-8493-3823-9 (v. 2)

Library of Congress Card Number 86-12937
Printed in the United States

EDITOR-IN-CHIEF

Angus A. Hanson, Ph.D., is Vice President-Research, W-L Research, Inc., Highland, Maryland, and has had broad experience in agricultural research and development. He is a graduate of the University of British Columbia, Vancouver, and McGill University, Quebec, and received the Ph.D. degree from the Pennsylvania State University, University Park, in 1951.

An employee of the U.S. Department of Agriculture from 1949 to 1979, Dr. Hanson worked as a Research Geneticist at University Park, Pa., 1949 to 1952, and at Beltsville, Md., serving successively as Research Leader for Grass and Turf Investigations, 1953 to 1965, Chief of the Forage and Range Research Branch, 1965 to 1972, and Director of the Beltsville Agricultural Research Center, 1972 to 1979. He has been appointed to a number of national and regional task forces charged with assessing research needs and priorities, and has participated in reviewing agricultural needs and research programs in various foreign countries. As Director at Beltsville, he was directly responsible for programs that included most dimensions of agricultural research.

In his personal research, he has emphasized the improvement of forage crops, breeding and management of turfgrasses, and the breeding of alfalfa for multiple pest resistance, persistence, quality, and sustained yield. He is the author of over 100 technical and popular articles on forage crops and turfgrasses, and has served as Editor of *Crop Science* and the *Journal of Environmental Quality*.

PREFACE

Plants are the ultimate source of food, fiber, fuel, and many other products of importance to man. Thousands of years ago, people in various parts of the world began to plant, cultivate, and harvest those species which could provide for their needs. From this early cultivation, a vast array of cultivated plants has developed. Some species, such as wheat, cotton, and alfalfa are grown on every continent of the world, while others are grown only within a small geographical area. In the *Handbook,* we have attempted to provide information on all economically important crops, except those grown for timber or as ornamentals.

The study of crop plants enabled early man to learn their cultivation and processing. In this century, there has been a vast amount of information accumulated, as the study of these plant species still retains a fascination for us. Information on crop plants, e.g., on their genetics, growth, morphology, physiology, cultivation, and processing, is accumulating daily, and is reported in a vast array of publications and conferences and in many different languages. To assemble all of that information in one place would require more than one book. Therefore the aim of this *Handbook* is to assemble, in a condensed form, as much information as possible. The contributors have attempted to collect and to present the most recent information available on their assigned topics. Not every topic is included, nor are all crops covered under every topic. Limitations of time and space necessitated some selection. Volume I presents information on the genetics, botany, and growth of crop plants, while Volume II covers the production of crops and their utilization.

This *Handbook* was produced to serve as a guide and reference for all those involved with and interested in crop plants, such as producers, processors, researchers, students, and teachers. It has been made possible through the time and effort provided by the members of the Advisory Board and by our various contributors. Any comments or suggestions for future editions would be welcomed.

I want to thank all those involved for their efforts, their cooperation, and for their patience. It has been a privilege to work with all of them.

B. R. Christie
Editor

THE EDITOR

Bertram R. Christie, Ph.D., is Professor of Crop Science, Ontario Agricultural College, University of Guelph, Guelph, Ontario.

Professor Christie obtained his B.S.A. and M.S.A. degrees from the Ontario Agricultural College (University of Toronto) in 1955 and 1956, respectively, and the Ph.D. Degree from Iowa State University, in 1959. Since then, he has been a member of the faculty of the Crop Science Department at the University of Guelph, where he is involved in undergraduate and graduate education and in research on forage crops. He was one of the early recipients of the O.A.C. Alumni Distinguished Teaching Award. In 1985, he was invited to the People's Republic of China to present a series of lectures on quantitative genetics and on forage crop breeding to Chinese scientists.

Professor Christie is a member of the Editorial Board of the *Canadian Journal of Genetics and Cytology* and has served in a similar capacity with the *Canadian Journal of Plant Science*. He is a Director of the American Forage and Grassland Council and the Canada Committee on Forage Crops. He has served as President of the Canadian Society of Agronomy and also of the Ontario Institute of Professional Agrologists. Professor Christie is also a member of the American Society of Agronomy, the Canadian Society of Genetics and Cytology, and the Agricultural Insitute of Canada.

Professor Christie has been the author or co-author of numerous articles and has developed several cultivars of forage crops.

ADVISORY BOARD

Charles O. Gardner, Ph.D.
Professor
Department of Agronomy
University of Nebraska
Lincoln, Nebraska

Donald G. Hanway, Ph.D.
Professor Emeritus
Department of Agronomy
University of Nebraska
Lincoln, Nebraska

Jules Janick, Ph.D.
Professor
Department of Horticulture
Purdue University
West Lafayette, Indiana

William C. Kennard, Ph.D.
Professor of Plant Physiology
Department of Plant Science
University of Connecticut
Storrs, Connecticut

Allan K. Stoner, Ph.D.
Director
Plant Genetics and Germplasm Institute
Agricultural Research Center
Beltsville, Maryland

CONTRIBUTORS

W. Powell Anderson
Associate Professor
Department of Agronomy
New Mexico State University
Las Cruces, New Mexico

K. C. Armstrong, Ph.D.
Research Cytogeneticist
Research Branch
Agriculture Canada
Ottawa, Ontario
Canada

Donald K. Barnes, Ph.D.
U.S.D.A., A.R.S. Plant Science
　Research Unit
Department of Agronomy and Plant
　Genetics
University of Minnesota
St. Paul, Minnesota

Stephen R. Bowley, Ph.D.
Assistant Professor
Department of Crop Science
University of Guelph
Guelph, Ontario
Canada

Calvin Chong, Ph.D.
Research Scientist
Horticultural Research Institute
Ontario Ministry of Agriculture and Food
Vineland Station, Ontario
Canada

Meryl N. Christiansen, Ph.D.
Director
Plant Physiology Institute
Agricultural Research Service
U.S. Department of Agriculture
Beltsville, Maryland

F. W. Cope, Ph.D.
Professor, Emeritus
Biological Science Department
University of the West Indies
St. Augustine
Trinidad

Bruce E. Coulman, Ph.D.
Associate Professor
Department of Plant Science
MacDonald College
McGill University
Ste. Anne de Bellevue, Quebec
Canada

S. K. A. Danso, Ph.D.
Joint FAO/IAEA Division
International Atomic Energy Agency
Vienna
Austria

Arnel R. Hallauer, Ph.D.
Research Geneticist, USDA, ARS
Department of Agronomy
Iowa State University
Ames, Iowa

Gudni Hardarson, Ph.D.
Soil Fertility Section
FAO/IAEA Program
Seibersdorf Laboratory
Seibersdorf
Austria

Jack R. Harlan, Ph.D.
Professor Emeritus, Plant Genetics
University of Illinois
Urbana, Illinois

S. B. Helgason, Ph.D.
Professor Emeritus
Department of Plant Science
Faculty of Agriculture
University of Manitoba
Winnipeg, Manitoba
Canada

Robert J. Hilton, Ph.D.
Professor Emeritus, Horticulture
University of Guelph
Guelph, Ontario
Canada

D. L. Jennings, Ph.D.
Scottish Crop Research Institute
Invergowrie, Dundee
Scotland

Joshua A. Lee
Department of Crop Science
North Carolina State University
Raleigh, North Carolina

Robert C. Leffel, Ph.D.
Research Agronomist
Agricultural Research Service
U.S. Department of Agriculture
Beltsville, Maryland

F. W. Liu, Ph.D.
Associate Professor
Department of Pomology
Cornell University
Ithaca, New York

E. V. Maas
Research Leader
U.S. Salinity Laboratory
U.S. Department of Agriculture
Riverside, California

David J. Major, Ph.D.
Senior Research Scientist
Plant Science Section
Agriculture Canada
Research Station
Lethbridge, Alberta
Canada

Beverley H. Marie
Research Associate
Department of Horticultural Science
University of Guelph
Guelph, Ontario
Canada

Douglas P. Ormrod, Ph.D.
Professor of Horticultural Science
University of Guelph
Guelph, Ontario
Canada

Craig C. Sheaffer, Ph.D.
Professor
Department of Agronomy and Plant
  Genetics
University of Minnesota
St. Paul, Minnesota

Alfred E. Slinkard, Ph.D.
Senior Research Scientist
Crop Development Centre
University of Saskatchewan
Saskatoon, Saskatchewan
Canada

Garry A. Smith, Ph.D.
Research Geneticist
Crops Research Laboratory
Agriculture Research Service
U.S. Department of Agriculture
Colorado State University
Fort Collins, Colorado

Thomas M. Starling, Ph.D.
Professor
Department of Agronomy
Virginia Polytechnic Institute and State
  University
Blacksburg, Virginia

Anna K. Storgaard, Ph.D.
Department of Plant Science
Faculty of Agriculture
University of Manitoba
Winnipeg, Manitoba
Canada

Norman L. Taylor, Ph.D.
Professor
Department of Agronomy
University of Kentucky
Lexington, Kentucky

M. Tollenaar, Ph.D.
Assistant Professor
Crop Science Department
University of Guelph
Guelph, Ontario
Canada

Dan Wiersma, Ph.D.
Water Resources Research Center
 (Retired)
Department of Agronomy
Purdue University
West Lafayette, Indiana

James R. Wilcox, Ph.D.
Research Geneticist, USDA, ARS
Department of Agronomy
Purdue University
West Lafayette, Indiana

B. Young
Research Assistant
Department of Horticultural Science
University of Guelph
Guelph, Ontario
Canada

F. Zapata, Ph.D.
International Atomic Energy Agency
Seibersdorf Laboratory
Seibersdorf
Austria

TABLE OF CONTENTS

Volume I

GENETICS OF CROPS
Chromosome Numbers of Crop Species 3
K. C. Armstrong

Centers of Origin 15
Jack R. Harlan

Introgressive Hybridization 23
S. R. Bowley and Norman L. Taylor

Breeding Systems 61
Arnel R. Hallauer

BOTANY OF CROPS
Plant Propagation 91
Calvin Chong

Life Cycles 115
S. B. Helgason and Anna K. Storgaard

Biological Nitrogen Fixation 165
Gudni Hardarson, S. K. A. Danso, and F. Zapata

ENVIRONMENTAL FACTORS AND PLANT GROWTH
Light and Photoperiod 195
David J. Major

Plant Temperature Stress 217
Meryl N. Christiansen

Sensitivity of Crop Plants to Gaseous Pollution Stress 225
Douglas P. Ormrod, B. Young, and Beverley Marie

INDEX 275

Volume II

CROP PRODUCTION
Water and Agricultural Productivity 3
Dan Wiersma and B. R. Christie

Salt Tolerance of Plants 57
E. V. Maas

World Production and Distribution 77
Bruce E. Coulman

Dry Matter Production from Crops 89
M. Tollenaar

Weed Science As It Relates to Crop Production 99
W. Powell Anderson

UTILIZATION OF CROPS
Cereals 117
Thomas M. Starling

Sugar Crops 125
Garry A. Smith

Starch Crops 137
D. L. Jennings

Oilseed Crops 145
James R. Wilcox and Robert C. Leffel

Protein Crops 167
Alfred E. Slinkard

Plant Fibers 173
Joshua A. Lee

Vegetable Crops 183
Robert J. Hilton

Fruit Crops 195
F. W. Liu

Drug Crops 209
F. W. Cope

Forage Crops 217
Craig C. Sheaffer and Donald K. Barnes

INDEX 253

Genetics of Crops

CHROMOSOME NUMBERS OF CROP SPECIES*

K. C. Armstrong

A knowledge of the chromosome number of a crop species is important for several reasons. The number of linkage groups (genes on the same chromosome are linked) will not exceed the gametic chromosome number of the species.[1] In diploids and alloploids the number of linkage groups equals the gametic number (sporophytic number divided by 2), but in autoploids it equals the sporophytic number divided by the level of ploidy. When developing or elucidating these linkage groups (chromosome gene maps), a prior knowledge of the number of linkage groups expected is useful.

The number of chromosomes is one of the factors that can have a powerful influence on recombination. An organism with the haploid number of n, and heterozygous for one gene pair on each chromosome pair, can produce 2^n genetically different gametes. Assuming that the haploid number is n = 7, then the number of possible chromosome combinations is 128, but if n = 14 the number of possible combinations is 16,384. The frequency of recombination between linked genes is controlled by the frequency of crossing-over that occurs between them. A low basic chromosome number is particularly restrictive on recombination frequency in outcrossing and highly heterozygous plant groups.[2]

Knowledge of the chromosome number of all the species of a genus or family can help determine the basic chromosome number of the genus or family. If the crop species deviates from this, it can be determined if the basic chromosome number has changed by aneuploidy or polyploidy.[3] If polyploidy appears to be involved, a knowledge of chromosome numbers can suggest that a species is an alloploid if the number suggests that the two parental species had different base numbers. However, distinguishing between alloploids and autoploids generally requires information from chromosome structure and chromosome pairing in parents, haploids, and interspecific hybrids and from various taxonomic comparisons.

Knowledge of the chromosome number is useful in interspecific hybridization programs. This will often indicate that cross-incompatibility, and hybrid inviability and sterility might be a consequence of different chromosome numbers in the parents or of their diploid nature, and these might be overcome by doubling the chromosome number of one or both parents and/or the F_1 hydrids.[3] In many cases, where differences in ploidy level are involved, crossing is more successful if the parent with the lowest chromosome number is used as the female parent.

Knowledge of the chromosome number makes the analysis of chromosome pairing much easier. In species with small chromosomes, and in particular polyploids and F_1 hybrids, it is almost essential. Complex pairing configurations can often not be interpreted accurately without prior knowledge of the number of somatic chromosomes.

The list of crop species presented in the tables was compiled from the species in the *World List of Plant Breeders*[4] and by consulting several crop production texts such as Klages[5] and Wilsie.[6] Chromosome numbers were taken from Darlington and Wylie[7] and other sources[8-19] were used for verification or when the number was not reported in Darlington.

---

* Tables follow text.

Table 1
CEREALS

Botanic name	Common name	2n	Ref.
Amaranthus paniculatus	Grain amaranth	32	7
Avena nuda (chinensis)	Chinese naked oat	42	7
A. sativa	Common oats	42	7
Eleusine coracana	African millet	36	7
Eragrostis tef	Tef grass	40	7
Fagopyrum esculentum	Buckwheat	16	7
F. tataricum	Rye buckwheat	16	7
Hordeum distichum	Two-row barley	14	7
H. vulgare	Barley	14	7
Oryza sativa	Rice	24	7
Panicum italicium	Italian millet	18	7
P. miliaceum	Broomcorn millet	36	7
Pennisetum glaucum	Spiked millet	14	7
P. spicatum (americanum)	Pearl millet	14	7
Secale cereale	Rye	14	7
Setaria sphacelata	Rhodesian timothy	36, 54	7
Sorghum vulgare	Grain sorghums	20	7
× *Triticosecale*	Triticale	42, 56	14
Triticum aestivum	Common wheat	42	7
T. dicoccum	Emmer	28	7
T. durum	Abyssinian hard wheat	28	7
T. monococcum	Einkorn wheat	14	7
Zea mays	Starchy, sweet corn	20	7

Table 2
PROTEIN CROPS

Botanical name	Common name	2n	Ref.
Arachis hypogea	Groundnut, peanut	40	7
Cajanus indicus	Pigeon pea	22, 44, 66	7
Cicer arietinum	Chickpea	16	7
Dolichos lablab	Hyacinth bean	22, 44	7
Glycine gracilis	Manchurian soya	40	7
G. max	Soyabean	40	7
G. ussuriensis	Wild soya	40	7
Lathyrus sativus	Grass pea	14	7
Lens culinaris	Lentil	14	12
Mucuna pruriens	Cowitch, Florida velvet bean	22	7
Phaseolus acutifolius	Tepary bean	22	7, 18
P. aureus	Mung, green gram	22	7, 18
P. lunatus	Lima bean, Sieva	22	7, 18
P. coccineus (multiflorus)	Scarlet runner bean	22	7, 18
P. mungo	Black gram	22, 24	7, 18
P. vulgaris	French, common, kidney, or dwarf bean	22	7, 18
Vicia faba	Horse bean, broad bean	12	7
Vigna angularis	Adzuki bean	22	7
V. sesquipedalis	Asparagus (yard-long bean)	24	7
V. sinensis	Cowpea	24	7, 18

Table 3
FRUIT CROPS

Botanical name	Common name	2n	Ref.
Amygdalus communis	Almond	16	7
Ananas comosa	Pineapple	50, 75, 100	7
Annona squamosa	Sweetsop	14	7
Artocarpus communis	Breadfruit	54, 56, ca 81(3x)	7
Carica candamarcensis	Heilborn	18	7
C. papaya	Papaya	18, 36	7
Citrullus vulgaris	Watermelon	22	7
Citrus grandis	Pummelo	18, 36	7
C. medica	Citron, lemon	18	7
C. nobilis	Tangerine	18	7
C. reticulata	Mandarin	18, 36	7
C. sinensis	Orange	18, 27, 36, 45	7
Cydonia oblonga	Quince	34	7
Diospyros kaki	Japanese persimmon	90	7
Fiscus carica	Fig	26	7
Fragaria chiloensis	Wild strawberry	56	7
F. virginiana	Strawberry	56	7
Garcinia mangostana	Mangosteen	ca 76	7
Litchi chinensis	Litchi	28, 30	7
Malus asiatica	Chinese apple	34	7
M. pumila	Apple	34, 51, 68	7, 12
Mangifera indica	Mango	40	7
Morus rubra	Red mulberry	28	7
Musa cavendishii	Banana	22, 33, 44, etc.	7
M. paradisiaca	Banana	22, 33, 44, etc.	7
Passiflora edulis	Passion fruit, purple granadilla	18	7
P. ligularis	Passion flower	18	7
Persea americana	Avocado pear	24	7
Phoenix dactylifera	Date palm	36	7
Prunus armeniaca	Apricot	16	7
P. avium	Sweet cherry, mazzard	16, 24, 32	7
P. cerasifera (divaricata)	Cherry plum	16, 17, 24	7
P. cerasus	Cherry	32	7
P. domestica	European plum or prune	48	7
P. maheleb	Maheleb cherry	16	7
P. persica	Peach and nectarine	16	7
P. pseudocerasus	Cherry	32	7
P. salicinia	Japanese peach	16	7
P. serotina	Wild black cherry	32	7
Psidium guajava	Guava	22, 33, 44	7, 13
Punica granatum	Pomegranate	16	13
Pyrus communis	Pear	34, 51, 68	7
P. ussuriensis	Chinese pear	34	7
Ribes grossularia	Gooseberry	16	7
R. nigrum	Black currant	16	7
R. rubrum	Red and white currant	16	7
Rubus idaeus	European raspberry	14, 21, 28	7
R. strigosus	American raspberry	14, 21	7
Tamarindus indica	Tamarind	24	7
Vaccinium angustifolium	Sugarberry	24	7
V. ashei	Rabbiteye blueberry	72	7
V. corymbosum	Highbush blueberry	48	7
V. myrtilloides	Canada blueberry	24	7
V. oxycoccus	Cranberry	48	12
Viburnum trilobum	American cranberry bush	18	7
Vitis labrusca	Fox grape	38	7
V. vinifera	Grape	38, 76	13

Table 4
VEGETABLES AND TUBERS

Botanical name	Common name	2n	Ref.
Allium escalonicum	Shallot, green onion	16	7, 15
A. cepa	Onion	16, 32	7, 15
A. chinense	Onion	32	12
A. fistulosum	Welsh onion	16	7, 15
A. porrum	Leek	32	7, 15
A. sativum	Garlic	16	7, 15
A. schoenoprasum	Chives	16, 32	7, 15
Amoracia lapathifolia	Horseradish	32	7
Apium graveolens	Celery	22	7
Arabidopsis thaliana	Common wall cress	10	7
Asparagus officinalis	Asparagus (var. *attilis*)	20	7, 15
Beta vulgaris	Sugar, mangold, or garden beet	18, 27	7
Brassica alboglabra	Chinese kale	18	7, 16
B. campestris	Turnip rape	20	7, 16
B. carinata	Abyssinian mustard	34	7, 16
B. cernea	Karashina	36	7, 16
B. chinensis	Chinese mustard or pak-choi	20	7, 16
B. hirta	White mustard	24	7, 16
B. juncea	Pai, brown mustard, Indian mustard	36	7, 16
B. kaber	Charlock	18	7, 16
B. napobrassica	Swede or rutabaga	38	7, 16
B. napus	Rape, oil rape, kale	38	7, 16
B. nigra	Black mustard	16	7, 16
B. oleracea	Cole, kale, collards brussels sprouts, cauliflower, cabbage, kohlrabi, broccoli	18	7, 16
B. pekinensis	Chinese cabbage or celery cabbage	20	7, 16
B. rapa	Turnip	20	7, 16
Cajanus cajan (indicus)	Pigeon pea	22, 44, 66	7
Canavalia ensiformis	Jack bean	22	7
Canna edulis	Edible canna	18, 27	7
Capsicum annuum	Common cultivated pepper	24	7, 19
C. chinense	Pepper	24	7, 19
C. frutescens	Cayenne pepper	24	7, 19
C. pendulum	Pepper	24	7, 19
C. pubescens	Pepper	24	7, 19
Cichorium endivia	Endive	18, 36	7
C. intybus	Chicory	18	7
Colocasia antiquorum	Taro	28, 36, 42, 48	7
Crambe maritima	Sea kale	60	7
Cucumis melo	Melon, cantaloupe	24	7
C. sativus	Cucumber	14	7
Cucurbita ficifolia	Malabar gourd	40	7
C. foetidissima	Calabazilla, buffalo gourd	40	7
C. maxima	Pumpkin	24, 40	7
C. moshata	Winter pumpkin	24, 40, 48	7
C. pepo	Summer squash, vegetable marrow	40	7
Cynara scolymus	Globe artichoke	34	7
Daucus carota	Carrot	18	7
Dioscorea alata	Yam	40, 60, 70, 80	12, 13
D. batatas	Chinese yam	ca 140	7
D. cayennensis	Attoto yam	ca 140	7
Eruca sativa Mill.	Rocket salad, rocket or roquette	22	7

Table 4 (continued)
VEGETABLES AND TUBERS

Botanical name	Common name	2n	Ref.
Ipomea batatas	Sweet potato	90	7
Lactuca sativa	Lettuce	18	7
Lepidium sativum	Garden cress	16, 32	7
Luffa cylindrica	Loofah	26	7
Lycopersicon cerasiforme	Cherry tomato	24	7
L. esculentum	Tomato	24	7
Manihot utilissima	Manioc, cassava	36	7
Nasturtium officinale	Water cress	32	7
Pastinaca sativa	Parsnip	22	7
Petroselinum crispum (sativum)	Parsley	22	7
Physalis peruviana	Ground cherry	24, 48	7
Pisum sativum	Pea	14	7
Raphanus sativus	Radish	18	7
Rheum officinale	Rhubarb	22, 44	7
R. rhaponticum	English rhubarb	44	7
Scorzonera hispanica	Black salsify	14	7
Sechium edule	Chayote	24	7
Solanum andigenum	Andean potato	48	7
S. melongena	Eggplant	24	7
S. muricatum	Pepino	24	7
S. phureja	Potato	24	7
S. tuberosum	Common potato	48	7
Spinacia oleracea	Spinach	75	7
Trichosanthes anguina	Snake gourd	22	7
Tropaeolum tuberosum	Edible nasturtium	42	7
Wikstroemia canescens	Akia	18	7
Zea mays	Sweet corn	18	7

Table 5
NUTS

Botanical name	Common name	2n	Ref.
Anacardium occidentale	Cashew	42	7
Arachis hypogaea	Peanut	40	7
Carya illinoensis	Pecan	32	7
C. ovata	Shagbark hickory	32	7
Castanea sativa	Chestnut	24	7
Cocos nucifera	Coconut palm	32	7
Corylus americana	American hazel, filbert	28	7
C. avellana	European hazel, cobnut	22	12, 13
Juglans nigra	Black walnut	32	7
J. regia	English walnut	32	7
J. sinensis	Walnut	32	7
Litchi chinensis	Litchi	28, 30	7
Macadamia integrifolia	Queensland nut	28	7
Pistacia vera	Pistachio nut	30	7
Prunus amygdalus	Almond	16	7

Table 6
OIL CROPS (COOKING OR VEGETABLE OILS)

Botanical name	Common name	2n	Ref.
Aleurites cordata	Tung, China wood oil	22	7
A. moluccana (triloba)	Candlenut	44	7
A. fordii	Tung	22	7
A. montana	Tung	22	7
Arachis hypogea	Ground nut, peanut	40	7
Brassica campestris	Rapeseed oil, canola oil	20	7
B. napus	Rapeseed oil, canola oil	38	7
Cannabis sativa	Hempseed oil	20	7
Carthamus tinctorius	Safflower oil	24	7
Cocos nucifera	Coconut palm	32	7
Elaeis guineensis	Oil palm	32	7
Glycine gracilus	Manch, soya	40	7
G. max	Soybean	40	7
G. ussuriensis	Wild soya	40	7
Gossypium hirsutum	Cottonseed oil	52	7
Helianthus annus	Sunflower	34	7
Linum usitatissimum	Flax	30, 32	7
Madia sativa	Madia, chili, tar weed	32	7
Olea europaea	Olive	46	7
Papaver somniferum	Poppy seed	22	7
Ricinus communis	Castor bean	20	7
Sesamum indicum	Sesame	26	7
Zea mays	Corn	20	7

Table 7
SUGAR CROPS

Botanical name	Common name	2n	Ref.
Acer saccharum	Sugar maple	26	7
Beta vulgaris	Sugar or mangold beet	18, 27	7
Saccharum officinarum (officinarum × spontaneum)	Sugarcane	64, 80	12
S. robustum		42, 60—144	7
S. sinense	Sugarcane	117—123	7
S. spontaneum	Kans grass	40—46, 54—128, 48—80, 96, 112, 48—128	7, 12
Sorghum dochna	Sorgo syrup, sugar sorghum	20	7

Table 8
FORAGE GRASSES

Botanical name	Common name	2n	Ref.
Agropyron cristatum	Crested wheatgrass	14	7
A. desertorum	Crested wheatgrass	28	10
A. elongatum	Tall wheatgrass	14, 28, 42, 56, 70	10
A. intermedium	Intermediate wheatgrass	42	7
A. smithii	Western wheatgrass	56	10
A. spicatum	Blue bunch wheatgrass	14, 28	7
A. trachycaulum	Slender wheatgrass	28	7
A. trichophorum	Pubescent wheatgrass	42	10
Alopecurus pratensis	Meadow foxtail	28, 42	7
Andropogon gerardi	Big bluestem	60, 70, 84—86	7
Arrhenatherum elatius	Tall oat grass	28	7
Bothriochloa barbinodis	Cane bluestem	c.180	7
B. ischaemum	King ranch bluestem	20, 40	7
B. saccharoides	Silver bluestem	60, 120	7
Bouteloua curtipendula	Side oats grama	20, 40	13
B. eriopoda	Black grama	20	13
B. gracilis	Blue grama	20	13
Bromus erectus	Upright brome	42, 56, 70	7, 9
B. inermis	Smooth brome	56	7
B. riparius	Meadow brome	70	9
Cenchrus ciliaris	Buffel grass	34, 36, 40, 44, 56	7, 12
Chloris gayana	Rhodes grass	20, 40	7
Cymbopogon nardus	Citronella grass	20	7
Cynodon dactylon	Bermuda grass	40	7
Dactylis glomerata	Cocksfoot	28	7
Dichanthium annulatum	Kleberg bluestem	20, 40, 60	9
D. aristatum	Angleton bluestem	20, 40	9
D. sericeum	Silky bluestem	20	9
Digitaria decumbens	Pangola grass	30	7
D. smutsii		18, 36	7
Elymus angustus	Altai wild rye	28, 42, 56	7
E. junceus	Russian wild rye	14	7
Eragrostis curvula	Weeping love grass	40, 50, 60, 80	13
Festuca arundinacea	Tall fescue	42, 56, 70	12
F. ovina	Sheeps fescue	14, 28, 42, 56, 70	7
F. pratensis	Meadow fescue	14	7
F. rubra	Red fescue	14, 28, 42, 56, 70	7
Lolium multiflorum	Italian ryegrass	14	7
L. perenne	Perennial ryegrass	14	7
Panicum maxima	Guinea grass	18, 32, 36, 48	7
P. purpurascens	Para grass	36	7
P. virgatum	Switchgrass	18, 36, 54, 72, 90, 108	7
Paspalum dilatatum	Dallis grass, large water grass	40, 50, 60	13
P. notatum	Bahia grass	20, 40	7
Pennisetum clandestinum	Kikuyu grass	36, 54	7, 12
P. purpureum	Dry napier grass	28, 56	7
Phalaris arundinacea	Reed canary grass	14, 28, 42	9
Phleum bertolinii	Timothy	14	12
P. pratense	Timothy	28, 42	13
Poa fendleriana	Mutton grass	56	13
P. nemoralis	Wood meadow grass	28—38, 42, 43, 47—49, c.70	7
P. pratensis	Kentucky bluegrass	36—123, 38—96	7
Setaria sphacelata	Rhodesian timothy	36, 54	7

Table 8 (continued)
FORAGE GRASSES

Botanical name	Common name	2n	Ref.
Sorghastrum nutans	Indian grass	20, 40	7
Sorghum halepense	Johnson grass	20, 40	13
S. sudanense	Sudan grass	20	7
Sorghum vulgare	Great millet	20	7

Table 9
FORAGE LEGUMES

Botanical name	Common name	2n	Ref.
Coronilla varia	Crown vetch	24	7
Crotalaria juncea	Sunn hemp	16	7
Lespedeza stipulacea	Korean lespedeza	20	7
L. striata	Common lespedeza	20	7
L. sericea		18	7
Lotus corniculatus	Birdsfoot trefoil	24	7
L. uliginosus		24	7
Lupinus albus	Wolf bean, white lupin	30, 40, 48, 50	7
L. angustifolius	Blue lupin	40, 48	7
L. luteus	European yellow lupin	48, 52	7
L. mutabilis	South American lupin	48	7
Medicago arabica	Spotted burr-clover	16	7
M. arborea	Tree alfalfa	32	7
M. falcata	Yellow lucerne	16, 32	7
M. hispida	California burr-clover	14, 16	7
M. sativa	Alfalfa	16, 32	7
M. tribuloides	Barrel medic	16	7
Melilotus alba	White sweet clover	16	7
M. officinalis	Yellow sweet clover	16	7
Onobrychis viciaefolia	Sainfoin	14, 28	7
Ornithopus sativus	Serradella	14, 16	7
Pueraria thunbergiana	Kudzu vine	22, 24	7
Trifolium alexandrinum	Egyptian clover	16	7
T. augustifolium	Fineleaf clover	14	7
T. diffusum	Rose clover	10	8
T. hybridum	Alsike clover	16	8
T. incarnatum	Crimson clover	14	8
T. ingrescens	Ball clover	16	8
T. pratense	Red clover	14	8
T. repens	White clover	32	8
T. resupinatum	Persian clover	16	8
T. subterraneum	Subterranean clover	16	8
Trigonella foenum-graecum	Fenugreek	16	7
Vicia sativa	Common vetch	12, 14	7
V. villosa	Hairy vetch	14	7
V. atropurpurea	Purple vetch	14	7
V. angustifolia	Narrowleaf vetch	12	7
V. dasycarpa	Woolly pod vetch	14	7
V. ervilia	Monantha (bitter) vetch	14	7
V. pannonica	Hungarian vetch	12	7
V. calcarata	Bard vetch	12, 14	7

Table 10
FIBER PLANTS AND WOODS

Botanical name	Common name	2n	Ref.
Agave fourcroydes	Henequen, Yucatan sisal	138	7
A. sisalana	Sisal hemp	138, c.149	7
Bambusa vulgaris	Bamboo	72	7
Boehmeria nivea	Rhea fiber, ramie	28	7
Cannabis indica	Hemp	20	7
C. sativa	Hemp	20	7
Ceiba pentandra	Silk cotton tree, kapok tree	72, 80	7
Corchorus capsularis	Jute	14	7
C. olitorius	Jute	14	7
Crotalaria juncea	Sunn hemp, crotalaria	16	7
C. mucronata (striata)	Hemp	16	7
Gossypium arboreum	Tree cotton	26	7
G. barbadense	Egyptian cotton	52	7
G. herbaceum	Cotton	26	7
G. hirsutum	Upland cotton	52	7
G. nanking	Oriental cotton	26	7
Hibiscus cannabinus	Kenaf	36	7
Linum angustifolium	Wild flax	28, 30, 32	7
L. usitatissimum	Flax	30, 32	7
Musa textiles	Manila hemp, abaca	20	7
Santalum album	Sandalwood	20	7
Urena lobata	Aramina, Congo jute	28, 56	7

Table 11
DYES, RUBBERS, HOPS (INDUSTRIAL CROPS)

Botanical name	Common name	2n	Ref.
Acacia arabica	Gum arabic	52	7
A. senegal	Gum arabic acacia	26	7
Hevea brasiliensis	Rubber tree	36	7
Humulus lupulus	Hop	20	7
Indigofera anil	Indigo	12	7
I. endecaphylla	Indigo	32, 36	7
I. tinctoria	Indigo	16	7
Parthenium argentatum	Mexican rubber guayule	36, 54, 72, 108—111	7

Table 12
DRUG CROPS

Botanical name	Common name	2n	Ref.
Althaea officinalis	Marshmallow	42	7
Angelica archangelica	Angelica	18	7
Anthemis nobilis	Chamomile	18	7
Camellia sinensis	Chinese tea	30	7
Chrysanthemum cinerariaefolium	Palm pyrethrum	18	7
C. coccineum	Pyrethrum	18	7
Cinchona ledgeriana	Quinine	34	7
C. succirunca	Cinchonidine	34	7
Cinnamonum camphora	Camphor	24	7
Coffea arabica	Coffee	44	7
C. canephora	Quillow coffee	22	7
Cola acuminata	Kola nut	40	7
Ilex paraguariensis	Mate, Paraguay tea	40	7
Nicotiana rustica	Tobacco	48	7
N. tabacum	Tobacco	48	7
Panax ginseng	Ginseng	22, 24	7
Papaver dubium	Poppy	28, 42	7
P. somniferum	Opium poppy	22	7
Ricinus communis	Castor bean	20	7
Theobroma cacao	Cacao	20	7
Valeriana officinalis	Valerian, all-heal	14, 28, 56	7

Table 13
SPICES AND FLAVORINGS

Botanical name	Common name	2n	Ref.
Artemisa dracunculus	Tarragon	18	7
Capsicum annuum	Red chili, pepper, paprika	24	7
C. frutescens	Cayenne pepper	24	7
Carum carvi	Caraway	20	7
Cinnamomum zeylanticum	Cinnamon tree	24	7
Hibiscus esculentus	Okra	72, 120, 130, 132	7
Maranta arundinacea	Arrowroot	18, 48	7
Myristica fragrans	Nutmeg	42	7
Piper nigrum	Black pepper	ca 128	7
Salvia officinalis	Sage	14	7
Sinapis alba	White mustard	24	7
Thymus vulgaris	Thyme	30	7
Glycyrrhiza glabra	Licorice root	16	7
Pimpinella anisum	Anise	18, 20	7
Trigonella foenum-graecum	Fenugreek	16	7
Vanilla fragrans	Vanilla	32	7
Mentha arvensis	Menthol	12, 54, 60, 72	7
M. piperita	Peppermint	36, 64, 66, 68, 70	7
M. viridis	Spearmint	36, 48, 84	7
Lavandula officinalis	Lavender	54	7
Coriandrum sativum	Coriander	22	7
Thymus serpyllum	Mother of thyme	24	7

REFERENCES

1. **Swanson, C. P.,** *Cytology and Cytogenetics,* Prentice-Hall, Englewood Cliffs, N.J., 1957, 98.
2. **Grant, V.,** *Genetics of Flowering Plants,* Columbia University Press, New York, 1975, chap. 23.
3. **Stebbins, G. L.,** *Variation and Evolution in Plants,* Columbia University Press, New York, 1950.
4. **Mao, Y. T.,** *World List of Plant Breeders,* Food and Agricultural Organization of the United Nations, Rome, 1961.
5. **Klages, K. H. W.,** *Ecological Crop Geography,* Macmillan, New York, 1942.
6. **Wilsie, C. P.,** *Crop Adaptation and Distribution,* W. H. Freeman, San Francisco, 1962, chap. 5.
7. **Darlington, C. D. and Wylie, A. P.,** *Chromosome Atlas of Flowering Plants,* George Allen and Unwin, London, 1945.
8. **Britten, E. J.,** Chromosome numbers in the genus *Trifolium, Cytologia,* 28, 428, 1963.
9. **Carnahan, H. L. and Hill, H. D.,** Cytology and genetics of forage grasses, *Bot. Rev.,* 27, 1, 1961.
10. **Dewey, D. R.,** Genomic and phylogenetic relationships among North American perennial Triticeae, in *Grasses and Grasslands: Systematics and Ecology,* Estes, J. R., Tyrl, R. J., and Brunken, J. N., Eds., University of Oklahoma Press, Norman, 1982, chap. 5.
11. **Myers, W. M.,** Cytology and genetics of forage grasses, *Bot. Rev.,* 13, 369, 1947.
12. **Moore, R. J., Ed.,** *Index to Plant Chromosome Numbers for 1973/74, Regnum Vegetabile 96,* Bohn, Scheltema and Holkema, Utrecht, 1977.
13. **Goldblatt, P., Ed.,** *Index to Plant Chromosome Numbers for 1975—1978, Monographs in Systematic Botany,* Missouri Botanical Garden, St. Louis, 1981, 5.
14. **Scoles, G. J. and Kaltsikes, P. J.,** The cytology and cytogenetics of *Triticale, Z. Pflanzenzuecht.,* 73, 13, 1974.
15. **Yarnell, S. H.,** Cytogenetics of the vegetable crops. I. Monocotyledons, *Bot. Rev.,* 20, 277, 1954.
16. **Yarnell, S. H.,** Cytogenetics of the vegetable crops. II. Crucifers, *Bot. Rev.,* 22, 81, 1956.
17. **Yarnell, S. H.,** Cytogenetics of the crucifers. III. Legumes. A. Garden peas, *Pisum sativum* L., *Bot. Rev.,* 28, 465, 1965.
18. **Yarnell, S. H.,** Cytogenetics of the vegetable crops. IV. Legumes, *Bot. Rev.,* 31, 247, 1965.
19. **Lippert, L. F., Smith, P. G., and Bergh, B. O.,** Cytogenetics of the vegetable crops. Garden pepper, *Capsicum* sp., *Bot. Rev.,* 32, 24, 1966.

CENTERS OF ORIGIN

Jack R. Harlan

The concept of "center of origin", when applied to cultivated plants, was developed by the Russian geneticist-agronomist Nikolai Ivanovich Vavilov (1887—1943). Other students of agricultural botany from Theophrastus and Pliny to Alphonse de Candolle and Sturtevant had attempted to locate geographic origins of cultivated plants, but the Vavilovian concept was more specialized and experimental. He proposed to locate the geographical region in which a plant was domesticated by analyses of variation of the plant in question. This required, first of all, extensive collections over the range of cultivation involving much plant exploration and field work. The collections were then assembled and grown at various experiment stations scattered over the U.S.S.R. Data were gathered on numerous characteristics including morphological variation, adaptation, cold and drought tolerance, daylength sensitivity, resistance to diseases and pests, and a limited number of chemical traits. Variation was then partitioned among species, subspecies, botanical varieties, forms, races, proles, or other formal taxa. The distributions of these taxa were plotted geographically, and the region of highest diversity was declared to be the center of origin.

The system was applied to many of the most important agricultural plants of the Soviet Union and was an enormously useful base for the national and regional plant-breeding programs of the country. The Institute, directed by Vavilov from 1920 to 1940, is now called the Vavilov Institute of Plant Industry (VIR). During the years of Vavilov's directorship, VIR conducted the most massive and systematic programs of plant exploration ever developed, and the many publications issued from the Institute were (and still are) invaluable contributions to our understanding of crop diversity. Studies of this kind can no longer be conducted for some crops because the ancient patterns of diversity have been destroyed by the adoption of modern cultivars. It is fortunate that the VIR research was conducted so early, and that Vavilov had the vision he did.

The concept of center of origin, however, has not stood the test of time so well. Vavilov recognized some of the problems himself. Some crops had too many centers. The chickpea was listed in five centers, pea in four. New World crops such as squash had centers in China and the Near East, and the African cowpea had centers in China and India. It was necessary to distinguish between "primary" and "secondary" centers. The primary centers were the regions of true origin and secondary centers developed later in places often far removed from the site of initial domestication. Primary centers could be identified by the presence of wild forms, the prevalence of primitive characters, and a higher frequency of dominant alleles. Secondary centers usually lack wild forms; variation is more specialized and derived, and recessive traits are more common.[1,2]

Vavilov also used the term "Genzentrum" or gene center for geographic areas rich in diversity of specified crops.[3] Centers of diversity are real phenomena and can be measured objectively, although few studies of this kind have ever been conducted. Centers of diversity are not necessarily centers of origin and, in fact, secondary centers are frequently richer in diversity than the primary ones.

In studying a broad spectrum of crops, Vavilov and his team of scientists noted that centers of diversity for a number of crops coincide, and this led him to propose eight basic "centers of origin" shown in Figure 1. The most complete list of crops by centers of origin is found in "The Phytogeographic Basis of Plant Breeding" (Vavilov, 1949 to 1950), translated by K. Starr Chester and published in *Chronica Botanica*.[2] It is relatively available and need not be repeated here. A synopsis, however, is given in Table 1 for the more important or characteristic crops of each region. It should be noted that Vavilov and his

FIGURE 1. The eight centers of origin, according to N. I. Vavilov.

staff actually used the Vavilovian method of analysis of variability on only some of the crops listed. There are some outright errors and some intuitive guesses. Southeast Asia and the South Pacific were not explored and Africa was touched only superficially. There is no evidence for his center VIIIa and very little for VIIIb. We know a good deal more today about crop origins, but for the time range of 1920 to 1940, the work was monumental and has not been equaled since.

The Ethiopian center, for example, is something of an artifact of exploration and sampling. It can readily be accepted as a center of diversity for a number of crops that originated elsewhere,[4] but a number of the African crops listed show far more diversity in other parts of Africa than in Ethiopia. A more complete analysis of African crops shows that activities of plant domestication took place right across sub-Sahara Africa from the Atlantic to the Indian Ocean. Some crops are clearly West African, a few are clearly Ethiopian in origin, and others clearly originated in between. There does not appear to be a real center of plant domestication in Africa; the pattern is diffused across the entire continent, south of the Sahara and north of the equator. Present evidence suggests equally diffuse patterns in Southeast Asia and the South Pacific, and probably in South America. I have termed such vast areas noncenters, in contrast to the well defined and more restricted centers found in the Near East, China, and Mesoamerica[5] (Figure 2).

Vavilov emphasized the fact that his centers were almost always in mountainous districts separated one from the other by vast deserts or regions of widely scattered cultivation. He felt that the diversity of habitat and environmental conditions in mountainous districts contributed to the diversity of crop plants. He was also keenly aware that the diversity of people in the mountains was influential in the generation of variation.[2] Yet, he paid less attention to evidence from enthography, linguistics, archaeology, etc., than did de Candolle.[6] He believed his method was more rigorous and scientific. It was, in fact, a very useful and pragmatic method. It described the variation available to plant breeders, indicated the sources of useful characteristics, and presented the breeders with the raw material needed to practice their arts and science. Furthermore, at the time, archaeological plant remains were largely confined to Egyptian tombs, Swiss Lake Dwellers, and coastal Peru. Modern linguistic analyses had hardly begun, and folk classifications were yet to be studied.

Some scholars have pointed out that the Vavilovian centers represent lines drawn about

Table 1
A SYNOPSIS OF CROP PLANTS LISTED BY CENTER OF ORIGIN ACCORDING TO VAVILOV[2]

I. Chinese center (136)[a]
Italian millet, proso, barnyard millet, kaoliang, naked oats, buckwheat, soybean, adzuki bean, velvet bean, several bamboos, giant radish, water chestnut, *Zizania* rice, taro, Chinese yam, Chinese cabbages, Chinese onions, udo, stem lettuce, Chinese pear, peach, jujube, ginkgo, Chinese walnut, Chinese hazelnut, oriental chestnut, oriental persimmon, several species of *Citrus*, litchi, loquat, tea, ginseng, lac tree, ramie, hemp, Chinese sugarcane

II. Indian center (117)
Rice, kodo millet, sorghum, chickpea, pigeonpea, mat bean, urd, mung bean, rice bean, guar, vegetable amaranths, egg plant, bitter gourd, cucumber, radish, mango, several species of *Citrus*, jambos, jack fruit, Indian sugarcane, sugar palm, sesame, coconut, sarson *(Brassica)*, tree cotton, jute, black pepper, betel palm, cardamon, cumin

IIa. Indo-Malayan center (55)
Job's tears (for grain), several yams, ginger, a few citrus, banana, plantain, mangosteen, jack fruit, durian, jambos, rambutan, coconut, sugarcane, sugar palm, nutmeg, cardamon, clove, black pepper, manila hemp

III. Central Asiatic center (42)
Common wheat, rye, pea, lentil, broad bean, chickpea, mustard, flax, sesame, safflower, Asian cotton, cantaloupe, carrot, turnip, radish, onion, garlic, spinach, pistachio, apricot, common pear, almond, grape, English walnut, apple, hazelnut

IV. Near Eastern center (82)
Common wheat, durum wheat, barley, rye, oats, chickpea, lentil, pea, lupin, alfalfa, sainfoin, several vetches, sesame, flax, rape, anise, coriander, opium poppy, cantaloupe, cucumber, turnip, garden beet, carrot, onion, leek, lettuce, fig, pomegranate, apple, pear, quince, cherry, almond, English walnut, hazelnut, chestnut, grape, apricot, pistachio

V. Mediterranean center (84)
Durum wheat, emmer, byzantine oats, barley, lentil, pea, broad bean, chickpea, berseem, crimson clover, common vetch, grasspea, flax, white mustard, black mustard, rape, olive, carob, garden beet, cabbage, artichoke, parsley, turnip, onion, garlic, leek, lettuce, celery, parsnip, fennel, caraway, anise, cumin, sage, fenugreek

VI. Abyssinian center (39)
Ethiopian durums, emmer, barley, sorghum, teff, finger millet, pearl millet, chickpea, lentil, pea, broad bean, fenugreek, cowpea, flax, noog *(Guizotia)*, safflower, sesame, castor bean, coriander, nigella, chat, coffee, onion, ensete, okra

VII. South Mexican and Central American center (49)
Indian corn, common bean, lima bean, tepary bean, jack bean, grain amaranth, several squashes, chayote, jícama, sweet potato, ground cherry, tomato, chili pepper, upland cotton, sisal, henequen, chirimoya, sapote, papaya, avocado, guava, cashews, cacao, rustica tobacco

VIII. South American center (45)
Potato (several species), oca, ulluco, año, maca, Andean lupin, quinoa, grain amaranth, Indian corn, lima bean, common bean, *Xanthosoma*, edible canna, pepiño, tomato, tree tomato, ground cherry, squash, hot pepper, coca, long staple cotton, passion fruit, papaya, guava, chirimoya, tobacco

VIIIa. Chiloe center (4)
Potato, Chilean strawberry, a tarweed for oil, a *Bromus* for grain

VIIIb. Brazilian-Paraguayan center (13)
Manioc, peanut, cacao, *Hevea* rubber, maté, pineapple, brazil nut, cashew, passion fruit

[a] The number in parentheses indicates the actual number of crop plants listed by Vavilov for each center.

regions in which ancient civilizations arose. A glance at the map (Figure 1) indicates this is quite true, and one may conclude that the interactions of time, space, and human history together are responsible for the development of centers of diversity regardless of crop origins.

P. M. Zhukovsky, a protege of Vavilov and his successor at VIR, was dissatisfied with the centers of origin outlined by Vavilov and attempted improved models. He enlarged some of them until they merged and even drew lines around whole continents[7] (Figure 3). The models might have been closer to the truth, but were no help to the concept of center of origin. E. N. Sinskaya, a distinguished scientist at VIR under Vavilov, also had difficulty

FIGURE 2. Centers and noncenters of agricultural origins. (A1) Near East center; (A2) African noncenter; (B1) North Chinese center; (B2) Southeast Asian and South Pacific noncenter; (C1) Mesoamerican center; and (C2) South American noncenter.

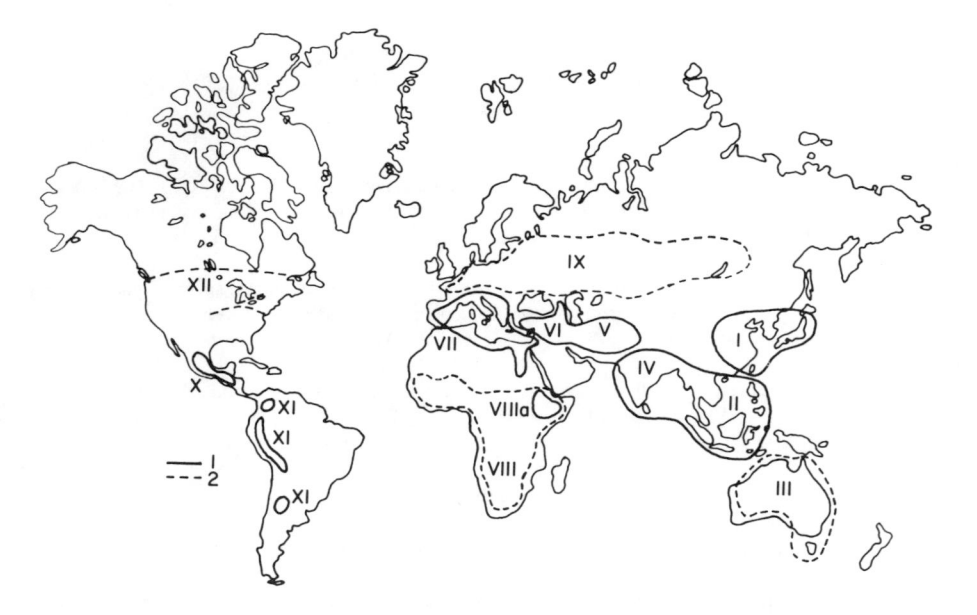

FIGURE 3. P. M. Zhukovsky's alterations (solid lines) and additions (broken lines) to Vavilov.

with "centers". She specialized in cruciferous crops and found that some, like *Brassica campestris* and the common radish *Raphanus sativus,* simply had no centers. They were domesticated repeatedly wherever they grew wild, the first right across Eurasia and the second along the sea coasts from western Europe to Japan (with a gap where Africa joins Asia).[8,9]

In recent years, more sources of information have come into play. Archaeobotanical information is now becoming massive in Europe, the Near East, Japan, and North America,

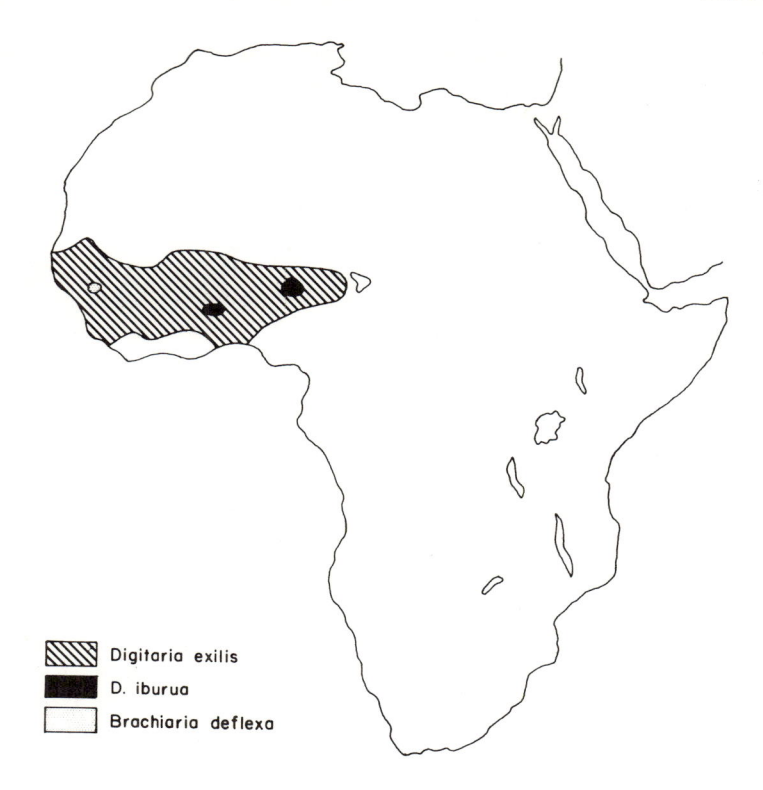

FIGURE 4. Distribution of three African millets. *Brachiaria deflexa* and *Digitaria iburua* are endemic crops; *D. exilis* is semiendemic.

with a yearly increment being added from tropical America and Africa. Ethnographic, ethnobotanical, linguistic, and other sources of information are accumulating. It is now clear that: (1) centers of origin and centers of diversity must be considered as distinct phenomena; (2) the two sometimes coincide, but most often do not; (3) a number of cultivated plants have multiple origins and noncentric variation patterns; (4) a single generalized model cannot characterize all cultivated plants, and (5) time, space, ecology, and human behavior must all be taken into account to explain the distribution of variability and the evolution of crops.

In an effort to bring these variables into perspective, this author has attempted a classification of variation patterns of cultivated plants. It is more complex than the Vavilovian system, but I believe more realistic.[10,11]

Endemic — A crop that originates in a small region and never spreads out of it appreciably may be said to have an endemic pattern of variation. In this case, the center of origin and the center of whatever diversity there may be coincide. A model is Guinea millet *(Brachiaria deflexa)* which is grown only in the Futa Djalon area of Guinea, West Africa. The crop is fully domesticated with large seeds, nonshattering habit, and considerable diversity. The wild races are widely distributed in Africa, but the crop is confined to a small district.

Semiendemic — Crops with similar variation patterns but with somewhat wider distributions may be called semiendemic. There may be one to several nodes of concentration of variability within the range of the crop which, however, is decidedly limited. Examples are African rice *(Oryza glaberrima)* and Fonio *(Digitaria exilis;* Figure 4). Despite the limited range of these crops, variation structures within the ranges can be detected with local concentrations of diversity.

Monocentric — Crops with a single well-defined center of origin, but often with wide distribution may be called monocentric. These are almost all new crops developed in late historical times, or crops that did not spread appreciably until recently. They are usually plantation crops such as coffee (*Coffea arabica*), cacao (*Theobroma cacao*), rubber (*Hevea braziliensis*), and African oil palm (*Elaeis guineensis*). Coffee may have been in use in Ethiopia and Yemen for a long period of time, but the coffee plantations of the world are a product of the last two centuries. Cacao was much used and highly regarded by the Mexican Indians but the germ plasm now deployed in cacao growing countries is derived recently and largely from other sources. Hevea rubber is primarily a product of the 20th century and the sources of many of the parental clones can be traced. Plantation exploitation of the oil palm is also a 20th-century development and sources of breeding materials are well known.

Oligocentric — Widespread crops with two to several centers of diversity may be called oligocentric. This is the most common pattern for ancient, widespread crops. For example, pea and lentil have centers in Central Asia, the Near East, the Mediterranean, and Ethiopia; common wheat has centers in the first three and barley in the last two.

Noncentric — Some crops do not seem to have discernable centers of origin, or have been repeatedly domesticated over a wide region. Sinskaya's studies have already been mentioned. Sorghum (*Sorghum bicolor*) and the common bean (*Phaseolus vulgaris*) are other examples. Various races of sorghum may show geographic affinities (e.g., guinea race in West Africa, kaffir race in southern Africa, caudatum race in the Chad-Sudan-Uganda region), but a region where sorghum itself was domesticated does not appear through standard analyses of variation patterns. Indirect evidence suggests the northeast quadrant of sub-Sahara Africa, but this region is not implicated by patterns of diversity.[12] It is incorrect to assume that a cultivated plant must *always* have a center of origin. The processes of domestication may have taken place over vast geographic expanses. Noncentric crops, of course, are confined to those with geographically widespread progenitors.

Nearly all crops will fit into this classification system in some fashion, but each must be taken as an individual case and the interplay of space, time, and history considered.

There is another problem concerning the origins of crops that Vavilov found necessary to treat: some crops were not domesticated in the region of the original wild sources. The wild races of the cultivated oat, for example, are found in the Mediterranean and Near Eastern regions, but the crop was domesticated in Northern Europe. The wild oats were carried from their homelands as weeds of emmer and barley and were domesticated outside of the native range of the species. Vavilov called such crops "secondary" as opposed to "primary" crops such as emmer and barley which were domesticated within the natural range of their progenitors. In a similar way, wild tomatoes are found along the coasts of Peru and Ecuador, but the plant is thought to have first been domesticated in Mexico.[13] Guar (*Cyamopsis tetragoaolobus*) is wild in Africa and cultivated in Pakistan and India. This example was called "transdomestication" by Hymowitz.[14]

From an historical perspective, it is now clear that the problem of crop origins, like agricultural origins, is far more complex and difficult than Vavilov visualized. Some plants originated in definable centers; some did not. Some were domesticated in the centers described by Vavilov, and some were domesticated in areas far removed from them. A number of crops have no definable centers at all. The value and applicability of the concept has declined accordingly. Brücher[15] found the patterns so ambiguous that he raised the question of whether centers really exist at all. The concept of concentration of diversity, however, has been immensely rewarding from the point of view of assembly of genetic diversity for plant breeding. Vavilov's emphasis on the value of the wild forms is only now beginning to be widely accepted. Recent international efforts in the field of plant genetic resources indicate that, at long last, there is acceptance of the idea that the world should be doing what Vavilov had tried to do for the U.S.S.R. during 1920 to 1940. We are too late in some cases, but the vision of Vavilov must be acknowledged.

REFERENCES

1. **Vavilov, N. I.**, *Studies on the Origin of Cultivated Plants* (in Russian and English), State Press, Leningrad, 1926, 248.
2. **Vavilov, N. I.**, The phytogeographic basis of plant breeding, in Chester, K. S., Trans., Selected writings of N. I. Vavilov: the origin, variation, immunity and breeding of cultivated plants, *Chron. Bot.*, 13 (1/6), 364, 1951.
3. **Vavilov, N. I.**, *Geographische Genzentren unsere Kulturpflanzen*, Verhandlungen V, Int. Kongr. Vererbungswiss (Berlin, 1927), Vol. 1, Borntraeger, Leipzig, 1928, 342.
4. **Harlan, J. R.**, Ethiopia: a center of diversity, *Econ. Bot.*, 23, 309, 1969.
5. **Harlan, J. R.**, Agricultural origins: centers and noncenters, *Science*, 174, 468, 1971.
6. **de Candolle, A.**, *Origin of Cultivated Plants*, 2nd ed., Kegan Paul, Trench & Trubner, London, 1909.
7. **Zhukovsky, P. M.**, New centres of the origin and new gene centres of cultivated plants including specifically endemic micro-centres of species closely allied to cultivated species, *Bot. Zh. (Leningrad)*, 53, 430, 1968.
8. **Sinskaya, E. N.**, The oleiferous plants and root crops of the family Crucifereae, *Bull. Appl. Bot. Genet. Plant Breeding*, 19(3), 1, 1928.
9. **Sinskaya, E. N.**, The wild radish from the sea-coast of Japan in connection with the problem of origin of the cultivated forms belonging to the genus *Raphanus, Bull. Appl. Bot. Genet. Plant Breeding*, 26(2), 3, 1931.
10. **Harlan, J. R.**, Geographic patterns of variation in some cultivated plants, *J. Hered.*, 66, 182, 1975.
11. **Harlan, J. R.**, Plant and animal distribution in relation to domestication, *Phil. Trans. R. Soc. London Series B*, 275, 13, 1976.
12. **Harlan, J. R. and Stemler, A. B. L.**, The races of sorghum in Africa, in *The Origins of African Plant Domestication*, Harlan, J. R., de Wet, J. M. J., and Stemler, A. B. L., Eds., Monton, The Hague, 1976, 465.
13. **Jenkins, J. A.**, The origin of the cultivated tomato, *Econ. Bot.*, 2, 379, 1948.
14. **Hymowitz, T.**, The trans-domestication concept as applied to guar, *Econ. Bot.*, 26, 9, 1972.
15. **Brücher, H.**, Gibt es Gen-Zentren?, *Naturwissenschaften*, 56, 77, 1969.

INTROGRESSIVE HYBRIDIZATION

S. R. Bowley and N. L. Taylor

Introgressive hybridization implies the transfer of genes from a donor to a recipient species that has some degree of genetic isolation. The genes that are under transfer are generally nuclear; however, they may be coded by the plasmon (cytoplasmic genome). Introduction of such alien genetic material into the gene pool of crop species furnishes a widened genetic base from which improved cultivars may be selected.

The successful incorporation of alien germplasm depends upon a number of factors. In general, the alien material is most easily incorporated when it is relatively small, the species are closely related, and the recipient species is polyploid.[1] The ease of transfer is greatest between species that have little chromosomal differentiation and show regular chromosomal pairing and chiasmata formation. This type of cross is not unlike an intraspecific hybrid, and backcross procedures are usually sufficient to allow recombination and introgression.

Crosses between species that have differing ploidies are facilitated by crossing at the same level of ploidy. For example, Wagenheim[2] found that *Solanum acaule* Bitt. (2n = 24) had increased crossability with *S. tuberosum* L. (2n = 48) when the ploidy of the former was artificially doubled. Treatment with colchicine and nitrous oxide are two commonly used methods whereby the chromosome number can be doubled.[3] Alternatively, the polyploid species may be reduced to the diploid level to facilitate crossing and selection of characters. This procedure has been used to cross *S. tuberosum* with a number of wild diploid species of potato.[4]

For species that have wider differentiation, the use of a natural or synthetic bridging species may be required. *Nicotiana sylvestris* Speg. and Comes (2n = 24) has been used as a bridge to obtain the cross between *N. repanda* Willd. (2n = 48) and *N. tabacum* L. (2n = 48).[5] The bridging species can also be used to bridge ploidy barriers. For example, Kerber and Dyck[6] used *Triticum turgidum* ssp. *durum* MacKey (2n = 28) as a bridge to transfer stem rust resistance from *T. monococcum* L. (2n = 14) to *T. aestivum* (L.) Thell. (2n = 42).

In many crosses, the F_1 is not sufficiently fertile to allow backcrossing to the recipient species. Artificial doubling of the chromosome number is often required to regain fertility. Johnston[7] used this procedure to transfer characters from *Brassica campestris* L. (2n = 20) to *B. napus* L. (2n = 36). In certain crosses, "automatic" doubling is obtained through nonreduction of gametes.[8] Additionally, suppression of homologous pairing in allopolyploid species may permit gene exchange between alien and homeologous chromosomes.[9]

To obtain introgression between species that have major chromosomal differentiation (i.e., little pairing, let alone crossing-over), induction of chromatin transfer is required. Although it is possible to incorporate intact chromosomes into a genome (substitution or addition lines), they usually carry too many detrimental characters to be of use. Breakage of chromosomes and transfer of small pieces have been used to lessen these linkage effects. Techniques employing ionizing radiation[10] and chemicals (e.g., 8-ethoxycaffeine)[11] have been used to induce chromosomal breaks and rearrangements.

Other problems are often encountered in obtaining the initial cross itself. Certain crosses are possible only when made in one direction. Crosses of *Lycopersicon esculentum* Mill. with species of the subgenus *Eriopersicon* are successful only if *L. esculentum* is used as the female parent,[12] and success varies depending on the parental genotype used.[13] To reduce incompatability, pretreatment of the stigma with immunosuppressants (e.g., E-amino caproic acid) has been tried with variable results.[14] Other methods employed to circumvent incompatibility include the application of growth regulators, shortening of the style, mentor pollen, irradiation of pollen or stigma, in vitro pollination, and perhaps cellular fusion.[15-17] Postzygotic failure may be avoided through embryo rescue, embryo transplants, grafting, and growth regulators.[16]

The introduction of alien genes often greatly disturbs the genetic system of the cultivated species. The wider the cross, the greater is the disturbance of the genome, plasmon, and genome/plasmon interactions. Hybrids may exhibit decreased vigor, chlorosis, or even necrosis.[18] Consequently, even with techniques such as embryo rescue, a viable hybrid may not be obtained. In other cases, a fertile F_1 may be obtained, but sterility or hybrid breakdown may be encountered in the F_2. In addition, certain genes may not be expressed or may act differently in the new genetic background created. Kerber and Dyck[6] found the degree of resistance of a dominant stem rust resistance gene declined when it was transferred from the diploid to the tetraploid and hexaploid wheat species. Considering the difficulties observed, introgressive hybridization should be explored as a breeding tool for those crops in which there is a lack of genetic variation or where specific genes are not found in the cultivated species but are available in the wild species.

The tables in this section are intended as a guide to wild germplasm that has been successfully introgressed with cultivated species. Successful introgression was defined as the production of a backcross, F_2 generation, or maintenance through reproductive apomixis. Those species in which one parent or the F_1 had to be artificially doubled to obtain the F_1 or backcross were indicated with an asterisk (*). In addition, the somatic chromosome number of the wild species was listed when it differed from that of the cultivated species. For ease of location, the species are separated into twelve groups or tables as follows:

Table 1.	Grain Crops
Table 2.	Seed Legumes
Table 3.	Forage Legumes
Table 4.	Forage Grasses
Table 5.	Root Crops
Table 6.	Stem and Leaf Crops
Table 7.	Fruit and Vegetable Crops
Table 8.	Nut Crops
Table 9.	Drug Crops
Table 10.	Fiber Crops
Table 11.	Latex, Rubber, Oil, and Wax Crops
Table 12.	Spice and Perfume Crops

Species that have dual purposes (e.g., flax oil and fiber), were classified based on their economic importance from a North American viewpoint. Similarly, in groups where the "wild" species are cultivated to some extent, the most commonly used species was designated as the cultivated species. Since many different techniques have been employed to hybridize wild and cultivated species, and certain crosses are possible only if one species is used as the female, referring to the original work is advised. Due to the extensive literature involved, the authors recognize the possibility of omissions and changes in nomenclature and apologize for any that inadvertently were not incorporated.

The literature search for compilation of these tables is intended to be complete through July 1982.

Table 1
INTROGRESSIVE HYBRIDIZATION IN GRAIN CROPS

Common name	Species	Species introgressed with the cultivated species[a]	Ref.
Amaranths	*Amaranthus caudatus* L. (2n = 32)	*A. edulis* Speg. *A. hybridus* L.	19
Oats	*Avena sativa* L. (2n = 42)	*A. powellii* S. Wats. (2n = 34) *A. abyssinica* Hochst. (2n = 28)* *A. barbata* Pott. (2n = 28) *A. byzantina* C. Koch *A. fatua* L. *A. hirtula* Lag. (2n = 14) *A. magna* Murphy et Terrell (2n = 28)* *A. murphyi* Ladiz. (2n = 28)* *A. sterilis* L. *A. strigosa* Schreb. (2n = 14)*	20—24
Buckwheat	*Fagopyrum esculentum* Moench (2n = 16)	*F. emarginatum* Meissn.	25
Barley	*Hordeum vulgare* L. (2n = 14)	*H. agriocrithon* E. Aberg *H. brachyantherum* L. × *H. bogdarii* Wilensky *H. distichon* L. *H. irregulare* Aberg et. Wiebe *H. jubatum* × *H. compressum* *H. spontaneum* C. Koch *Triticum aestivum* (L.) Thell. (2n = 42)	26, 27
Rice	*Oryza sativa* L. (2n = 24)	*O. australiensis* Domin *O. glaberrima* Steud. *O. nivara* Sharma et Shastry *O. officinalis* Wall. *O. perennis* ssp. *balunga* Moench. *O. perennis* ssp. *barthii* Moench.	22, 28—32
Pearl millet	*Pennisetum americanum* (L.) Leeke (2n = 14)	*P. fallax* Stapf & Hubb. *P. mollissimum* Hochst. *P. purpureum* Schumach. (2n = 28)* *P. setaceum* (Forsk.) Choiv. (2n = 27) *P. violaceum* (Lam.) L. Rich	22, 33—37

Table 1 (continued)
INTROGRESSIVE HYBRIDIZATION IN GRAIN CROPS

Common name	Species	Species introgressed with the cultivated species[a]	Ref.
Rye	*Secale cereale* L. (2n = 14)	*Hordeum jubatum* L. *S. africanum* Stapf. *S. anatolicum* Boiss. *S. ancestrale* Zhuk. *S. dalmaticum* Vis. *S. fragile* Bieb. *S. kuprijanovii* Grossh. *S. montanum* Guss. *S. segetale* (Zhuk) Roshev *S. silvestre* Host. *S. turkestanicum* Bensin *S. vavilovii* Grossh.	38—41
Sorghum	*Sorghum bicolor* (L.) Moench (2n = 20)	*S. aethiopicum* (Hack) Rupr. ex Stapf. *S. almum* Parodi (2n = 40) *S. arundinaceum* (Desf.) Stapf. *S. halepense* (L.) Pers. *S. propinquum* (Kunth) Hitchc. *S. sudanense* (Piper) Stapf. *S. verticilliflorum* (Steud.) Stapf. *S. virgatum* (Hack) Stapf. *Saccharum officinarum* L. (2n = 80)	42, 43
Bread wheat	*Triticum aestivum* (L.) Thell. (2n = 42)[b]	*T. monococcum* L. (2n = 14)[c] *T. timopheevi* Zhuk. (2n = 28)[d] *T. turgidum* (L.) Thell. (2n = 28)[e] *T. zhukovskyi* Men. et Er. (2n = 28) *Aegilops aucheri* Boiss. (2n = 14) *Ae. bicornis* (Forsk.) Jaub. et Spach (2n = 14) *Ae. caudata* L. (2n = 14) *Ae. columnaris* Zhuk. (2n = 28) *Ae. comosa* Sibth. et Sm. (2n = 14) *Ae. cylindrica* Host. (2n = 28)	1, 20, 22, 44—60

		Ae. heldereichii Holzm. (2n = 14)	
		Ae. juvanalis (Thell.) Eig (2n = 42)	
		Ae. kotshyi Boiss. (2n = 28)	
		Ae. longissima Schw. et Muschl. (2n = 14)	
		Ae. mutica Boiss. (2n = 14)	
		Ae. ovata L. (2n = 28)	
		Ae. sharonensis Eig. (2n = 14)	
		Ae. speltoides Tausch. (2n = 14)	
		Ae. squarrosa L. (2n = 14)	
		Ae. triaristata Willd. (2n = 28)	
		Ae. triuncialis L. (2n = 28)	
		Ae. umbellulata Zhk. (2n = 14)	
		Ae. variabilis Eig. (2n = 28)	
		Ae. ventricosa Tausch. (2n = 28)	
		Agropyron distichum (Thunb.) Beauv. (2n = 28)*	
		Ag. elongatiforme (Host.) Beauv. (2n = 56)	
		Ag. elongatum (Host.) Beauv. (2n = 14,56,70)	
		Ag. glaucum (2n = 42)	
		Ag. intermedium (Host.) Beauv. (2n = 42)	
		Ag. junceum (L.) Beauv. (2n = 28)	
		Ag. repens L. (2n = 42)	
		Ag. trichophorum (Link) Richt. (2n = 42)	
		Elymus giganteus L. (2n = 28)	
		Haynaldia villosa (L.) Schur. (2n = 14)	
		Hordeum vulgare L. (2n = 14)	
		Secale cereale L. (2n = 14)	
		S. montanum Guss. (2n = 14)	
		× Triticosecale Wittmack (2n = 42 – 56)	
		T. aestivum (2n = 42)[b]	1, 20, 22,
		T. monococcum (2n = 14)[c]	44—60
		T. timopheevi[d]	
		Aegilops bicornis (2n = 14)	
		Ae. bincialis Vis. (2n = 28)	
		Ae. caudata (2n = 14)	
		Ae. cylindrica	
		Ae. heldereichii (2n = 14)	
		Ae. kotschyi	
Durum wheat	Triticum turgidum ssp. durum (Desf.) Mackey (2n = 28)[e]		

Table 1 (continued)
INTROGRESSIVE HYBRIDIZATION IN GRAIN CROPS

Common name	Species	Species introgressed with the cultivated species[a]	Ref.
Durum wheat (cont.)		Ae. longissima (2n = 14)	
		Ae. ovata	
		Ae. sharonensis (2n = 14)	
		Ae. speltoides (2n = 14)	
		Ae. squarrosa (2n = 14)	
		Ae. triaristata	
		Ae. triuncialis	
		Ae. umbellulata (2n = 14)	
		Ae. uniaristata Vis (2n = 14)	
		Ae. variabilis	
		Ae. ventricosa	
		Agropyron distschum	
		Ag. elongatum	
		Ag. intermedium (2n = 42)	
		Ag. trichophorum (2n = 42)	
		Haynaldia villara (2n = 14)	
		Hordeum vulgare L. (2n = 14)	
		Secale cereale (2n = 14)	
		× Triticosecale (2n = 42 – 56)	
Triticale	× Triticosecale Wittmack (2n = 42,56)	Triticum aestivum (L.) Thell. (2n = 42)	61, 62
		T. turgidum ssp. durum (Desf.) Mackey (2n = 28)	
		Secale cereale L. (2n = 14)	
Corn	Zea mays L. (2n = 20)	Z. mexicana (Schrad.) Küntze	16, 22, 63—65
		Z. perennis (Hitchc.) Reeves & Mangelsdorf (2n = 40)*	
		Sorghum bicolor (L.) Moench (2n = 20)	
		Tripsacum australe Cutler & Anderson	
		T. dactyloides L. (2n = 36,72)	
		T. floridanum Porter	
Wild rice	Zizania aquatica L. (2n = 30)	Z. palustris Dore	66
		Z. texana Hitchc.	

a Introgression means the production of a backcross, F_2, or maintenance through reproductive apomixis. Species crosses in which one parent or the F_1 had to be artificially doubled are indicated with an asterisk (*). The somatic number of the wild species is given only when it differs from the cultivated.
b *Triticum aestivum* includes ssp. *compactum*, *macha*, *spelta*, *sphaerococcum*, *vavilovi*, and *vulgare*.[67]
c *T. monococcum* includes ssp. *bocoticum* and *monococcum*.[67]
d *T. timopheevi* includes ssp. *araraticum* and *timopheevi*.[67]
e *T. turgidum* includes ssp. *carthilicum*, *dicoccoides*, *dicoccum*, *durum*, *paleocolchicum*, *polonicum*, and *turanicum*.[67]

Table 2
INTROGRESSIVE HYBRIDIZATION IN SEED LEGUMES

Common name	Species	Species introgressed with the cultivated species[a]	Ref.
Peanuts	*Arachis hypogaea* L. (2n = 40)	*A. batizocoi* Krap. et Greg. (2n = 20)*	68—76
		A. cardenasii Krap. et Greg. nom. nudum (2n = 20)*	
		A. chacoense Krap. et Greg. nom. nudum	
		A. correntina (Burk.) Krap. et Greg. nom. nudum (2n = 20)*	
		A. diogoi Hoehne (2n = 20)*	
		A. duranensis Krap. et Greg. nom. nudum (2n = 20)*	
		A. monticola Krap. et Rig.	
		A. spegazzinii Krap. et Greg. nom. nudum (2n = 20)*	
		A. stenosperma Krap. et Greg. nom. nudum (2n = 20)*	
		A. villosa Benth. (2n = 20)*	
Pigeon pea	*Cajanus cajan* (L.) Millsp. (2n = 22)	*Atylosia cajanifolia*	73, 77
		A. lineata Wight et Arn.	
		A. scarabaeiodes (L.) Benth.	
		A. sericea Benth.	
Chickpea	*Cicer arietinum* L. (2n = 16)	*C. echinospermum* Davis	73, 78—81
		C. reticulatum Ladiz.	
Soybean	*Glycine max* (L.) Merrill (2n = 40)	*G. soja* Sieb. et Zucc.	73, 82
Lentils	*Lens culinaris* Medik (2n = 14)	*L. nigricans* Beib.	79, 83, 84
		L. orientalis (Boiss.) Popov.	
Field bean	*Phaseolus vulgaris* L. (2n = 22)	*P. acutifolius* Gray	73, 85—87
		P. coccineus Lam.	
		P. ritensis Jones*	
Peas	*Pisum sativum* L. (2n = 14)	*P. abyssinicum* Braun	73, 88
		P. arvense L.	
		P. aucheri Jaubert and Spach	
		P. elatius Bieb.	
		P. formosum Alefeld	
		P. fulvum Sibth et Smith	
		P. humile Boiss. et Noe	
		P. jomardi Schrank	
		P. transcaucasicum (Gov.) Stankov	
Green gram	*Vigna radiata* (L.) Wilczek (2n = 14)	*V. angularis* (Willd.) Ohwi & Ohashi	73, 89, 90
		V. mungo (L.) Hepper	
		V. sublobatus Roxb.	
		V. trilobata	
		V. umbellata (Thunb.) Ohwi & Ohashi	
Cowpea	*Vigna unguiculata* (L.) Walp. (2n = 22)	*V. cylindrica* (L.) Skeels	73, 91
		V. sesquipedalis (L.) Fruw.	

[a] Introgression means the production of a backcross, F_2, or maintenance through reproductive apomixis. Species crosses in which one parent or the F_1 had to be artificially doubled are indicated with an asterisk (*). The somatic number of the wild species is given only when it differs from the cultivated.

Table 3
INTROGRESSIVE HYBRIDIZATION IN FORAGE LEGUMES

Common name	Species	Species introgressed with the cultivated species[a]	Ref.
Centrosema	*Centrosema pubescens* Benth. (2n = 20)	*C. brasilianum* (L.) Benth.	92
Lespedeza	*Lespedeza cuneata* (D. Don) G. Don (2n = 20)	*L. hedysaroides*	93, 94
		L. inschanica (Maxim) Schindl.	
		L. latissima Nakai	
Trefoil	*Lotus corniculatus* L. (2n = 24)	*L. alpinus* (DC) Schleich. (2n = 12)*	94—97
		L. coimbrensis Willd. (2n = 12)*	
		L. filicaulis Dur. (2n = 12)*	
		L. japonicus (Regel) Larsen (2n = 12)*	
		L. krylovii Sch. and Serg. (2n = 12)*	
		L. palustris Willd. (2n = 12)*	
		L. pedunculatus Cav. (2n = 12)	
		L. schoelleri Schweinf. (2n = 12)*	
		L. tenuis Waldst. et Kit. (2n = 12)	
		L. uliginosus Schk. (2n = 12)*	
Blue lupines	*Lupinus angustifolius* L. (2n = 40)	*L. linifolius* Roth.	98—100
		L. opsianthus	
Yellow lupines	*Lupinus luteus* Kell. (2n = 52)	*L. rothmaleri* Klink.	98—100
White lupines	*Lupinus albus* L. (2n = 50)	*L. jugoslavicus* Kazim et Now.	98—100
Alfalfa	*Medicago sativa* L. (2n = 16,32)	*M. cancellata* Bieb. (2n = 48)	101—105
		M. coerulea Less. (2n = 16)*[b]	
		M. dzhawakhetica Bordz (2n = 32)*	
		M. erecta (2n = 16)*	
		M. falcata L. (2n = 16,32)[c]	
		M. lupulina L. (2n = 16)	
		M. rhodopoea Velen (2n = 16)*	
		M. saxatilis Bieb. (2n = 48)	
Yellow sweet clover	*Melilotus officinalis* (L.) Lam. (2n = 16)	*M. alba* Desr.	101, 106—108
		M. hirsuta Lipsky	
		M. polonica (L.) Desr.	

Table 3 (continued)
INTROGRESSIVE HYBRIDIZATION IN FORAGE LEGUMES

Common name	Species	Species introgressed with the cultivated species[a]	Ref.
White sweet clover	*Melilotus alba* Desr. (2n = 16)	*M. dentata* (W. K.) Pers. *M. hirsuta* Lipsky *M. polonica* (L.) Desr. *M. suaveolens* Ledeb. *M. taurica* (M. B.) Ser.	101, 106—108
Red clover	*Trifolium pratense* L. (2n = 14)	*T. diffusum* Ehrh. (2n = 16)*	109—111
White clover	*Trifolium repens* L. (2n = 32)	*T. ambiguum* M. Bieb. (2n = 32) *T. ishmocarpum* *T. nigrescens* Viv. (2n = 16)* *T. occidentale* Coombe (2n = 16) *T. uniflorum* L. (2n = 32) *T. xerocephalum* Fenzl.	109, 112—115
Sub. clover	*Trifolium subterraneum* L. (2n = 16)	*T. eriosphaerum* Boiss. *T. pilulare* Boiss.	116
Common vetch	*Vicia sativa* L. (2n = 12)	*V. angustifolia* L. (2n = 12)* *V. cordata* Wulf. (2n = 10) *V. macrocarpa* Mor. *V. narbonensis* Hermann *V. pannonica* Crantz *V. serratifolia* Jacq. (2n = 14)	94, 117—112

[a] Introgression means the production of a backcross, F_2, or maintenance through reproductive apomixis. Species crosses in which one parent or the F_1 had to be artificially doubled are indicated with an asterisk (*). The somatic number of the wild species is given only when it differs from the cultivated.

[b] Syn. *M. sativa*.

[c] *M. falcata* includes *M. glomerata* Balk (2n = 16), *M. glutinosa* Bieb (2n = 32), *M. prostrata* Jacq. (2n = 16, 32), and *M. quasifalcata* Sinsk (2n = 16).

Table 4
INTROGRESSIVE HYBRIDIZATION IN FORAGE GRASSES

Common name	Species	Species introgressed with the cultivated species[a]	Ref.
Wheat grasses			
Crested	*Agropyron cristatum* (L.) Gaertn. (2n = 14)	*A. desertorum* Fisch. ex Link (2n = 28)*	123, 124
Intermediate	*Agropyron intermedium* (Host.) Beauv. (2n = 28,42)	*A. podperae* Nabel *A. trichophorum* (Link) Richt *Triticum turgidum* ssp. *durum* (Desf.) Mackey (2n = 28)	125, 126
Pubescent	*Agropyron trichophorum* (Link) Richt.	*A. intermedium* (Host.) Beauv. *Triticum turgidum* spp. *durum* (2n = 28)*	126
Quackgrass	*A. repens* L. (2n = 42)	*A. desertorum* Fisch. ex Link (2n = 28)* *A. elongatiforme* (Host.) Beauv. (2n = 56) *A. spicatum* (Pursh) Schribn. et Smith (2n = 28) *A. stipifolium*	127—129
Meadow foxtail	*Alopecurus pratensis* L. (2n = 28)	*A. geniculatus* L.	130
Big bluestem	*Andropogon gerardi* Vitman (2n = 60)	*A. halli* Hack.	131
Bluestem	*Andropogon gayanus* Kunth (2n = 40)	*A. tectorum* Schum et Thonn	131
Bothriochloa	*Bothriochloa intermedia* Camus (2n = 40)	*B. ischaemum* (L.) Keng. *Capillipedium parviflorum* (R. Br.) Stapf *C. spicigerum* Blake *Dichanthium annulatum* (Forsk.) Stapf	132
Bromegrass	*Bromus inermis* Leyss (2n = 56)	*B. erectus* Huds. (2n = 28,70)* *B. pumpellianus* Scribn. (2n = 56) *B. tytholepsis* Nevski (2n = 70)	101, 133—136
Buffel grass	*Cenchrus ciliaris* L. (2n = 34)	*C. setigerus* Vah.	137
Orchard grass	*Dactylis glomerata* L. (2n = 48)	*D. aschersoniana* Graebn. (2n = 14)* *D. hispanica* (Roth) Koch (2n = 28) *D. marina* Borill. (2n = 28)	101, 138
Digitalis	*Digitalis purpurea* L. (2n = 56)	*D. ambigua* Murr.	139
Wild rye	*Elymus riparius* Wiegard (2n = 28)	*E. canadensis* L. *E. wiegandii*	140, 141
Wild rye	*Elymus angustus* Trin. (2n = 84)	*E. cinereus* Schribn. et Merr. (2n = 56) *E. giganteus* Vahl. (2n = 28)	140, 141
Fescues			
Tall	*Festuca arundinacea* Schreb. (2n = 42)	*F. gigantea* (L.) Vill. (2n = 42)*	140, 142

Table 4 (continued)
INTROGRESSIVE HYBRIDIZATION IN FORAGE GRASSES

Common name	Species	Species introgressed with the cultivated species[a]	Ref.
Fescues (cont.)			
Sheep	*F. ovina* L. (2n = 14)	*F. pratensis* Huds. (2n = 14)*	140, 142
Meadow	*F. pratensis* Huds. (2n = 14)	*Lolium perenne* L. (2n = 14)	140, 142
		L. multiflorum Lam. (2n = 14)	
		F. rubra L. (2n = 42)	
		F. gigantea (L.) Vill. (2n = 42)	
		Lolium perenne L.	
Ryegrasses			
Annual	*Lolium multiflorum* Lam. (2n = 14)	*Festuca arundinacea* Schreb. (2n = 42)	143
		L. perenne L.	
Perennial	*Lolium perenne* L. (2n = 14)	*Festuca arundinacea* Schreb. (2n = 42)	143
		F. pratensis Huds.	
		L. loliaceum (Bory et Chaub.) Hand-Mazz.	
		L. remotum Schrank	
		L. rigidum Gaud.	
		L. temulentum L.	
Dallisgrass	*Paspalum dilatatum* Poir.	*P. malacophyllum* Trin. (2n = 40)	144
Pennisetum	*Pennisetum ciliare* (L.) Link (2n = 36 – 54)	*Cenchrus setigerus* Vahl.	137
Reed canary grass	*Phalaris arundinacea* L. (2n = 28)	*P. aquatica* L. (2n = 28)	145
Timothy	*Phleum pratense* L. (2n = 42)	*P. alpinum* L. (2n = 28)*	146
		P. nodosum L. (2n = 14)	
		P. subulatum (Savi) Aschers. et Graebn. (2n = 14)	
Kentucky bluegrass	*Poa pratensis* L. (2n = 22 to 147)	*P. alpina* L.	101, 147—150
		P. ampla Merr.	
		P. chaixii Vill.	
		P. compressa L.	
		P. longifolia Trin.	
		P. nemoralis	
		P. scabrella (Thurb) Benth.	
Johnson grass	*Sorghum halepense* (L.) Pers. (2n = 40)	*S. sudanense* (Piper) Stapf (2n = 20)	151
Zoysia grass	*Zoysia japonica* Steud. (2n = 40)	*Z. matrella* (L.) Merr.	152
		Z. tenuifolia Willd.	

[a] Introgression means the production of a backcross, F_2, or maintenance through reproductive apomixis. Species crosses in which one parent or the F_1 had to be artificially doubled are indicated by an asterisk (*). The somatic number of the wild species is given only when it differs from the cultivated.

Table 5
INTROGRESSIVE HYBRIDIZATION IN ROOT CROPS

Common name	Species	Species introgressed with the cultivated species[a]	Ref.
Onion	*Allium cepa* L. (2n = 16)	*A. drobovii* Vved.	153, 154
		A. fistulosum L.	
		A. galanthum Kar. et Kir	
		A. pskemense B. Fedtsch.	
Beets	*Beta vulgaris* L. (2n = 18)	*B. atriplicifolia* Rouy	1, 155—159
		B. corolliflora Zoss. (2n = 36)	
		B. intermedia Bunge (2n = 36)*	
		B. lomatogona Fisch. et Mey.	
		B. macrocarpa Guss.	
		B. maritima L.	
		B. patellaris Moq. (2n = 36)*	
		B. patula Ait.	
		B. procumbens Chr. Sm. (2n = 18)*	
Carrot	*Daucus carota* subsp. *sativus* L. (2n = 18)	*D. capillifolius* Gilli	160
		D. gingidium L.	
Yam	*Dioscorea rotundata* Poir. (2n = 40)	*D. cayenensis* Lam (2n = 36,54)	161
Jerusalum artichoke	*Helianthus tuberosus* L. (2n = 102)	*H. annus* L. (2n = 34)	162—164
		H. hirsutus Rafin. (2n = 68)	
		H. resinosus	
		H. rigidus (Cass.) Desf.	
		H. schweinitzii	
		H. strumosus	
Sweet potato	*Ipomea batatas* (L.) Lam. (2n = 90)	*T. trifida* (H.B.K.) G. Don	165
Cassava	*Manihot esculenta* Crantz (2n = 36)	*M. anomala* Poh.	166, 167
		M. catingae	
		M. dichotoma Ule.	
		M. glaziovii Muell. Arg.	
		M. gracilis Pax	
		M. melanobasis	
		M. oligantha subsp. *nesteli* Pax emend. Nassar	
		M. saxicola	
		M. zehntneri Ule.	

Table 5 (continued)
INTROGRESSIVE HYBRIDIZATION IN ROOT CROPS

Common name	Species	Species introgressed with the cultivated species[a]	Ref.
Radish	*Raphanus sativus* L. (2n = 18)	*R. maritimus* Smith.	168—170
		R. raphanistrum L.	
		Brassica oleracea L. (2n = 18)	
		B. napus L. (2n = 38)	
		B. kaber L.	
		B. campestris (D.C.) L.C. Wheeler (2n = 20)	
		B. rapa L. (2n = 20)	
		B. chinensis Jusl. (2n = 20)	
		B. carinata A. Br. (2n = 34)	
Potato	*Solanum tuberosum* L. (2n = 48)	*S. acuale* Bitt. (2n = 48)*	1, 45, 171—173
		S. ajanhuire Juz. et Buk. (2n = 24)	
		S. andraeanum Baker (2n = 24)	
		S. camarguense Hawkes (2n = 24)	
		S. chacoense Bitt. (2n = 24)	
		S. demissum Lindl. (2n = 72)	
		S. goniocalyx Juz. et Buk. (2n = 24)	
		S. kurtzianum	
		S. multidissectum Hawkes (2n = 24)	
		S. oplocense Hawkes (2n = 24)	
		S. phureja Juz. et. Buk. (2n = 24)	
		S. polytrichum (2n = 24)	
		S. raphanifolium Card. et Hawkes (2n = 24)	
		S. semidemissum Juz. (2n = 24)	
		S. spegazzinii (2n = 24)	
		S. stenotomum Juz. et Buk. (2n = 24)	
		S. stoloniferum Schlecht. et Bche (2n = 48)	
		S. sucrense Hawkes (2n = 24)	
		S. vernei Bitt. et Wittm. (2n = 24)	
		S. verrucosum Schlechtend. (2n = 24)*	
Salsify	*Tragopogon porrifolius* L. (2n = 12)	*T. pratensis* L.	174

[a] Introgression means the production of a backcross, F_2, or maintenance through reproductive apomixis. Species crosses in which one parent or the F_1 had to be artificially doubled are indicated with an asterisk (*). The somatic number of the wild species is given only when it differs from the cultivated.

Table 6
INTROGRESSIVE HYBRIDIZATION IN STEM AND LEAF CROPS

Common name	Species	Species introgressed with the cultivated species[a]	Ref.
Celery	*Apium graveolens* L. (2n = 22)	*Petroselinum hortense* Hoffm.	175, 176
Asparagus	*Asparagus officinalis* L. (2n = 20)	*A. tenuifolius* Lam.	177
Brassica[b]			
Cabbage, etc.	*Brassica oleracea* L. (2n = 18)	*B. alboglabra* Bailey (2n = 18)	178—184
		B. campestris L. (2n = 20)*	
		B. chinensis Jusl. (2n = 20)	
		B. carinata A. Br (2n = 34)	
		B. hirta Moench (2n = 24)	
		B. nigra (L.) Koch. (2n = 16)	
		B. pekinesis (Lour.) Rupr. (2n = 20)	
		B. rapa L. (2n = 20)	
Rape, etc.	*Brassica campestris* L. (2n = 20)	*B. rapa* L. (2n = 20)	178—184
		B. parachinesis Bailey (2n = 20)	
		B. chinensis Jusl. (2n = 20)	
		B. pekinesis (Lour.) Rupr. (2n = 20)	
		B. nigra (L.) Koch (2n = 16)	
		B. kaber (D.C.) L.C. Wheeler (2n = 16)	
		B. juncea (L.) Coss. (2n = 16)	
		B. napobrassica (L.) Mill. (2n = 16)	
		B. corinata A. Br. (2n = 34)	
		B. cernua Forbes & Hensl. (2n = 34)	
		B. napus L. (2n = 38)	
		Diplotaxis muralis (L.) D.C. (2n = 42)	
Turnip, etc.	*Brassica rapa* L. (2n = 20)	*B. pekinesis* (Lour.) Rupr. (2n = 20)	178—184
		B. chinensis Jusl. (2n = 20)	
		B. trilocularis (Roxb.) Hook f. & Thoms. (2n = 20)	
		B. kaber (D.C.) L.C. Wheeler (2n = 24)	
		B. carinata A. Br. (2n = 34)	
		B. cernua Forbes & Hensl. (2n = 34)	
		B. nigra (L.) Koch (2n = 16)	
		B. napobrassica (L.) Mill. (2n = 16)	

Table 6 (continued)
INTROGRESSIVE HYBRIDIZATION IN STEM AND LEAF CROPS

Common name	Species	Species introgressed with the cultivated species[a]	Ref.
Black mustard	Brassica nigra (L.) Koch. (2n = 16)	B. kaber (D.C.) L.C. Wheeler B. carinata A. Br. (2n = 34) B. juncea (L.) Coss. (2n = 16)	178—184
Charlock	Brassica kaber (D.C.) L.C. Wheeler (2n = 16)	B. pekinesis (Lour.) Rupr. (2n = 20)*	178—184
Brown mustard	Brassica juncea (L.) Coss. (2n = 36)	B. hirta Moench (2n = 24) B. carinata A. Br. (2n = 34) B. pekinensis (Lour.) Rupr. (2n = 20)* B. carinata A. Br. (2n = 34) B. napus L. (2n = 38)	178—184
Swedes	Brassica napus L. (2n = 38)	B. campestris L. (2n = 20) B. cernua Forbes & Hemsl. (2n = 20) B. chinensis Jusl. (2n = 20) B. pekinesis (Lour.) Rupr. (2n = 20) B. nipposinica Bailey (2n = 20) B. napobrassica (L.) Mill. (2n = 38) B. oleracea L. (2n = 18) B. carinata A. Br. (2n = 34)	178—184
Chinese mustard	Brassica chinensis Jusl. (2n = 20)	B. carinata A. Br. (2n = 34) B. alboglabra Bailey (2n = 18) B. cernua Forbes & Hemsl. (2n = 20)	178—184
Celery cabbage	Brassica pekinesis (Lour.) Rupr. (2n = 20)	B. carinata A. Br. (2n = 34) Raphanus sativus L. (2n = 18)	178—184
Endive	Cichorium endiva L. (2n = 18)	C. intybus L.	185
Artichoke	Cynara scolymus L. (2n = 34)	C. cardunculus L. C. syriaca Boiss.	186
Lettuce	Lactuca sativa L. (2n = 18)	L. serriola L. L. saligna L.* L. virosa L.*	187

Parsley	*Petroselinum hortense* Hoffm. (2n = 22)	
	Apium graveolens L. (2n = 22)	175, 176
Sugarcane	*Saccharum officinarum* L. (2n = 80)	
	Miscanthus sinensis Anderss. (2n = 35 to 41)	16, 42, 188, 189
	S. barberi Jesweit	
	S. robustum Brandis (2n = 60,80)	
	S. sinense Roxb.	
	S. spontaneum L. (2n = 64 to 124)	
	Sorghum bicolor (L.) Moench (2n = 20)	

a Introgression means the production of a backcross, F_2, or maintenance through reproductive apomixis. Species crosses in which one parent or the F_1 had to be artificially doubled are indicated with an asterisk (*). The somatic number of the wild species is given only when it differs from the cultivated.

b Species names used here are recognized as subspecies by certain authorities.

Table 7
INTROGRESSIVE HYBRIDIZATION IN FRUIT AND VEGETABLE CROPS

Common name	Species	Species introgressed with the cultivated species[a]	Ref.
Pineapple	*Ananas comosus* (L.) Merr. (2n = 50)	*A. ananassoides* (Bak.) Smith *A. bracteatus* (Lindl.) Schultes *A. erectifolius* L. B. Smith *Pseudananas macrodontes* Morr. (2n = 100)	190—192
American pawpaw	*Asimina triloba* (L.) Dunal (2n = 18)	*A. longifolia* Kral. *A. obovata* (Wild.) Nash	193
Sweet pepper	*Capsicum annuum* L. (2n = 24)	*C. baccatum* H.B.K.* *C. chinense* Jacq. *C. frutescens* L. *C. pendulum* Willd.*	185, 194—196
Watermelon	*Citrullus lanatus* (Thunb.) Mansf.	*C. colocynthis* Ludw.	197
Citrus	*Citrus* spp. (2n = 18, 36)	*Ponicirus trifoliata* (L.) Raf.	198, 199
Cantelope	*Cucumis melo* L. (2n = 24)	*C. humifructus* Stent. *C. metuliferus* E. Mey. *C. sagittatus* Peyr	200, 201
Cucumber	*Cucumis sativus* L. (2n = 14)	*C. sativus* L. *C. dinteri* Cogn. *C. hardwickii* Royle *C. melo* L. *C. sagittatus* Peyr	200, 201
Buffalo gourd	*Cucurbita foetidissima* H.B.K. (2n = 40)	*C. moschata* Duch.	202
Squash	*Cucurbita maxima* Duch. (2n = 40)	*C. lundelliana* Bailey *C. moshata* Duch.	45, 203—206
Cushaw	*Cucurbita moschata* Duch. (2n = 40)	*C. lundelliana* Bailey *C. maxima* Duch.	45, 203—206
Pumpkin	*Cucurbita pepo* L. (2n = 40)	*C. ecuadorensis* *C. maxima* Duch. *C. moschata* Duch.	45, 203—206
Fig	*Ficus carica* L. (2n = 26)	*F. afghanistanica* *F. palmata* Forsk. *F. pseudo-carica* Miq.	207, 208

Strawberry	*Fragaria* × *ananassa* Duch. (2n = 56)	
	F. pumila L.	
	F. chiloensis (L.) Duch.	209—214
	F. moschata Duch. (2n = 42)*	
	F. moupinensis (Franck) Card. (2n = 28)	
	F. nubicola Los. (2n = 28)	
	F. orientalis A. Los. (2n = 28)	
	F. ovalis (Lehm.) Rydb.*	
	F. vesca L. (2n = 14)*	
	F. viridis Duch. (2n = 14)*	
	F. virginiana Duch.	
	Potentilla palustris (2n = 98)*	
Okra	*Hibiscus esculentus* L. (2n = 72)	215
Tomato	*Lycopersicon esculentum* Mill. (2n = 24)	1, 12, 45
	L. cheesmannii Riley	216—218
	L. chilense	
	L. chmielewskii	
	L. hirsutum Humb. et Bonpl.	
	L. parviflorum	
	L. peruvianum Mill.	
	L. pimpinellifolium Dun.	
	Solanum lycopersicoides Dun. (2n = 24)	
	S. pennelli (2n = 24)	
Apple	*Malus pumila* Mill. (2n = 34,51,68)	219, 220
	M. astrosanguinea Schneid.	
	M. baccata (L.) Borkh.	
	M. floribunda Sieb.	
	M. micromalus Mak.	
	M. prunifolia Borkh.	
	M. robusta Rehd.	
	M. zumi Rehd.	
Banana	*Musa paradisiaca* L. (2n = 33)	221, 222
	M. acuminatan Colla (2n = 22)	
	M. balbisiana Colla (2n = 22)	
	M. chiliocarpa (2n = 22)	
	M. rosacea Jacq. (2n = 22)	
Maracuja	*Passiflora alata* Dryand. (2n = 18)	223, 224
	P. coerulea L.	
	P. edulis Sims	
	P. quadrangularis L.	
	P. racemosa Brot.	
	P. raddiana	

Table 7 (continued)
INTROGRESSIVE HYBRIDIZATION IN FRUIT AND VEGETABLE CROPS

Common name	Species	Species introgressed with the cultivated species[a]	Ref.
Passiflora	Passiflora ligularis Juss. (2n = 18)	P. laurifolia L.	223, 224
		P. maliformis L.	
Passiflora	Passiflora edulis Sims (2n = 18)	P. alata Dryand.	223, 224
		P. cincinnata Mast.	
		P. incarnata L.	
Avocado	Persea americana Mill. (2n = 24)	P. floccosa Mez.	225
Sweet cherry	Prunus avium L. (2n = 16)	P. cerasus L. (2n = 32)	226
		P. fruticosa Pall. (2n = 32)	
Sour cherry	Prunus cerasus L. (2n = 32)	P. avium L. (2n = 16)	226
		P. fruticosa Pall. (2n = 32)	
Plum	Prunus domestica L. (2n = 48)	P. insititia L. (2n = 48)	227, 228
		P. spinosa L.	
Peach	Prunus persica (L.) Batsh. (2n = 16)	P. amygdalus Stokes	229
		P. besseyi Bailey	
		P. davidiana (Carr) French	
		P. kansuensis Rehd.	
		P. mira Koehne	
		P. salicina Lindl.	
		P. spinosa L.	
		P. tenella Batsh.[b]	
Pear	Pyrus communis L. (2n = 34)	P. pyrifolia (Burm) Nak.	230, 231
		P. serotina Ehrh.	
		P. ussuriensis Max.	
Black currant	Ribes nigrum L. (2n = 16)	R. bracteosum Dougl.	1, 232, 233
		R. cereum Dougl.	
		R. dicuscha Fisch.	
		R. glutinosum Benth.	
		R. hudsonianum Rich.	
		R. niveum Lindl.	
		R. petiolare Dougl.	
		R. sanguineum Pursh	
		R. ussuriense Jancz	

Red currant	*Ribes sativum* Syme (2n = 16)	*R. multiflorum* Kit.	1, 232, 233
		R. petraeum Wulf.	
Gooseberry	*Ribes grossularia* L. (2n = 16)	*R. aciculare* Sm.	1, 232, 233
		R. alpestre Dcne.	
		R. cynosbati L.	
		R. divaricatum Dougl.	
		R. gracile	
		R. hirtellum Michx.	
		R. inermis	
		R. irriguum Dougl.	
		R. missouriense Nutt.	
		R. niveum Lindl.*	
		R. non-scriptum	
		R. oxyancanthoides L.	
		R. pinetorum Greene	
		R. rotundifolium	
		R. sanguineum Pursh.	
		R. stenocarpum Maxim.	
Raspberries	*Rubus idaeus* L. (2n = 14)	*R. biflorus* Buch-Ham.	234
		R. coreanus Miq.	
		R. kuntzeanus Hemsl.	
		R. occidentalis L.	
		R. parvifolius Nutt.	
		R. phoenicolasius Maxim.	
Blackberries	*Rubus occidentalis* L. (2n = 14)	*R. albescens*	234
		R. crataegifolius Bunge.	
		R. idaeus L.	
		R. leucodermis Dougl.	
Eggplant	*Solanum melongena* L. (2n = 24)	*S. incanum* L.	235, 236, 237
		S. indicum L.	
		S. integrifolium Poir.	
		S. khasianum Clarke	
		S. macrocarpon L.	
		S. sodomaeum L.	

Table 7 (continued)
INTROGRESSIVE HYBRIDIZATION IN FRUIT AND VEGETABLE CROPS

Common name	Species	Species introgressed with the cultivated species[a]	Ref.
Highbush blueberry	*Vaccinium corymbosum* L. (2n = 48)	*V. angustifolium* (2n = 48) *V. ashei* Reade (2n = 72)* *V. atrococcum* Heller (2n = 24)* *V. darrowi* Camp (2n = 24) *V. tenellum* Ait. (2n = 24)*	238, 239
Muscadine grape	*Muscadinia rotundifolia* Small (2n = 40)	*Vitis vinifera* L. (2n = 38)*	240—242
Grape	*Vitis vinifera* L. (2n = 38)	*Muscadinia rotundifolia* Small (2n = 40) *V. aestivalis* Michx. *V. amurensis* Rupr. *V. berlandieri* Planch. *V. bicolor* Leconte *V. californica* *V. candicans* Engelm. *V. cinerea* Engelm. *V. cordifolia* Michx. *V. labrusca* L. *V. lincecumii* Buckl. *V. longii* *V. monticola* Buckl. *V. riparia* Michx. *V. rupestris* Scheele	240—242

[a] Introgression means the production of a backcross, F_2, or maintenance through reproductive apomixis. Species crosses in which one parent or the F_1 had to be artificially doubled are indicated with an asterisk (*). The somatic number of the wild species is given only when it differs from the cultivated.

[b] Syn. *P. nana* Stokes.

Table 8
INTROGRESSIVE HYBRIDIZATION IN NUT CROPS

Common name	Species	Species introgressed with the cultivated species[a]	Ref.
Pecan	*Carya illinoensis* (Wang) Koch (2n = 32)	*C. cordiformis* (Wang) Koch	243
		C. laciniosa Loud.	
		C. ovata Koch	
Chestnut	*Castanea dentata* Borkh. (2n = 24)	*C. ashei* Sudw.	244, 245
		C. crenata Sieb. et Zucc.	
		C. henryi Rehd. & Wils.	
		C. mollissima	
		C. orzarkensis Ashe.	
		C. pumila Mill.	
		C. sativa Mill.	
		C. seguinii Dode	
Filbert	*Corylus avellana* L. (2n = 28)	*C. colurna* L.	246—248
		C. jacquemontii	
Black walnut	*Juglans nigra* L. (2n = 32)	*J. californica* Wats.	249
		J. cineria L.	
		J. hindsii Jeps.	
		J. major Heller	
		J. microcarpa Berlander	
		J. regia L.	
		J. sieboldiana Maxim.	
Macadamia	*Macadamia tetraphylla* (L.) Johnson (2n = 28)	*M. integrifolia* Maid. et Betche	250
Almond	*Prunus amygdalus* Batsch. (2n = 16)	*P. argentea* (Lam) Rehd.	251
		P. bucharia Fed.	
		P. fenzliana Fritch	
		P. persica (L.) Batsch.	
		P. tangutica Batal.	
		P. webbi Spach.	
Oak	*Quercus macranthera* Fisch. et May (2n = 24)	*Q. alba* L.	252
		Q. macrocarpa Michx.	
		Q. robur L.	
		Q. rubra L.	

[a] Introgression means the production of a backcross, F_2, or maintenance through reproductive apomixis. Species crosses in which one parent or the F_1 has to be artificially doubled are indicated by an asterisk (*). The somatic number of the wild species is given only when it differs from that of the cultivated.

Table 9
INTROGRESSIVE HYBRIDIZATION IN DRUG CROPS

Common name	Species	Species introgressed with the cultivated species[a]	Ref.
Belladonna	*Atropa belladonna* L. (2n = 72)	*A. acuminata* Royle ex Lindl.	253, 254
		A. boetica Willk.	
		A. lutea	
Coffee	*Coffea arabica* L. (2n = 44)	*C. canephora* Pierre (2n = 88)	255—261
		C. liberica Bull.	
		C. racemosa	
		C. resinosa (2n = 22)	
		C. robusta Linden	
		C. stenophylla G. Don	
Coffee	*Coffea canephora* Pierre (2n = 88)	*C. arabica* L.	255—261
		C. eugenioides (2n = 22)	
		C. kianjavatensisi	
Tobacco	*Nicotiana tabacum* L. (2n = 48)	*N. alata* Link et Otto (2n = 18)	1, 16, 262—269
		N. attenuata Torr. (2n = 24)	
		N. benavidesii Goodsp. (2n = 24)	
		N. benthamiana Domin. (2n = 38)	
		N. bigelovi (Torr.) Wats.	
		N. debneyi Domin.	
		N. glauca Grah. (2n = 24)	
		N. glutinosa L. (2n = 24)	
		N. goodspeedii Wheeler (2n = 40)	
		N. gossei Domin. (2n = 36)*	
		N. langsdorffi Wein. (2n = 18)	
		N. longiflora Cav. (2n = 20)	
		N. megalosiphon Heurck et Muell. (2n = 40)	
		N. nesophila Johnst.	
		N. occidentalis Wheller (2n = 42)*	
		N. otophora Griseb. (2n = 24)*	
		N. paniculata L. (2n = 24)	
		N. plumbaginifolia Viv. (2n = 20)*	
		N. repanda Willd.	
		N. rustica L.	
		N. setchellii Goodsp. (2n = 24)	
		N. stocktonii Brandg.	
		N. suaveolens Lehm. (2n = 32)	
		N. sylvestris Speg. et Comes (2n = 24)	
		N. tomentosa Ruiz et Pav. (2n = 24)	
		N. tomentosiformis Goodsp. (2n = 24)*	
		N. trigonphylla Dun. (2n = 24)	
		N. undulata Vent. (2n = 24)	
Opium	*Papaver somniferum* L. (2n = 22)	*P. orientale* L. (2n = 14)*	270—271
		P. setigerum D.C. (2n = 44)	

[a] Introgression means the production of a backcross, F_2, or maintenance through reproductive apomixis. Species crosses in which one parent or the F_1 had to be artificially doubled are indicated by an asterisk (*). The somatic number of the wild species is given only when it differs from the cultivated.

Table 10
INTROGRESSIVE HYBRIDIZATION IN FIBER CROPS

Common name	Species	Species introgressed with the cultivated species[a]	Ref.
Sisal	*Agave sisalana* Perrine (2n = 138)	*A. cantala* Roxburgh (2n = 90)	272
Sisal	*A. fourcroydes* Lem. (2n = 140)	*A. sisalana* Perrine (2n = 138)	272
Sisal	*A. amaniensis* Trelease & Nowell (2n = 60)	*A. angustifolia* Haw. (2n = 60)	272
Hemp	*Cannabis sativa* L. (2n = 20)	*C. ruderalis* Janisch.	273
Jute	*Corchorus olitorius* L. (2n = 14)	*C. aestuans* L.	274—277
		C. capsularis L.	
		C. depressus Christensen	
Cotton	*Gossypium hirsutum* L. (2n = 52)	*G. anomalum* Wawr & Peyr (2n = 26)*	278—283
		G. arboreum L. (2n = 26)*	
		G. amourianum Kearney (2n = 26)*	
		G. barbadense L.	
		G. davidsonii Kell. (2n = 26)	
		G. harknessii Brandg. (2n = 26)*	
		G. herbaceum L. (2n = 26)*	
		G. longicalyx Hutch. & Lee (2n = 26)*	
		G. raimondii Ulb (2n = 26)*	
		G. stocksii Mast (2n = 26)*	
		G. sturtianum J. H. Willis (2n = 26)*	
		G. thurberi Tod. (2n = 26)*	
		G. tomentosum Nutt.	
		Hibiscus rosasinensis L. (2n = 36,46,72)	
Kenaf	*Hibiscus cannabinus* L. (2n = 36,72)	*H. radiatus* (2n = 72)	284, 285
		H. sabdariffa L. (2n = 36,72)	
Musa	*Musa textilis* Nee (2n = 20)	*M. balbisiana* Colla (2n = 22)	286
Sansevieria	*Sansevieria trifasciata* Prain (2n = 36,40)	*S. deserti* N.E. Brown	287

[a] Introgression means the production of a backcross, F_2, or maintenance through reproductive apomixis. Species crosses in which one parent or the F_1 had to be artificially doubled are indicated with an asterisk (*). The somatic number of the wild species is given only when it differs from the cultivated one.

Table 11
INTROGRESSIVE HYBRIDIZATION IN LATEX, RUBBER, OIL, AND WAX CROPS

Common name	Species	Species introgressed with the cultivated species[a]	Ref.
Tung	*Aleurites fordii* Hemsl. (2n = 22)	*A. montana* (Lour.) Wils.	288
Safflower	*Carthamus tinctorius* L. (2n = 24)	*C. curdicus* Hanelt	289
		C. gypsicolur Il.	
		C. oxyacanthus Bieb.	
		C. palaestinus Eig.	
		C. persicus Willd.	
Crambe	*Crambe abyssinica* Hoch. (2n = 90)	*C. hispanica* L. (2n = 30,60)	290
Crepis	*Crepis alpina* L. (2n = 10)	*C. syriaca* Bornm.	291
Oil palm	*Elaeis guineensis* Jacq. (2n = 32)	*E. melanococca* Gaertner	292—294
		E. oleifera (HBK) Cortes	
Sunflower	*Helianthus annuus* L. (2n = 34)	*H. agrophyllus* Torr. et A. Gray	45, 162
		H. bolanderi	295—297
		H. debilis	
		H. decapetalus L.	
		H. giganteus L.*	
		H. grosseserratus Martens	
		H. hirsutus Rafin. (2n = 68)	
		H. maximilianii Schrad.*	
		H. paradoxus	
		H. petiolaris Nutt.	
		H. rigidus (Cass.) Desf. (2n = 102)	
		H. ruderalis Wenzl.	
		H. tuberosus L. (2n = 102)	
Rubber	*Hevea brasiliensis* Muell. Arg. (2n = 36)	*H. benthamiana* Muell. Arg.	298
		H. pauciflora Spruce ex Benth.	
Flax	*Linum usitatissimum* L. (2n = 30)	*L. africanum* L.	299, 300
		L. angustifolium Huds.	
		L. corymbiferum Desf.	
		L. decumbens Desf.	
		L. hispanicum Mill.	
		L. humile Mill.	
		L. marginale A.Cunn.	
		L. nervosum Waldst.	
		L. pallescens Ledeb.	
		L. tenue Mmnby.	
Sesame	*Sesamum indicum* L. (2n = 26)[b]	*S. malabaricum**	301
		*S. schenckii**	
Dandelion	*Taraxacum officinale* Web. (2n = 24)	*T. kok-saghyz* Rodin (2n = 16)	302

[a] Introgression means the production of a backcross, F_2, or the maintenance through reproductive apomixis. Species crosses in which one parent or the F_1 had to be artificially doubled are indicated with an asterisk (*). The somatic number of the wild species is given only when it differs from the cultivated.

[b] Syn. *S. alatum* and *S. capense*.

Table 12
INTROGRESSIVE HYBRIDIZATION IN SPICE AND PERFUME CROPS

Common name	Species	Species introgressed with the cultivated species[a]	Ref.
Mint	*Mentha arvensis* L. (2n = 72,96)	*M. piperita* L.	303, 304
Mint	*Mentha longifolia* (L.) Huds. (2n = 48)	*M. spicata* (L.) Huds.	303, 304
Basil	*Ocimum basilicum* L. (2n = 16)	*O. sanctum* L.	305
Oregano	*Origanum vulgare* L. (2n = 30)	*O. majorana*[b]	306
Rose oil	*Rosa damascena* L. (2n = 14)	*R. gallica* L.	307
		R. moschata	
		R. phoenicea Boiss.	
		R. rugosa Thunb.	
Vanilla	*Vanilla fragrans* (Salisb.) Ames (2n = 32)	*V. haapape* (2n = 64)	308

[a] Introgression means the production of a backcross, F_2, or maintenance through reproductive apomixis. Species crosses in which one parent or the F_1 had to be artificially doubled are indicated with an asterisk (*). The somatic number of the wild species is given only when it differs from the cultivated.

[b] Syn. *Majorana hortensis* Moench.

REFERENCES

1. **Knott, D. R. and Dvořák, J.**, Alien germplasm as a source of resistance to disease, *Annu. Rev. Phytopathol.*, 14, 211, 1976.
2. **Wagenheim, K.-H. F. von**, Zur ursache der Kreazungsschwiergkeiten zwischen *Solanum tuberosum* L. und *S. acaule* Bitt. bzw. *S. stoloniferum* Schlechtd. et. Bouche, *Z. Pflanzenzuecht.*, 34, 7, 1955.
3. **Taylor, N. L., Anderson, M. K., Quesenberry, K. H., and Watson, L.**, Doubling the chromosome number of *Trifolium* species using nitrous oxide, *Crop Sci.*, 16, 516, 1976.
4. **Hougas, R. W. and Peloquin, S. J.**, Exploitation of *Solanum* germplasm, in *The Potato and Its Wild Relatives*, Correll, D. S., Ed., Texas Research Foundation, Renner, Texas, 1962, 21.
5. **Burk, L. G.**, An interspecific bridge cross. *Nicotiana repanda* through *N. sylvestris* to *N. tabacum*, *J. Hered.*, 58, 215, 1967.
6. **Kerber, E. R. and Dyck, P. L.**, Inheritance of stem rust resistance transferred from diploid wheat *(Triticum monococcum)* to tetraploid and hexaploid wheat and chromosome location of the gene involved, *Can. J. Genet. Cytol.*, 15, 397, 1973.
7. **Johnston, T. D.**, Transfer of disease resistance from *Brassica campestris* L. to rape *(B. napus* L.), *Euphytica*, 23, 681, 1974.
8. **Harlan, J. R. and de Wet, J. M. J.**, On O. Winge and a prayer: the origins of polyploidy, *Bot. Rev.*, 41, 361, 1975.
9. **Riley, R., Chapman, V., and Johnson, R.**, The incorporation of alien disease resistance in wheat by genetic interference with the regulation of meiotic chromosome synapsis, *Genet. Res.*, 12, 199, 1968.
10. **Sears, E. R.**, The transfer of leaf-rust resistance from *Aegilops umbellulata* to wheat, *Brookhaven Symp. Biol.*, 9, 1, 1956.
11. **Auerbach, C.**, *Mutation Research*, Chapman and Hall, New York, 1975, 1.
12. **Rick, C. M. and Butler, L.**, Cytogenetics of the tomato, *Adv. Genet.*, 8, 267, 1956.
13. **Kuriyama, T., Kuniyasu, K., and Mochizuki, H.**, Studies on the breeding of disease-resistant tomato by interspecific hybridization, *Bull. Hortic. Res. Stn. Jpn. Ser. B.*, 11, 33, 1971.
14. **Anon.**, Radical research and CIMMYT's role, in *CIMMYT Review 1974*, Mexico City, Mexico, 1974, 76.
15. **Matzk, F.**, Attempts to overcoming of incompatibilities in interspecific and intergeneric crosses with grasses, *Incompatibility Newslett.*, 7, 65, 1976.
16. **Stalker, H. T.**, Utilization of wild species for crop improvement, *Adv. Agron.*, 33, 111, 1981.
17. **Yeung, E. C., Thorpe, T. A., and Jensen, C. J.**, *In vitro* fertilization and embryo culture, in *Plant Tissue Culture, Methods and Applications in Agriculture*, Thorpe, T. A., Ed., Academic Press, New York, 1981, 253.
18. **Chu, Y. E. and Oka, H. I.**, The distribution and effects of genes causing F_1 weakness in *Oryza breviligulata* and *O. glaberrima*, *Genetics*, 70, 163, 1972.

19. **Mohinder, P. and Khoshov, T. N.**, Evolution and improvement of cultivated amaranths, *Theor. Appl. Genet.*, 43, 242, 1973.
20. **Lacadena, J.-R.**, Interspecific gene transfer in plant breeding, in *Interspecific Hybridization in Plant Breeding*, Proc. 8th. Congr. Eucarpia, Sanches-Monge, E. and Garcia-Olmedo, F., Eds., Escuela Técnica Superior de Ingenieros Agronómos, Madrid, 1977, 45.
21. **Ladizinsky, G. and Fainstein, R.**, Introgression between the cultivated hexaploid oat *A. sativa* and the tetraploid wild *A. magna* and *A. murphyi*, *Can. J. Genet. Cytol.*, 19, 59, 1977.
22. **Harlan, J. R.**, *Crops and Man*, American Society of Agronomy, Madison, 1975, 107.
23. **Rajhathy, T. and Thomas, H.**, *Cytogenetics of Oats (Avena L.)*, Misc. Pub. No. 2, Genetic Society of Canada, Ottawa, 1974, 1.
24. **Thomas, H., Haki, J. M., and Arangzeb, S.**, The introgression of characters of the wild oat *Avena magna* (2n = 4X = 28) into the cultivated oat *A. sativa* (2n = 6X = 42), *Euphytica*, 29, 391, 1980.
25. **Marshall, H. G.**, Buckwheat, in *Hybridization of Crop Plants*, Fehr, W. R. and Hadley, H. H., Eds., American Society of Agronomy, Madison, 1980, 215.
26. **Schooler, A. B. and Anderson, M. K.**, Interspecific hybrids between *(Hordeum brachyantherum* L. × *H. bogdanii* Wilensky) × *H. vulgare* L., *J. Hered.*, 70, 70, 1979.
27. **Starling, T. M.**, Barley, in *Hybridization of Crop Plants*, Fehr, W. R. and Hadley, H. H., Eds., American Society of Agronomy, Madison, 1980, 189.
28. **Coffman, W. R. and Herrera, R. M.**, Rice, in *Hybridization of Crop Plants*, Fehr, W. R. and Hadley, H. H., Eds., American Society of Agronomy, Madison, 1980, 511.
29. **Khush, G. S.**, Rice, in *Handbook of Genetics*, Vol. 2, King, R. C., Ed., Plenum Press, New York, 1974, 31.
30. **Nayar, N. M.**, Origin and cytogenetics of rice, *Adv. Genet.*, 17, 153, 1973.
31. **Shin, Y-B. and Katayama, T.**, Cytogenetical studies on the genus *Oryza*. XI. Alien addition lines of *O. sativa* with single chromosomes of *O. officinalis*, *Jpn. J. Genet.*, 54, 1, 1979.
32. **Yabuno, T.**, Genetic studies on the interspecific cytoplasm substitution lines of Japonica varieties of *Oryza sativa* L. and *O. glaberrima* Steud., *Euphytica*, 26, 451, 1977.
33. **Brunken, J., de Wet, J. M. J., and Harlan, J. R.**, The morphology and domestication of pearl millet, *Econ. Bot.*, 31, 163, 1977.
34. **Burton, G. W.**, Pearl millet, in *Hybridization of Crop Plants*, Fehr, W. R. and Hadley, H. H., Eds., American Society of Agronomy, Madison, 1980, 457.
35. **Hanna, W. W.**, Interspecific hybrids between pearl millet and fountain-grass, *J. Hered.*, 70, 425, 1979.
36. **Muldoon, D. K. and Pearson, C. J.**, The hybrid between *Pennisetum americanum* and *Pennisetum purpureum*, *Herb. Abstr.*, 49, 189, 1979.
37. **Pernês, J., Nguyen, É., Beninga, M., and Belliard, J.**, Analyse des relations génétiques entre formes spontanées et cultivées chez le Mil à chandelle *(Pennisetum americanum* (L.) Leeke, *P. mollissimum* Hochst), *Ann. Ameloir. Plant.*, 30, 253, 1980.
38. **Deodikar, G. B.**, *Rye Secale cereale L.*, Indian Council of Agricultural Research, New Delhi, 1963, 70.
39. **Evans, L. E. and Scoles, G. J.**, Cytogenetics, plant breeding and agronomy, in *Rye: Production, Chemistry and Technology*, Bushuk, W., Ed., American Association of Cereal Chemists, St. Paul, Minnesota, 1976, 13.
40. **Morey, D. D. and Barnett, R. D.**, Rye, in *Hybridization of Crop Plants*, Fehr, W. R. and Hadley, H. H., Eds., American Society of Agronomy, Madison, 1980, 249.
41. **Wojciechowska, B.**, Hybrid between *Hordeum jubatum* L. × *Secale cereale* and its backcross generations with rye. III. Meiosis in BC_1, *Genet. Pol.*, 22, 25, 1981.
42. **Gupta, S. C., Harlan, J. R., de Wet, J. M., and Grassl, C. O.**, Cytology of four individuals derived from a *Saccharum-Sorghum* hybrid, *Caryologia*, 29, 351, 1976.
43. **Schertz, K. F. and Dalton, L. G.**, Sorghum, in *Hybridization of Crop Plants*, Fehr, W. R. and Hadley, H. H., Eds., American Society of Agronomy, Madison, 1980, 577.
44. **Alderov, A. A.**, The genetics of plant height in some tetraploid wheat species, *Plant Breeding Abstr.*, 51, 9466, 1981.
45. **Brezhnev, D. D.**, The utilization of world plant gene pool of the USSR in distant hybridization, in *Interspecific Hybridization in Plant Breeding*, Proc. 8th. Congr. Eucarpia, Sanches-Monge, E. and Garcia-Olmedo, F., Eds., Escuela Técnica Superior de Ingenieros Agronómos, Madrid, 1977, 23.
46. **Fedak, G.**, Barley-wheat hybrids, in *Interspecific Hybridization in Plant Breeding*, Proc. 8th. Congr. Eucarpia, Sanches-Monge, E. and Garcia-Olmedo, F., Eds., Escuela Técnica Superior de Ingenieros Agronómos, Madrid, 1977, 261.
47. **Fedak, G. and Armstrong, K. C.**, Production of trigeneric (Barley × wheat) × rye hybrids, *Theor. Appl. Genet.*, 56, 221, 1980.
48. **Gerechter-Amitai, Z. K., Wahl, I., Vardi, A., and Zohary, D.**, Transfer of stem rust seedling resistance from wild diploid Einkorn to tetraploid *Durum* wheat by means of a triploid hybrid bridge, *Euphytica*, 20, 281, 1971.

49. **Kihara, H.,** Nucleus and chromosome substitution in wheat and *Aegilops*. I. Nucleus substitution, in *Proc. 2nd Int. Wheat Genet. Symp.,* MacKey, J., Ed., Berlingska Boktryckerict, Lund, Sweden, 1963, 313.
50. **Law, C. N., Worland, A. J., Chapman, V., and Miller, T. E.,** Homeologous chromosome transfers into hexaploid wheat and their influence on grain protein amounts, in *Interspecific Hybridization in Plant Breeding,* Proc. 8th. Congr. Eucarpia, Sanches-Monge, E. and Garcia-Olmedo, F., Eds., Escuela Técnica Superior de Ingenieros Agronomós, Madrid, 1977, 73.
51. **Maan, S. S.,** Cytoplasmic variability in *Triticinae,* in *Proc. 4th Int. Wheat Genet. Symp.,* Sears, E. R. and Sears, L. M. S., Eds., University of Missouri Press, Columbia, 1973, 367.
52. **Mujeeb-Kazi, A.,** Apomictic progeny derived from intergeneric *Hordeum-Triticum* hybrids, *J. Hered.,* 72, 284, 1981.
53. **Mujeeb-Kazi, A. and Rodriguez, R.,** An intergeneric hybrid of *Triticum aestivum* L. × *Elymus giganteus, J. Hered.,* 72, 253, 1981.
54. **Panagotov, I. and Gotsov, K.,** Interaction between nucleus of *Triticum aestivum* L. and cytoplasms of certain species of *Triticum* and *Aegilops,* in *Proc. 4th Int. Wheat Genet. Symp.,* Sears, E. R. and Sears, L. M. S., Eds., University of Missouri Press, Columbia, 1973, 381.
55. **Pienaar, R. de V.,** Genome relationships in wheat × *Agropyron distichum* (Thunb.) Beauv. hybrids, *Z. Pflanzenzuecht.,* 87, 193, 1981.
56. **Sears, E. R.,** The cytology and genetics of wheats and their relatives, *Adv. Genet.,* 2, 240, 1948.
57. **Sears, E. R.,** The wheats and their relatives, in *Handbook of Genetics,* Vol. 2, King, R. C., Ed., Plenum Press, New York, 1974, 59.
58. **Suemoto, H.,** The origin of the cytoplasm of tetraploid wheats, in *Proc. 3rd Int. Wheat Genet. Symp.,* Finlay, K. W. and Shepherd, K. W., Eds., Plenum Press, New York, 1968, 141.
59. **Tsuji, S.,** Cytoplasmic relationships among the *Aegilops* species having the D genome, *Plant Breeding Abstr.,* 51, 10394, 1981.
60. **Zohary, D., Harlan, J., and Vardi, A.,** The wild diploid progenitors of wheat and their breeding value, *Euphytica,* 18, 58, 1969.
61. **Larter, E. N. and Gustafson, J. P.,** Triticale, in *Hybridization of Crop Plants,* Fehr, W. R. and Hadley, H. H., Eds., American Society of Agronomy, Madison, 1980, 681.
62. **Pienaar, R. de V.,** Methods to improve the gene flow from rye and wheat to triticale, in *Proc. 4th Int. Wheat Genet. Symp.,* Sears, E. R. and Sears, L. M. S., Eds., University of Missouri Press, Columbia, 1973, 253.
63. **De Wet, J. M. J., Engle, L. M., Grant, C. A., and Tanaka, S. T.,** Cytology of maize-tripsacum introgression, *Am. J. Bot.,* 59, 1026, 1972.
64. **Mangelsdorf, P. C. and Reeves, R. G.,** Hybridization of maize, *Tripsacum* and *Euchlaena, J. Hered.,* 22, 329, 1931.
65. **Russell, W. A. and Hallauer, A. R.,** Corn, in *Hybridization of Crop Plants,* Fehr, W. R. and Hadley, H. H., Eds., American Society of Agronomy, Madison, 1980, 299.
66. **Elliott, W. A.,** Wild rice, in *Hybridization of Crop Plants,* Fehr, W. R. and Hadley, H. H., Eds., American Society of Agronomy, Madison, 1980, 721.
67. **MacKey, J.,** Species relationship in *Triticum,* in *Proc. 2nd Int. Wheat Genet. Symp.,* MacKey, J., Ed., Berlingska Boktryckerict, Lund, Sweden, 1963, 237.
68. **Hammons, R. O.,** Genetics of *Arachis hypogaea,* in *Peanuts — Culture and Uses,* American Peanut Research and Education Assoc., Inc., Stillwater, Okla., 1973, 135.
69. **Jay, P. R.,** Introgression of wild *Arachis* species into three botanical varieties of *Arachis hypogaea* L., *Plant Breeding Abstr.,* 52, 3263, 1982.
70. **Norden, A. J.,** Peanut, in *Hybridization of Crop Plants,* Fehr, W. R. and Hadley, H. H., Eds., American Society of Agronomy, Madison, 1980, 443.
71. **Peters, R. J., Simpson, C. E., and Smith, O. D.,** Fertility, cytological stainability and phenotypic traits of a backcross progeny of an *Arachis* amphiploid, *Crop Sci.,* 22, 357, 1982.
72. **Singh, A. K., Sastri, D. C., and Moss, J. P.,** Utilization of wild *Arachis* species at ICRISAT, in *Proc. Int. Workshop on Groundnuts,* Gibbons, R. W. and Merton, J. V., Eds., International Crops Research Institute for the Semi-Arid Tropics, Patancheru, India, 1980, 82.
73. **Smartt, J.,** Interspecific hybridization in the grain legumes — a review, *Econ. Bot.,* 33, 329, 1979.
74. **Spielman, I. V., Burge, A. P., and Moss, J. P.,** Chromosome loss and meiotic behaviour in interspecific hybrids in the genus *Arachis* L. and their implications in breeding for disease resistance, *Z. Pflanzenzuecht.,* 83, 236, 1979.
75. **Stalker, H. T., Wynne, J. C., and Company, M.,** Variation in progenies of an *Arachis hypogaea* × diploid wild species hybrid, *Euphytica,* 28, 675, 1979.
76. **Stalker, H. T.,** Department of Crop Science, North Carolina State University, Raleigh, personal communication, 1982.
77. **Sharma, D. and Green, J. M.,** Pigeonpea, in *Hybridization of Crop Plants,* Fehr, W. R. and Hadley, H. H., Eds., American Society of Agronomy, Madison, 1980, 471.

78. **Auckland, A. K. and van der Maesen, L. J. G.,** Chickpea, in *Hybridization of Crop Plants,* Fehr, W. R. and Hadley, H. H., Eds., American Society of Agronomy, Madison, 1980, 249.
79. **Ladizinsky, G.,** The origin of lentil and its wild genepool, *Euphytica,* 28, 179, 1979.
80. **Ladizinsky, G. and Alder, A.,** The origin of chickpea *Cicer arientinum* L., *Euphytica,* 25, 211, 1976.
81. **Ladizinsky, G. and Adler, A.,** Genetic relationships among the annual species of *Cicer* L., *Theor. Appl. Genet.,* 48, 197, 1976.
82. **Fehr, W. R.,** Soybean, in *Hybridization of Crop Plants,* Fehr, W. R. and Hadley, H. H., Eds., American Society of Agronomy, Madison, 1980, 589.
83. **Goshen, D., Ladizinsky, G., and Muehlbaur, F. J.,** Restoration of meiotic regularity and fertility among derivatives of *Lens culinaris* × *L. nigricans* hybrids, *Euphytica,* 31, 795, 1982.
84. **Muehlbaur, F. J., Slinkard, A. E., and Wilson, V. E.,** Lentil, in *Hybridization of Crop Plants,* Fehr, W. R. and Hadley, H. H., Eds., American Society of Agronomy, Madison, 1980, 417.
85. **Bliss, F. A.,** Common bean, in *Hybridization of Crop Plants,* Fehr, W. R. and Hadley, H. H., Eds., American Society of Agronomy, Madison, 1980, 273.
86. **Braak, J. P. and Kooistra, E.,** A successful cross between *Phaseolus vulgaris* L. and *Phaseolus ritensis* Jones with the aid of embryo culture, *Euphytica,* 24, 669, 1975.
87. **Rutger, J. N. and Beckham, L. S.,** Natural hybridization of *Phaseolus vulgaris* L. × *Phaseolus coccineus* L., *J. Am. Soc. Hortic. Sci.,* 95, 659, 1970.
88. **Gritton, E. T.,** Field pea, in *Hybridization of Crop Plants,* Fehr, W. R. and Hadley, H. H., Eds., American Society of Agronomy, Madison, 1980, 347.
89. **Ahn, C. S. and Hartmann, R. W.,** Interspecific hybridization between rice bean *[Vigna umbellata* (Thunb.) Ohwi and Ohashi] and adzuki bean [*Vigna angularis* (Willd.) Ohwi and Ohashi], *J. Am. Soc. Hortic. Sci.,* 103, 435, 1978.
90. **Singh, D. P.,** Breeding for resistance to diseases in greengram and blackgram, *Theor. Appl. Genet.,* 59, 1, 1981.
91. **Blackhurst, H. T. and Miller, J. C., Jr.,** Cowpea, in *Hybridization of Crop Plants,* Fehr, W. R. and Hadley, H. H., Eds., American Society of Agronomy, Madison, 1980, 327.
92. **Grof, B.,** Interspecific hybridization in *Centrosema:* hybrids between *C. brasilianum, C. virginianum* and *C. pubescens, Q. J. Agric. Anim. Sci.,* 27, 385, 1970.
93. **Taylor, N. L., Harlan, J. R., Buckner, R. C., Cope, W. A., and Craigmiles, J. P.,** Some approaches to breeding difficult to handle species, in *Rep. 16th South. Past. Imp. Conf.,* Mississippi State, Mississippi, 1959, 3.
94. **Townsend, C. E.,** Forage legumes, in *Hybridization of Crop Plants,* Fehr, W. R. and Hadley, H. H., Eds., American Society of Agronomy, Madison, 1980, 367.
95. **Bent, F. C.,** Interspecific hybridization in the genus *Lotus, Can. J. Genet. Cytol.,* 4, 151, 1962.
96. **Seaney, R. R. and Henson, P. R.,** Birdsfoot trefoil, *Adv. Agron.,* 22, 120, 1970.
97. **Somaroo, B. H. and Grant, W. F.,** Crossing relationships between synthetic *Lotus* amphiploids and *L. corniculatus, Crop Sci.,* 12, 103, 1972.
98. **Kazimierski, T.,** An interspecific hybrid in the genus *Lupinus (Lupinus albus* L. × *Lupinus jugoslavicus* Kazim.), *Genet. Pol.,* 1, 3, 1960.
99. **Majsurjan, N. A.,** The theoretical basis of plant industry, *Plant Breeding Abstr.,* 38, 1871, 1968.
100. **Pukhalskaya, N. F.,** Inheritance of alkaloid content in hybrids of *Lupinus angustifolius, Plant Breeding Abstr.,* 46, 2538, 1976.
101. **Atwood, S. S.,** Cytogenetics and breeding of forage crops, *Adv. Genet.,* 1, 1, 1947.
102. **Bingham, E. T.,** Department of Agronomy, University of Wisconsin, Madison, personal communication, 1982.
103. **Chernenko, E. G.,** Polyploidy in interspecific crosses of lucerne species differing in ploidy, *Plant Breeding Abstr.,* 48, 11740, 1978.
104. **Lesins, K.,** Interspecific crosses involving alfalfa. VII. *Medicago sativa* × *M. rhodopea, Can. J. Genet. Cytol.,* 14, 221, 1972.
105. **Lesins, K. and Gillies, C. B.,** Toxonomy and cytogenetics of *Medicago,* in *Alfalfa Science and Technology,* Hanson, C. H., Ed., American Society of Agronomy, Madison, 1972, 53.
106. **Sano, Y. and Kita, F.,** Reproductive barriers distributed in *Melilotus* species and their genetic bases, *Can. J. Genet. Cytol.,* 20, 275, 1978.
107. **Smith, W. K.,** Transfer from *Melilotus denta* to *M. alba* of the genes for reduction in coumarin content, *Genetics,* 33, 124, 1948.
108. **Smith, W. K. and Gorz, H. J.,** Sweetclover improvement, *Adv. Agron.,* 17, 161, 1965.
109. **Taylor, N. L.,** Clovers, in *Hybridization of Crop Plants,* Fehr, W. R. and Hadley, H. H., Eds., American Society of Agronomy, Madison, 1980, 261.
110. **Taylor, N. L., Stroube, W. H., Collins, G. B., and Kendall, W. A.,** Interspecific hybridization of red clover *(Trifolium pratense* L.), *Crop Sci.,* 3, 549, 1963.

111. **Schwer, J. F. and Cleveland, R. W.,** Diploid interspecific hybrids of *Trifolium pratense* L., *T. diffusum* Ehrh. and some related species, *Crop Sci.,* 12, 321, 1972.
112. **Evans, P. T. and Rupert, E. A.,** Department of Agronomy, Clemson University, Clemson, S.C., personal communication, 1982.
113. **Gibson, P. B. and Beinhart, G.,** Hybridization of *Trifolium occidentale* with two other species of clover, *J. Hered.,* 60, 93, 1969.
114. **Hovin, A. W.,** Interspecific hybridization between *Trifolium repens* L. and *T. nigrescens* Viv. and analysis of hybrid meiosis, *Crop Sci.,* 2, 251, 1962.
115. **Williams, E. G. and Verry, I. M.,** A partially fertile hybrid between *Trifolium repens* and *T. ambiguum, N. Z. J. Bot.,* 19, 1, 1981.
116. **Katznelson, J.,** Interspecific hybridization in *Trifolium, Crop Sci.,* 7, 307, 1967.
117. **Donnelly, E. D.,** Breeding hard seeded vetch using interspecific hybridization, *Crop Sci.,* 11, 721, 1971.
118. **Donnelly, E. D.,** Selecting lines of vetch that breed true for hard seed, *Crop Sci.,* 20, 259, 1980.
119. **Donnelly, E. D. and Clark, E. M.,** Hybridization in the genus *Vicia, Crop Sci.,* 2, 141, 1962.
120. **Donnelly, E. D. and Hoveland, C. S.,** Interspecific reseeding *Vicia* hybrids for use on summer perennial grass sods in southeastern U.S.A., *Proc. X Int. Grasslands Congr.,* Helsinki, 1966, 679.
121. **Ionushite, R.,** The influence of pollen irradiation on cross compatibility of some vetch species, *Plant Breeding Abstr.,* 44, 117, 1974.
122. **Yamamoto, K.,** On the hybrid between *Vicia sativa* (2n = 12) and *V. macrocarpa* (2n = 12), *Jpn. J. Breeding,* 18, 156, 1968.
123. **Asay, K. H. and Dewey, D. R.,** Bridging ploidy differences in crested wheatgrass with hexaploid × diploid hybrids, *Crop Sci.,* 19, 519, 1979.
124. **Dewey, D. R. and Pendse, P. C.,** Hybrids between *Agropyron deserotum* and induced-tetraploid *Agropyron cristatum, Crop Sci.,* 8, 607, 1968.
125. **Dewey, D. R.,** Morphology, cytology and fertility of *Agropyron podperae* and its hybrids with *A. intermedium, Crop Sci.,* 18, 315, 1978.
126. **Schulz-Schaeffer, J.,** A possible source of cytoplasmic male sterility in intermediate wheatgrass, *Agropyron intermedium* (Host) Beauv., *Crop Sci.,* 10, 204, 1970.
127. **Dewey, D. R.,** Derivation of a new forage grass from *Agropyron repens* × *Agropyron spicatum* hybrids, *Crop Sci.,* 16, 175, 1976.
128. **Dewey, D. R.,** Morphological, cytological and taxonomic relationships between *Agropyron repens* and *A. elongatiforme* (Gramineae), *Syst. Bot.,* 5, 61, 1980.
129. **Schaeffer, J. R.,** A source of male sterility in crested wheatgrass, *Plant Breeding Abstr.,* 51, 6234, 1981.
130. **Sieber, V. K. and Murray, B. G.,** The cytology of the genus *Alopecurus* (Gramineae), *Bot. J. Linn. Soc.,* 70, 343, 1979.
131. **Peters, L. V. and Newell, L. C.,** Hybridization between divergent types of big bluestem, *Andropogon gerardi* Vitman and Sand bluestem, *Andropogon halli* Hack., *Crop Sci.,* 1, 359, 1961.
132. **Harlan, J. R., Chheda, H. R., and Richardson, W. L.,** Range of hybridization with *Bothriochloa intermedia* (R. Br.) A. Camus, *Crop Sci.,* 2, 480, 1962.
133. **Armstrong, K. C.,** Chromosome pairing in hexaploid hybrids from *B. erectus* (2n = 28) × *B. inermis* (2n = 56), *Can. J. Genet. Cytol.,* 15, 427, 1973.
134. **Armstrong, K. C.,** A and B genome homologies in tetraploid and octoploid cytotypes of *Bromus inermis, Can. J. Genet. Cytol.,* 21, 65, 1979.
135. **Elliott, F. C.,** The cytology and fertility relations of *Bromus inermis* and some of its relatives, *Agron. J.,* 41, 298, 1949.
136. **Nielsen, E. L., Drolsom, P. N., and Jalal, S. M.,** Analysis of F_2 progeny from *Bromus* species hybrids, *Crop Sci.,* 2, 459, 1962.
137. **Read, J. C. and Bashaw, E. C.,** Cytoplasmic relationships and the role of apomixis in speciation in Buffelgrass and Birdwoodgrass, *Crop Sci.,* 9, 805, 1969.
138. **Breese, E. L. and Davies, W. E.,** Herbage breeding, in *Welsh Plant Breeding Station Report, 1972,* Aberystwyth, 1972, 25.
139. **Buxton, B. H. and Newton, F. C.,** Hybrids of *Digitalis ambigua* and *Digitalis purpurea,* their fertility and cytology, *J. Genet.,* 19, 269, 1928.
140. **Carnahan, H. L. and Hill, H. D.,** Cytology and genetics of forage grasses, *Bot. Rev.,* 27, 1, 1961.
141. **Dewey, D. R.,** Cytogenetics of *Elymus angustus* and its hybrids with *Elymus giganteus, Elymus cinereus* and *Agropyron repens., Bot. Gaz.,* 133, 57, 1972.
142. **Berg, C. C., Webster, G. T., and Jauhar, P. P.,** Cytogenetics and genetics, in *Tall Fescue,* Buckner, R. C. and Bush, L. P., Eds., American Society of Agronomy, Madison, 1979, 93.
143. **Terrell, E. E.,** Taxonomic implications of genetics in ryegrass *(Lolium), Bot. Rev.,* 32, 138, 1966.
144. **Bennett, H. W. and Bashaw, E. C.,** An interspecific hybrid in *Paspalum, J. Hered.,* 51, 81, 1960.
145. **McWilliam, J. R.,** Interspecific hybridization in *Phalaris:* hybrids between *Phalaris tuberosa* and the hexaploid race of *Phalaris arundinacea, Aust. J. Agric. Res.,* 13, 585, 1962.

146. **Nath, J.,** Cytogenetical and related studies in the genus *Phleum* L., *Euphytica,* 16, 267, 1967.
147. **Clausen, J.,** Introgression facilitated by apomixis in polyploid *Poa, Euphytica,* 10, 87, 1961.
148. **Dijk, G. E., van**, Wild species for the breeding of grasses, in *Proc. Conf. Broadening Genet. Base Crops,* Zeven, A. C. and Harten, A. M. van, Eds., Center for Agricultural Publishing and Documentation, Wageningen, Netherlands, 1978, 211.
149. **Williamson, C. J.,** Scottish Crop Research Inst., Pentlandfield, personal communication, 1982.
150. **Williamson, C. J. and Watson, P. J.,** Production and description of interspecific hybrids between *Poa pratensis* and *P. longifolia, Euphytica,* 29, 715, 1980.
151. **Casady, A. J. and Anderson, K. L.,** Hybridization, cytological and inheritance studies of a sorghum cross-autotetraploid sudangrass × (Johnsongrass × 4n sudangrass), *Agron. J.,* 44, 189, 1952.
152. **Forbes, I. F.,** Chromosome numbers and hybrids in *Zoysia, Agron. J.,* 44, 194, 1952.
153. **Dolezel, J., Novak, F. J., and Luzny, J.,** Embryo development and *in vitro* culture of *Allium cepa* and its interspecific hybrids, *Z. Pflanzenzuecht.,* 85, 177, 1980.
154. **Saini, S. S. and Davis, G. N.,** Male sterility in *Allium cepa* and some species hybrids, *Econ. Bot.,* 23, 37, 1969.
155. **Cleij, G., De Bock, T. S. M., and Lekkerkerker, B.,** Crosses between *Beta intermedia* Bunge and *B. vulgaris* L., *Euphytica,* 17, 11, 1968.
156. **Cleij, G., De Bock, T. S. M., and Lekkerkerker, B.,** Crosses between *Beta vulgaris* L. and *Beta lomatogona* F. et M., *Euphytica,* 25, 539, 1976.
158. **Dalke, C.,** Interspecific hybrids between sugar beet and *Beta corolliflora* of the *Corollinae* section, in *Interspecific Hybridization in Plant Breeding,* Proc. 8th. Congr. Eucarpia, Sanches-Monge, E. and Garcia-Olmedo, F., Eds., Escuela Técnica Superior de Ingenieros Agronómes, Madrid, 1977, 113.
159. **Smith, G. A.,** Sugarbeet, in *Hybridization of Crop Plants,* Fehr, W. R. and Hadley, H. H., Eds., American Society of Agronomy, Madison, 1980, 601.
160. **McCollum, G. D.,** Hybrids of *Daucus gingidium* with cultivated carrots *(D. carota* subsp. *sativus)* and *D. capillifolius, Bot. Gaz.,* 138, 56, 1977.
161. **Anon.,** Nigeria International Institute of Tropical Agriculture, Annual Report, 1974, *Plant Breeding Abstr.,* 47, 10088, 1977.
162. **Dedio, W. and Putt, E. D.,** Sunflower, in *Hybridization of Crop Plants,* Fehr, W. R. and Hadley, H. H., Eds., American Society of Agronomy, Madison, 1980, 631.
163. **Heiser, C. B. and Smith, D. M.,** Species crosses in *Helianthus.* II. Polyploid species, *Rhodora,* 66, 344, 1964.
164. **Schilling, E. E. and Heiser, C. B.,** Infrageneric classification of *Helianthus (Compositae), Taxon,* 30, 393, 1981.
165. **Austin, D. F.,** Hybrid polyploids in *Ipomea* section *batatas, J. Hered.,* 68, 259, 1977.
166. **Kawano, K.,** Cassava, in *Hybridization of Crop Plants,* Fehr, W. R. and Hadley, H. H., Eds., American Society of Agronomy, Madison, 1980, 225.
167. **Nassar, N. M. A.,** Attempts to hybridize wild *Manihot* species with cassava, *Econ. Bot.,* 34, 13, 1980.
168. **Cauderon, Y.,** Allopolyploidy, in *Interspecific Hybridization in Plant Breeding,* Proc. 8th. Congr. Eucarpia, Sanches-Monge, E. and Garcia-Olmedo, F., Eds., Escuela Técnica Superior de Ingenieros Agronómes, Madrid, 1977, 131.
169. **Kato, M. and Tokumasu, S.,** Nucleus substitution of *Brassica japonica* Sieb with *Raphanus sativus* L. and its resultant chlorophyll deficiency, *Euphytica,* 29, 97, 1980.
170. **Kobabe, G.,** Natural crossing of wild radish *(Raphanus raphanistrum* L.) with cultivated radish *(R. sativus* var. *radicula* D.C.) and the behavior of tuber form and colour in the following F and BC generations, *Z. Pflanzenzuecht.,* 42, 1, 1959.
171. **Astley, D. and Hawkes, J. G.,** The nature of the Bolivian weed potato species *Solanum sucrense* Hawkes, *Euphytica,* 28, 685, 1979.
172. **Ramanna, M. S. and Hermsen, J. H. T. H.,** Unilateral 'eclipse sterility' in reciprocal crosses between *Solanum verrucosum* Schlechtd. and diploid *S. tuberosum* L., *Euphytica,* 23, 417, 1974.
173. **Ross, H.,** The use of wild *Solanum* species in German potato breeding of the past and today, *Am. Potato J.,* 43, 63, 1966.
174. **Fahselt, D., Ownbey, M., and Borton, M.,** Seed fertility in *Tragopogon* hybrids Compositae, *Am. J. Bot.,* 63, 1109, 1976.
175. **Honma, S. and Lacy, M. L.,** Hybridization between Pascal celery and parsley, *Euphytica,* 29, 801, 1980.
176. **Ryder, E. J.,** *Leafy Salad Vegetables,* AVI, Westport, Conn., 1979, 95 and 248.
177. **Bozzini, A.,** Interspecific hybridization and experimental mutagenesis in breeding *Asparagus, Genet. Agrar.,* 16, 212, 1963.
178. **Beversdorf, W. D., Weiss-Lerman, J., Erickson, L. R., and Souza-Machado, V.,** Transfer of cytoplasmically inherited triazine resistance from Birds rape to cultivated oilseed rape *(Brassica campestris* and *B. napus), Can. J. Genet. Cytol.,* 22, 167, 1980.

179. **Cai, Y. X. and Jiang, Z. Q.,** Synthesis of new species and cytogenetic studies in *Brassica*. IV. Studies on genome substitution engineering in *Brassica napus* and its two constituent species, *Plant Breeding Abstr.*, 52, 5282, 1982.
180. **Downey, R. K., Klassen, A. J., and Stringam, G. R.,** Rapeseed and mustard, in *Hybridization of Crop Plants,* Fehr, W. R. and Hadley, H. H., Eds., American Society of Agronomy, Madison, 1980, 495.
181. **Gland, A.,** Doubling chromosomes in interspecific hybrids by colchicine treatment, *Cruciferae Newslett.*, 6, 20, 1981.
182. **Hinata, K. and Konno, N.,** Studies on a male sterile strain having the *Brassica campestris* nucleus and the *Diplotaxis muralis* cytoplasm, *Jpn. J. Breeding,* 29, 305, 1979.
183. **Johnston, T. D.,.** Transfer of disease resistance from *Brassica campestris* L. to rape *(B. napus* L.), *Euphytica,* 23, 681, 1974.
184. **Yarnell, S. H.,** Cytogenetics of the vegetable crops. II. Crucifers, *Bot. Rev.,* 22, 81, 1956.
185. **Anon.,** Annual Report of the Institute for Horticulture Plant Breeding, 1979, Wageningen, Netherlands, *Plant Breeding Abstr.,* 51, 10968, 1981.
186. **Zohary, D. and Basnizky, J.,** The cultivated artichoke — *Cynara scolymus.* Its probable wild ancestors, *Econ. Bot.,* 29, 233, 1975.
187. **Whitaker, T. W.,** Salads for everyone — a look at the lettuce plant, *Econ. Bot.,* 23, 261, 1969.
188. **Grassl, C. O.,** Breeding *Andropogoneae* at the generic level for biomass, *Sugarcane Breeders Newslett.,* 43, 41, 1980.
189. **James, N. I.,** Sugarcane, in *Hybridization of Crop Plants,* Fehr, W. R. and Hadley, H. H., Eds., American Society of Agronomy, Madison, 1980, 617.
190. **Collins, J. L.,** History, taxonomy and culture of the pineapple, *Econ. Bot.,* 3, 335, 1949.
191. **Collins, J. L.,** *The Pineapple, Botany, Cultivation and Utilization,* Interscience, New York, 1960, 78.
192. **Pickersgill, B.,** Pineapple, in *Evolution of Crop Plants,* Simmonds, N. W., Ed., Longman, New York, 1976, 14.
193. **Zimmerman, G. A.,** Hybrids of the american pawpaw, *J. Hered.,* 32, 83, 1941.
194. **Boukema, I. W.,** Allelism of genes controlling resistance to TMV in *Capsicum* L., *Euphytica,* 29, 433, 1980.
195. **Dumas de Vaulx, D. and Pitrat, M.,** Realization of the interspecific cross between *Capsicum annuum* L. and *Capsicum baccatum,* in *Interspecific Hybridization in Plant Breeding,* Proc. 8th. Congr. Eucarpia, Sanches-Monge, E. and Garcia-Olmedo, F., Eds., Escuela Técnica Superior de Ingenieros Agronómos, Madrid, 1977, 327.
196. **Lippert, L. F., Smith, P. G., and Bergh, B. O.,** Cytogenetics of the vegetable crops. Garden pepper, *Capsicum sp., Bot. Rev.,* 32, 24, 1966.
197. **Fulks, B. K., Scheerens, J. C., and Bemis, W. P.,** Natural hybridization of two *Citrullus* species, *J. Hered.,* 70, 214, 1979.
198. **Cameron, J. W. and Soost, R. K.,** Leaf types of F_1 hybrids and backcrosses involving unifoliate *Citrus* and trifoliate *Poncirus, J. Am. Soc. Hortic. Sci.,* 105, 517, 1980.
199. **Soost, R. K. and Cameron, J. W.,** Citrus, in *Advances in Fruit Breeding,* Janick, J. and Moore, J. N., Eds., Purdue University Press, Layfayette, Ind., 1975, 507.
200. **Deakin, J. R., Bohn, G. W., and Whitaker, T. H.,** Interspecific hybridization in *Cucumis, Econ. Bot.,* 25, 195, 1971.
201. **Robinson, R. W. and Whitaker, T. W.,** *Cucumis,* in *Handbook of Genetics,* Vol. 2, King, R. C., Ed., Plenum Press, New York, 1974, 145.
202. **Bemis, W. P., Curtis, L. D., Weber, C. W., and Berry, J.,** The feral buffalo gourd, *Cucurbita foetidissima, Econ. Bot.,* 32, 87, 1978.
203. **Vaulx, R. D. de and Pitrat, M.,** Realization of the interspecific hybridization (F_1 and BC_1) between *Cucurbita pepo* and *C. ecuadorensis, Plant Breeding Abstr.,* 51, 11015, 1981.
204. **Whitaker, T. W.,** An interspecific cross in *Cucurbita: C. lundelliana* Bailey × *C. maxima* Duch., *Euphytica,* 11, 173, 1962.
205. **Whitaker, T. W.,** *Cucurbita,* in *Handbook of Genetics,* Vol. 2, King, R. C., Ed., Plenum Press, New York, 1974, 135.
206. **Whitaker, T. H. and Bohn, G. W.,** The taxonomy, genetics, production, and uses of the cultivated species of *Cucurbita, Econ. Bot.,* 4, 52, 1950.
207. **Arendt, N. L.,** Variability of characters in interspecific hybrids of fig, *Plant Breeding Abstr.,* 50, 5541, 1980.
208. **Storey, W. B.,** Figs, in *Advances in Fruit Breeding,* Janick, J. and Moore, J. N., Eds., Purdue University Press, Lafayette, Ind., 1975, 568.
209. **Asker, S.,** Some viewpoints on *Fragaria* × *Potentilla* intergeneric hybridization, *Hereditas,* 67, 181, 1971.
210. **Evans, W. D.,** The use of synthetic octoploids in strawberry breeding, *Euphytica,* 26, 497, 1977.

211. **Evans, W. D.,** Improving vegetatively propagated polyploids using induced polyploids of interspecific hybrids, in *Proc. Int. Symp. New Genetical Approaches to Crop Improvement,* Tadhojam, Sind, Pakistan, 1982, in press.
212. **Evans, W. D.,** University of Guelph, Ontario, Canada, personal communication, 1982.
213. **Scott, D. H. and Lawrence, F. J.,** Strawberries, in *Advances in Fruit Breeding,* Janick, J. and Moore, J. N., Eds., Purdue University Press, Lafayette, Ind., 1975, 71.
214. **Sukharevan, N. B. and Kipko, V. P.,** Cytogenetic characteristics of seedlings obtained from backcrosses of octoploid *Fragaria ananassa* × *F. orientalis* hybrids with *F. ananassa, Plant Breeding Abstr.,* 51, 10923, 1981.
215. **Arumugam, A. and Muthukrishnan, C. R.,** Gene effects on some quantitative characters in okra, *Indian J. Agric. Sci.,* 49, 602, 1979.
216. **Ponti, O. M. B., de and Hogenboom, N. G.,** Breeding tomato *(Lycopersicon esculentum)* for resistance to the greenhouse whitefly *(Trialeurodes vaporariorum),* in *Integrated Control of Insect Pests in the Netherlands,* Minks, A. K. and Gruys, P., Eds., Pudoc, Wageningen, Netherlands, 1980, 187.
217. **Rick, C. M.,** Tomato, in *Hybridization of Crop Plants,* Fehr, W. R. and Hadley, H. H., Eds., American Society of Agronomy, Madison, 1980, 669.
218. **Rick, C. M.,** University of California, Davis, personal communication, 1982.
219. **Brown, A. G.,** Apples, in *Advances in Fruit Breeding,* Janick, J. and Moore, J. N., Eds., Purdue University Press, Lafayette, Ind., 1975, 3.
220. **Shay, J. R., Dayton, D. F., and Hough, L. F.,** Apple scab resistance from a number of *Malus* species, *Proc. Am. Soc. Hortic. Sci.,* 62, 348, 1953.
221. **Naik, K. C., Muthuswamy, S., and Raman, V. S.,** Recent achievements in tropical fruit improvements in south India, *S. Indian Hortic.,* 7, 3, 1959.
222. **Raman, V. S., Alikhan, W. M., Manimekalai, G., and Bhakthavathsalu, C. M.,** A study of the cytomorphology of some banana hybrids, *Madras Agric. J.,* 58, 55, 1971.
223. **Martin, F. W. and Nakasone, H. Y.,** The edible species of *Passiflora, Econ. Bot.,* 24, 333, 1970.
224. **Ruberte-Torres, R. and Martin, F. W.,** First generation hybrids of edible protein fruit species, *Euphytica,* 23, 61, 1974.
225. **Bergh, B. O.,** Avocados, in *Advances in Fruit Breeding,* Janick, J. and Moore, J. N., Eds., Purdue University Press, Lafayette, Ind., 1975, 541.
226. **Fogle, H. W.,** Cherries, in *Advances in Fruit Breeding,* Janick, J. and Moore, J. N., Eds., Purdue University Press, Lafayette, Ind., 1975, 348.
227. **Tamassy, I.,** New ways of horticultural plant breeding, *Plant Breeding Abstr.,* 47, 5748, 1977.
228. **Weinberger, J. H.,** Plums, in *Advances in Fruit Breeding,* Janick, J. and Moore, J. N., Eds., Purdue University Press, Lafayette, Ind., 1975, 336.
229. **Hesse, C. O.,** Peaches, in *Advances in Fruit Breeding,* Janick, J. and Moore, J. N., Eds., Purdue University Press, Lafayette, Ind., 1975, 285.
230. **Cociu, V. and Thiesz, R.,** Studies on the production of some varieties of fruit crops resistant to unfavorable environmental conditions, *Plant Breeding Abstr.,* 49, 3032, 1979.
231. **Layne, R. E. C. and Quamme, H. A.,** Pears, in *Advances in Fruit Breeding,* Janick, J. and Moore, J. N., Eds., Purdue University Press, Lafayette, Ind., 1975, 38.
232. **Keep, E.,** Currants and gooseberries, in *Advances in Fruit Breeding,* Janick, J. and Moore, J. N., Eds., Purdue University Press, Lafayette, Ind., 1975, 197.
233. **Melekhina, A. A., Yankelevich, B. B., and Eglite, M. A.,** Self fertility in the progeny from crosses with *Ribes petiolare* Dougl., *Plant Breeding Abstr.,* 51, 746, 1981.
234. **Ourecky, D. K.,** Brambles, in *Advances in Fruit Breeding,* Janick, J. and Moore, J. N., Eds., Purdue University Press, Lafayette, Ind., 1975, 98.
235. **Schaff, D. A., Boyer, C. H., and Pollack, B. L.,** Interspecific hybridization of *Solanum melongena* L. × *S. macrocarpon, Hortic. Sci.,* 15, 419, 1980.
236. **Sharma, D. R., Chowdhury, J. B., Ahuja, U., and Dhankhar, B. S.,** Interspecific hybridization in genus *Solanum* : a cross between *S. melongena* and *S. khasianum* through embryo culture, *Z. Pflanzenzuecht.,* 85, 248, 1980.
237. **Yamakawa, K. and Mochizuki, H.,** Nature and inheritance of *Fusarium* wilt resistance in egg plant cultivars and related wild *Solanum* species, *Plant Breeding Abstr.,* 51, 6673, 1981.
238. **Draper, A. D.,** USDA, Beltsville Agricultural Research Center, Beltsville, Md., personal communication, 1982.
239. **Galletta, G. J.,** Blueberries and cranberries, in *Advances in Fruit Breeding,* Janick, J. and Moore, J. N., Eds., Purdue University Press, Lafayette, Ind., 1975, 154.
240. **Einset, J. and Pratt, C.,** Grapes, in *Advances in Fruit Breeding,* Janick, J. and Moore, J. N., Eds., Purdue University Press, Lafayette, Ind., 1975, 130.
241. **Olmo, H. P.,** Genetic problems and general methodology of breeding, in *Proc. II Symp. Int. sur l'Amelioration de la Vigne,* Institut National de la Recherche Agronomique, Paris, 1978, 1.

242. **Reisch, B.,** Department of Pomology and Viticulture, New York State Agricultural Experiment Station, Geneva, N.Y., personal communication, 1982.
243. **Madden, G. D. and Malstrom, H. L.,** Pecans and hickories, in *Advances in Fruit Breeding,* Janick, J. and Moore, J. N., Eds., Purdue University Press, Lafayette, Ind., 1975, 420.
244. **Anon.,** Chestnut blight and resistant chestnuts, *USDA Farmers Bull.* No. 2068, 1965, 1.
245. **Jaynes, R. A.,** Chestnuts, in *Advances in Fruit Breeding,* Janick, J. and Moore, J. N., Eds., Purdue University Press, Lafayette, Ind., 1975, 490.
246. **Farris, C. W.,** The trazels, in *North. Nut Growers Assoc. 59th Annu. Rep.,* East Lansing, Mich., 1968, 32.
247. **Farris, C. W.,** Inheritance of parental characteristics in filbert hybrids, *North. Nut Growers Assoc. 61st Annu. Rep.,* Ithaca, N.Y., 1970, 54.
248. **Lagerstedt, H. B.,** Filberts, in *Advances in Fruit Breeding,* Janick, J. and Moore, J. N., Eds., Purdue University Press, Lafayette, Ind., 1975, 456.
249. **Forde, H. I.,** Walnuts, in *Advances in Fruit Breeding,* Janick, J. and Moore, J. N., Eds., Purdue University Press, Lafayette, Ind., 1975, 439.
250. **Saleeb, W. F., Yermanos, D. M., Huszar, C. K., Storey, W. B., and Labanauskar, C. K.,** The oil and protein in nuts of *Macadamia tetraphylla* (L.) Johson, *Macadamia integrifolia* Maiden and Betche, and their F_1 hybrid, *J. Am. Soc. Hortic. Sci.,* 98, 453, 1973.
251. **Kester, D. E. and Asay, R. A.,** Almonds, in *Advances in Fruit Breeding,* Janick, J. and Moore, J. N., Eds., Purdue University Press, Lafayette, Ind., 1975, 387.
252. **Khmaldaze, S. I.,** Segregation characteristics of second generation hybrid forms of oak bred by S. S. Pyatnitskii, *Plant Breeding Abstr.,* 51, 10947, 1981.
253. **Albors Yodi, E.,** Contribution to the study and production of alkaloids in the genus *Atropa, Plant Breeding Abstr.,* 30, 1880, 1960.
254. **Dhar, A. K.,** Studies on chiasma frequency of *Atropa* species, *Sci. Cult.,* 36, 473, 1970.
255. **Costa, W. M.,** The relationship between degree of resistance to *Hemilea vastatrix* and yield in Icatu coffee, *Plant Breeding Abstr.,* 49, 8275, 1979.
256. **Ferwerda, F. P.,** Coffee breeding in Java, *Econ. Bot.,* 2, 258, 1948.
257. **Lanaud, C.,** Study of genetical problems posed in coffee by the introgression of characters of a wild species *(C. kianjavatensisi* Mascarocoffea) into the cultivated species *C. caneophora* (Eucoffea), *Plant Breeding Abstr.,* 50, 1413, 1980.
258. **Lanaud, C. and Zickler, D.,** Premieres informations sur la fertilité des hybrides pentaploides et hexaploides entre *C. arabica* (Eucoffea) et *C. resinosa* (Mascarocoffea), *Cafe, Cacao, The,* 24, 169, 1980.
259. **Medina Filho, H. P., Carvalho, A., and Medina, D. M.,** *Coffea racemosa* germplasm and its potential for coffee breeding, *Plant Breeding Abstr.,* 48, 11908, 1978.
260. **Medina Filho, H. P., Carvalho, A., and Monaco, L. C.,** Coffee breeding. XXXVII. Observations on the resistance of coffee trees to the leaf miner, *Plant Breeding Abstr.,* 48, 11912, 1978.
261. **Srinivasan, C. S., Suryakantha, R. K., and Vishweshwara, S.,** Pattern of fruit growth and development in interspecific hybrids of *Coffee caneophora* × *C. arabica, Plant Breeding Abstr.,* 50, 10582, 1980.
262. **Berbec, A.,** Chromosome substitution and gene mutation concurring to produce a white flowered derivative of the interspecific hybrid *Nicotiana tabacum* L. × *Nicotiana benavidesii* Goodspeed, *Genet. Pol.,* 21, 283, 1980.
263. **Burk, L. G. and Dean, C. E.,** Hybrid fertility and aphid resistance in the cross *Nicotiana tabacum* × *N. gossei, Euphytica,* 24, 59, 1975.
264. **Clayton, E. E.,** A wildfire resistant tobacco, *J. Hered.,* 38, 35, 1947.
265. **Gerstel, D. U.,** Cytoplasmic male sterility in *Nicotiana* (a review), *North Carolina Agr. Res. Ser. Tech. Bull. 263,* Raleigh, N.C., 1980.
266. **Reed, S. M. and Collins, G. B.,** Chromosome pairing relationships and black shank resistance in three *Nicotiana* interspecific hybrids, *J. Hered.,* 71, 423, 1980.
267. **Smith, H. H.,** Recent cytogenetic studies in the genus *Nicotiana, Adv. Genet.,* 14, 1, 1968.
268. **Ternovskii, M. F., Moiseeva, M. E., and Grebenkin, A. P.,** Experimentally produced new type of cytoplasmic male sterility in interspecies *Nicotiana* hybrids, *Sov. Genet.,* 9, 693, 1973.
269. **Wernsman, E. A. and Matzinger, D. F.,** Tobacco, in *Hybridization of Crop Plants,* Fehr, W. R. and Hadley, H. H., Eds., American Society of Agronomy, Madison, 1980, 657.
270. **Belyaeva, R. G. and Nevkrytaya, N. V.,** A phenological and genetical analysis of mutants with altered flower structure in *Papaver somniferum* L., *Plant Breeding Abstr.,* 49, 10360, 1979.
271. **Jonsson, R. and Loof, B.,** Poppy hybrid *(Papaver somniferum* L. × *P. orientale* L.), *Plant Breeding Abstr.,* 44, 3986, 1974.
272. **Salgado, A. L. de B., Ciaramello, D., and Azzini, A.,** [Breeding agave by hybridization] Melhoramento de Agave por hibridacao, *Bragantia,* 38, 1, 1979.
273. **Emboden, W. A.,** *Cannabis* — a polytypic genus, *Econ. Bot.,* 28, 304, 1974.

274. **Das, P. K.,** Trisomic analysis in jute — cytological behavior of trisomic hybrids from F_3 and F_4 progenies of the cross between *Corchorus olitorius* L. and *C. capsularis* L., *Nucleus,* 15, 163, 1972.
275. **Islam, A. S. and Haque, M.,** A natural species of hybrid of jute and its possible use in the evolution of desired plant types, *Plant Breeding Abstr.,* 43, 482, 1973.
276. **Islam, A. S., Haque, M., and Dewan, M. B.,** An attempt to produce a photo-neutral strain of jute through interspecific hybridization, *Jpn. J. Breeding,* 25, 349, 1975.
277. **Singh, D. P.,** Jute, in *Hybridization of Crop Plants,* Fehr, W. R. and Hadley, H. H., Eds., American Society of Agronomy, Madison, 1980, 407.
278. **Lee, J. A.,** Cotton, in *Hybridization of Crop Plants,* Fehr, W. R. and Hadley, H. H., Eds., American Society of Agronomy, Madison, 1980, 313.
279. **Lee, J. A.,** A new linkage relationship in cotton, *Crop Sci.,* 21, 346, 1981.
280. **Lee, J. A.,** Genetics of D_3 complementary lethality in *Gossypium hirsutum* and *G. barbadense, J. Hered.,* 72, 299, 1981.
281. **Mehetre, S. S., Thembre, M. V., and Tyyab, M. A.,** Cytomorphological studies in an intergeneric hybrid between *Gossypium hirsutum* L. (2n = 52) and *Hibiscus panduraeformis* Burm., *Euphytica,* 29, 323, 1980.
282. **Meyer, V. G.,** Interspecific cotton breeding, *Econ. Bot.,* 28, 56, 1974.
283. **Phillips, L. L.,** Cotton *(Gossypium),* in *Handbook of Genetics,* Vol. 2, King, R. C., Ed., Plenum Press, New York, 1974, 111.
284. **Anon.,** Annual Report of the Jute Agricultural Research Institute (1956—57), *Plant Breeding Abstr.,* 29, 4154, 1959.
285. **Wilson, F. D. and Adamson, W. C.,** Reaction to the cotton root-knot nematodes and the pollen and seed fertility of Kenaf-Roselle *(Hibiscus cannabinus* × *H. sabdariffa)* allohexaploids, *Euphytica,* 19, 349, 1970.
286. **Bernardo, F. A., Villareal, R. L., and Garcia, M. U.,** Reaction of *Musa balbisiana - Musa textilis* BC_1 and BC_2 to abaca mosaic virus, *Plant Breeding Abstr.,* 36, 6599, 1966.
287. **Wilson, F. D., Joyner, J. F., and Fishler, D. W.,** Fibre yields in *Sansieveria* interspecific hybrids, *Econ. Bot.,* 23, 148, 1969.
288. **Draper, A. D.,** Cytological irregularities in interspecific hybrids of *Aleurites, Proc. Am. Soc. Hortic. Sci.,* 89, 157, 1966.
289. **Knowles, P. F.,** Safflower, in *Hybridization of Crop Plants,* Fehr, W. R. and Hadley, H. H., Eds., American Society of Agronomy, Madison, 1980, 535.
290. **Lessman, K. J. and Anderson, W. P.,** Crambe, in *Hybridization of Crop Plants,* Fehr, W. R. and Hadley, H. H., Eds., American Society of Agronomy, Madison, 1980, 339.
291. **White, G. A., Willingham, B. C., and Wheeler, C.,** Agronomic evaluation of prospective new crop species. III. *Crepis alpina* — source of crepenynic acid, *Econ. Bot.,* 27, 320, 1973.
292. **Hardon, J. J. and Tan, G. Y.,** Interspecific hybrids in the genus *Elaeis.* I. Crossability, cytogenetics and fertility of F_1 hybrids of *E. guineensis* × *E. oleifera, Euphytica,* 18, 372, 1969.
293. **Mashindano, B. M.,** Contribution to the study of the hybrid from the backcross of *Elaeis melanococca* Gaertner × *E. guineensis* Jacquin to *E. guineensis* Jacquin in the selection nursery at Yangambi, *Plant Breeding Abstr.,* 48, 3700, 1978.
294. **Zeven, A. C.,** The partial and complete domestication of the oil palm *(Elaeis guineensis), Econ. Bot.,* 26, 274, 1972.
295. **Heiser, C. B.,** Study in the evolution of the sunflower species *Helianthus annus* and *H. bolanderi, Univ. Calif. Berkeley Publ. Bot.,* 23, 157, 1949.
296. **Heiser, C. B.,** Hybridization in the annual sunflowers: *Helianthus annus* × *H. debilis* var. *cucumerifolius, Evolution,* 5, 42, 1951.
297. **Heiser, C. B.,** The North American sunflowers, *Helianthus, Mem. Torrey Bot. Club,* 22(3), 1, 1969.
298. **Imle, E. P.,** *Hevea* rubber — past and future, *Econ. Bot.,* 32, 264, 1978.
299. **Beard, B. H. and Comstock, V. E.,** Flax, in *Hybridization of Crop Plants,* Fehr, W. R. and Hadley, H. H., Eds., American Society of Agronomy, Madison, 1980, 357.
300. **Wicks, Z.,** A search for rust genes in related species, *Flax Inst. Proc.,* 15, 1975.
301. **Nayar, N. M. and Mehra, K. L.,** Sesame: its uses, botany, cytogenetics and origin, *Econ. Bot.,* 24, 20, 1970.
302. **Malecka, J.,** Cytotaxonomical and embryological investigations on a natural hybrid between *Taraxacum kok-saghyz* Rodin and *T. officinale* Web. and their putative parental species, *Genet. Pol.,* 13(3), 59, 1972.
303. **Belyaeva, R. G. and Kovineva, V. M.,** A study of second generation interspecific hybrids of *Mentha* L., *Plant Breeding Abstr.,* 45, 799, 1975.
304. **Hefendehl, F. W.,** Monoterpene composition of a carvone containing polyploid strain of *Mentha longifolia* (L.) Huds., *Plant Breeding Abstr.,* 48, 3971, 1978.
305. **Darrah, H. H.,** Investigation of the cultivars of the basils *(Ocimum), Econ. Bot.,* 28, 63, 1974.

306. **Dzevaltov'skii, A. K. and Polishchuk, V. S.,** Experiments on the artificial hybridization of *Majorana hortensis* Moench with *Origanum vulgare, Plant Breeding Abstr.,* 45, 7899, 1975.
307. **Widrlenchner, M. P.,** History and utilization of *Rosa damascena, Econ. Bot.,* 35, 42, 1981.
308. **Anon.,** Institut de Recherches Agronomiques Tropicales et des cultures vivrieres, Annu. Rep. 1970, *Plant Breeding Abstr.,* 43, 1693, 1973.

BREEDING SYSTEMS

Arnel R. Hallauer

INTRODUCTION

Plant breeding is an activity practiced by humans for systematic improvement of preferred plant types. Plant breeding is older than recorded history and has been practiced since humans made the transition from nomadic hunter gatherers to gardening and farming. The level of plant breeding, however, has changed dramatically during the development of human civilization, particularly during the past 200 years. Initially, plant breeding included simple phenotypic (or mass) selection among wild species of plants available at particular localities. Selection was for types that had particular appeal because of quantity of product produced, flavor, attractiveness, and availability for use as food, fiber, or fuel. Although the methods of selection used by the early plant breeders would be considered crude by present-day standards, they were very successful in transforming wild species of plants into highly productive cultivated species that provided for the needs of settlements. The impact of the early plant breeders was greater, however, in the development of cultivated crop species than the recent efforts for improving the productivity, quality, and pest resistance of our major crop plant species.

Plant breeding is often defined as the art and the science for the improvement of crop plant species. Until about 200 years ago, plant breeding was restricted to the farmers and gardeners who selected seed from favored plant types for propagation of the next crop generation. Selection was largely an art (or eye appeal) for what the farmer or gardener considered to be the visualized ideal plant type; selection, therefore, was based on the appearance of the plant, or what we refer to as the phenotype. An example of early plant breeding methods was the use of the corn show card in the late 19th century. The corn show card included a list of traits to use in selecting for ear types that were considered to be the ideal type for highly productive corn. It was found, however, that the esthetic traits of ear conformation were not necessarily indicative of relative productivity; i.e., ear samples that conformed to the show card standards were no more productive than those samples that rated low relative to the show card standards. It soon became apparent that other methods of evaluation were needed to determine the productivity of the different selections.

Vilmorin in 1850 suggested the practice of progeny testing as an adjunct to the previously used methods of individual plant selection.[1] Progeny testing included evaluating a group (or family) of individuals that had descended from common individuals. In this manner, evaluations could be made for selected individuals that could be repeated in time and space. Progeny testing, therefore, was an important factor in changing the emphasis of selection, based only on phenotype of individuals, to selection based on progenies of individuals. Two other developments also had an important impact in changing the emphasis of plant breeding methods: (1) rediscovery of Mendel's laws of inheritance in 1901, and (2) development of the statistical concepts of replication and randomization for making valid comparisons. These three developments (progeny testing, Mendelian genetics, and experimental design) were instrumental in changing the emphasis of plant breeding from an activity based on art to one based on science.

Although plant breeding during the 20th century has developed into an important science, the art of plant breeding is still considered by many to be an important facet in successful plant breeding. But it is generally agreed that the art, or "eye", of plant breeding must be accompanied by use of reliable data collected from replicated trials conducted over time and space. The art of plant breeding is based on an idealized mental concept of the breeder; that is, an ideotype that is considered desirable and acceptable to the farmer and gardener with regard to appearance, esthetic qualities, color, stature, etc. The idealized concept, however, must be supported by performance, which may include yield, ease of handling, resistance to pests, drought tolerance, etc. Hence, modern plant breeding includes both the art and science, with more emphasis in recent years based on the science rather than the art. Subsequent discussion will outline some of the breeding systems used for the improvement of multicellular plant species.

PLANT REPRODUCTION

Although the principles of plant breeding are similar for nearly all plant species, some differences occur in their application because of the manner in which crop species reproduce. In contrast to large-animal breeding where we have separate sexes in different individuals, plant species can be self-fertilizing (male gametes fertilize female gametes of the same individuals), cross-fertilizing (male gametes fertilize female gametes of different individuals, but in some instances they can fertilize the same individuals), dioecious (male and female individuals cross-fertilize, which is similar to large animals), reproduce vegetatively (clones, tubers, and bulbs), and by apomixis (seed produced asexually in sexually reproducing plants). Additionally, breeding methods may be adjusted because of self-incompatibility mechanisms, male or female sterility, and different levels of polyploidy (more than two sets of chromosomes). None of these variations in reproduction among plant species, however, alters the breeding methods used based on the laws of Mendelian genetics. Specific techniques have to be developed for the different plant species to efficiently and effectively improve them. In most instances, breeding techniques for crossing and selfing to produce progenies have been developed for the different types of plant reproduction.[1a]

Table 1 includes some examples of the different plant species included in the different categories. There is no consistent pattern in the occurrence of ploidy levels and whether plant species are self- or cross-fertilized, dioecious, or vegetatively propagated. The perennial species tend to have a greater frequency of polyploidy, and with the occurrence of vegetative propagation, the perennial growth habit may favor polyploidy. With respect to plant breeding and breeding methods used, however, the differences in plant propagation may be either an advantage or disadvantage. A plant species that is an obligate self-fertilizer is an advantage to the breeder for developing pure lines, but it would be a disadvantage for producing hybrids, because of the difficulties in making the crosses and obtaining sufficient quantities of seed for extensive testing. In contrast, to produce pure lines in corn, a cross-fertilizing species, it is necessary to manually control the gametes in self-fertilization, but it is relatively easy to produce both sufficient quantities of self- and cross-seed. Sorghum is mainly self-fertilized, with instances of cross-fertilization ranging from 2 to 35% and averaging about 6%; hence, manual methods of pollen control similar to corn are required although sorghum is primarily a self-fertilizer.

Many vegetatively propagated legumes and grasses can be either self- or cross-fertilized, and often have different levels of polyploidy. Breeding methods have to be adapted for effective breeding strategies to develop improved cultivars. But after a superior cultivar has been identified, it can be vegetatively propagated to increase available supplies of the superior cultivar. Although the techniques and consequences of self- and cross-fertilization may be undesirable, use of vegetative propagation would be an advantage in many instances.

The method of plant propagation has to be well understood before effective breeding strategies can be developed. Once the method of plant propagation has been determined, the plant breeder can select the breeding system that would most effectively develop the desired ideotype. Further refinements of the breeding system will evolve with experience in breeding and a better understanding of the advantages and disadvantages of the unique

Table 1
EXAMPLES OF PLANTS HAVING DIFFERENT METHODS OF PROPAGATION

Crop plant	Genus and species	Method of propagation	Ploidy level
Cereal grasses			
Barley	*Hordeum vulgare*	Self-fertilizing	2x
Corn	*Zea mays*	Cross-fertilizing	2x
Oats	*Avena sativa*	Self-fertilizing	6x
Rye	*Secale cereale*	Cross-fertilizing	2x
Wheat	*Triticum* spp.	Self-fertilizing	2x—6x
Legumes			
Alfalfa	*Medicago sativa*	Cross-fertilizing	4x
Common bean	*Phaseolus vulgaris*	Self-fertilizing	2x
Lima bean	*Phaseolus lunatus*	Self-fertilizing	2x
Pea	*Pisum sativa*	Self-fertilizing	2x
Soybean	*Glycine max*	Self-fertilizing	2x
Forage grasses			
Tall fescue	*Festuca arundinacea*	Cross-fertilizing	2x—10x
Annual ryegrass	*Lolium multiflorum*	Cross-fertilizing	2x
Tall wheatgrass	*Agropyron elongatum*	Self-fertilizing	2x—5x
Smooth bromegrass	*Bromus inermis*	Cross-fertilizing	2x—4x
Reed canary grass	*Phalaris arundinacea*	Cross-fertilizing	2x—4x
Others			
Cotton	*Gossypium barbadense*	Mostly self-fertilizing	4x
Flax	*Linum usitatissimum*	Self-fertilizing	2x
Hop	*Humulus lupulus*	Cross-fertilizing	2x—3x
Sorghum	*Sorghum bicolor*	Mostly self-fertilizing	4x
Sugar beet	*Beta vulgaris*	Cross-fertilizing	2x
Tobacco	*Nicotiana tabacum*	Self-fertilizing	4x
Vegetables			
Asparagus	*Asparagus*	Cross-fertilizing (dioecious)	2x
Eggplant	*Solanum melongena*	Self-fertilizing	2x
Onion	*Allium cepa*	Cross-fertilizing	2x
Potato	*Solanum tuberosum*	Cross-fertilizing	2x—5x
Tomato	*Lycopersicon esculentum*	Self-fertilizing	2x

methods of plant propagation for the particular plant species. The breeding systems to be discussed are applicable to most plant species. The facility in their use, however, will vary among plant species because of their different methods of propagation.

Genetic variation, regardless of type of plant propagation, must exist for selection to be effective. Without genetic variation, the plant breeder has no opportunity to utilize his talents for plant improvement. The most common method for generating genetic variation is by hybridization. For species that are cross-fertilized, a reservoir of genetic variation is maintained by the nature of plant propagation. Hybridization between plants that are normally self-fertilized can be accomplished (both naturally and by artificial means) to provide genetic variation for selection. If genetic variation is not considered adequate or if favorable combinations of genetic variation are not available, planned hybridization would be the most common mode for creating genetic variation. Genetic variation also can be generated by mutations, polyploidy, and by crossing with distant relatives. All these forms of creating genetic variation are used in modern plant-breeding programs.

BREEDING SYSTEMS

Plant breeders measure or observe phenotypes of individuals. It is important, however, to be able to partition the phenotypic values of the traits into components attributable to the effects of the factors inherited (genes) and to the influence of the environment. Some traits (e.g., flower color) are inherited in a qualitative manner (e.g., one or two genes), permitting the breeder to classify the phenotypes of individuals into discrete classes. Environmental effects, therefore, usually do not mask the expression of the genes, and the breeder can monitor the inheritance of the genes. Most of our economically important traits, however, are determined by a large, and unknown, number of genes, each individually having a small effect on the total expression of the trait, and whose effects are modified by the environment. These types of traits have a complex inheritance and are determined in a quantitative manner. For traits inherited in a quantitative manner, it is not possible to classify individuals in discrete classes, and the breeders use mating and experimental designs to partition the phenotypic values into the effects due to genetic and environmental forces. The parameters estimated and used in selection are means and variances.

In the very simplest case, where the genotype (G) and environment (E) do not interact, the phenotypic value (P) can be expressed as P = G + E. For traits whose expressions are simply inherited and can be classified in discrete classes, P = G and the environmental component does not play an important role. Environmental effects, however, can have a significant impact on the expression of quantitative traits. For instance, yield, an important economic trait, is determined by a large number of genes whose effects are influenced by the environmental conditions throughout the growing season. It is very important, therefore, to be able to determine the magnitude of the G and E components of P. Further difficulties occur if there are interations of G and E; that is, P = G + E + GE. Interactions occur when there are different responses of genotypes, either magnitudinal or ranking, in different environments. It becomes necessary in these instances to repeat the evaluation of genotypes over time and space to partition P into G, E, and GE.

Heritability is an important concept in plant breeding that relates the proportion of the genetic variance (σ_G^2) to the total phenotypic variance ($\sigma_P^2 = \sigma_G^2 + \sigma_E^2 + \sigma_{GE}^2$). It is a ratio of the σ_G^2 and σ_P^2 that provides for the breeder an estimate of the relative importance of genetic effects in the expression of the phenotype. Or, heritability is the relative expression of the genes transmitted from the parents to their offspring. Heritability is not an absolute value. It varies among traits for the same individuals, among the same traits for different groups of individuals, among types of progenies tested, and among environments in which individuals are tested.

Heritability (H), in its simplest form, can be expressed as H = σ_G^2/σ_P^2, where $\sigma_P^2 = \sigma_G^2 + \sigma_E^2 + \sigma_{GE}^2$. For simply inherited traits, $\sigma_G^2 = \sigma_P^2$ and H is 1.00 or 100%. For yield, however, H can vary from 0.10 to 0.80, depending on types of progenies evaluated and the G, E, and GE effects. In corn, for example, H for yield may be 10% or less for selection based on nonreplicated plants in one environment to 80% or more based on pure-line progenies tested in replicated trials repeated over environments. Heritability estimates can be further refined if the genetic variance can be partitioned into that due to additive (σ_A^2), or fixable, and nonadditive (σ_N^2) genetic effects. For these instances H = σ_A^2/σ_P^2, where $\sigma_P^2 = \sigma_G^2 + \sigma_E^2 + \sigma_{GE}^2$ and $\sigma_G^2 = \sigma_A^2 + \sigma_N^2$. Heritability estimates for quantitatively inherited traits provide useful guidelines in planning breeding strategies, but they are only estimates restricted in their interpretation to the individuals and environments sampled. Hanson[2] has discussed the use, estimation, and interpretation of heritability estimates in plant breeding.

Early Methods of Selection

Humans started practicing plant selection when they developed cultures for specific sites. To sustain their societies it became critical to develop reliable food, fiber, and fuel supplies within a restricted area. Hence, plant species adapted to their localities were chosen that provided for their needs. Plant breeding was initiated firstly to domesticate wild, weedy plant species and secondly to improve the strains that had been domesticated. These phases of plant breeding were initiated at the dawn of human civilization and continue in a very limited degree today. Many of our major crop species were domesticated by 5000 B.C., and the time required to convert wild, weedy plant species to forms of the domesticates included in present-day breeding programs undoubtedly required several millennia.[3] Subsequent breeding efforts have been to affect subtle improvements and refinements of the domesticates for different, and quite often minor, traits.

After the major changes had occurred from wild, weedy species to a domesticate, continuous improvements have been made to extend the range of the domesticated plants. Migration of people, exchange of seeds, and the looting by conquering armies all contributed to the wide distribution of useful plant species. In each instance, the peasant farmer or gardener would select for types that were most adapted and useful for their purposes. Different strains within the cultivated species were developed that had resistance to prevalent pests, day-length response, and tolerance to drought, cool growing conditions, pH differences of soils, and salt and mineral elements in the soils. Often, the selections retained would be those unique individuals that survived the immediate environment.

Early plant breeding systems were restricted to the selection, in the eyes of the breeder, of the most desirable individuals. Selection was based on the phenotype (or visual appearance), which in present-day terminology is referred to as mass selection. The genetics for the transmission of traits from parents to offspring (heritability) were not available or understood until the early part of the 20th century. Selection emphasized, therefore, that like begets like. The extent to which selection was effective depended on how the plant species was propagated and the confounding effects of the environment in which it was grown. Selection, however, was obviously effective, as evidenced by the many different cultivars and strains of the same plant species that are distributed over large areas. Many cultivars and strains have only small differences (e.g., color of flower or grain), but the early plant breeders were aware of the differences between closely related plants. Identifiable traits were fixed to develop recognizable cultivars and strains that played a prominent role in the early societies for useful, esthetic, and ceremonial purposes.

Progeny Selection

Selection based on progeny performance was suggested to enhance selection among superior individuals. The nature of an individual plant is determined by its genotype (genetic constitution) whose expression is modified by the environment in which it is grown to produce the phenotype (physical appearance). Two plants may have the same genotype but different phenotypes because of the environment, or, conversely, two plants may have similar phenotypes but different genotypes. Except for certain situations, each plant will have a unique genotype. The purpose of progeny selection, therefore, is to separate the genetic and environmental effects in the expression of the phenotype of the plants. Vilmorin and Hopkins[4] were the first to suggest some form of progeny testing. Their reasoning for use of progeny testing was clearly demonstrated by Johannsen's (1903) research on seed size of the common bean; he demonstrated that inheritance of phenotypic variation was only possible when it was a result of genotypic differences.[4a]

Pedigree

Pedigree selection is probably the most widely used breeding system in plant improvement. It has wide application regardless of the method of plant propagation. Basically, the pedigree method of breeding is a system of selection that is practiced within segregating populations (based on individuals and progenies); the segregating populations are derived either from crosses of pure lines or broad, genetically based cultivars (landrace varieties, synthetic varieties, and composites).

Table 2
ILLUSTRATION OF ALLELE SEGREGATION IN AN F_2 POPULATION PRODUCED BY CROSSING TWO PURE-LINE PARENTS DIFFERING FOR ONE FACTOR PAIR

Generation		One factor pair		Homozygosity (%)
Parents	A_1A_1	mult	A_2A_2	100
F_1		A_1A_2		0
$F_2 (S_0)$	$¼A_1A_1$	$+ ½A_1A_2$	$+ ¼A_2A_2$	50
	↓	↓	↓	
S_1	$¼A_1A_1 + ½[¼A_1A_1$	$+ ½A_1A_2 + ¼A_2A_2]$	$+ ¼A_2A_2$	
	$= ⅜A_1A_1$	$+ ¼A_1A_2$	$+ ⅜A_2A_2$	75
	↓	↓	↓	
S_2	$⅜A_1A_1 + ¼[¼A_1A_1$	$+ ½A_1A_2 + ¼A_2A_2]$	$+ ⅜A_2A_2$	
	$= ⁷⁄_{16}A_1A_1$	$+ ⅛A_1A_2$	$+ ⁷⁄_{16}A_2A_2$	87.5
	↓	↓	↓	
S_3	$⁷⁄_{16}A_1A_1 + ⅛[¼A_1A_1$	$+ ½A_1A_2 + ¼A_2A_2]$	$+ ⁷⁄_{16}A_2A_2$	
	$= ¹⁵⁄_{32}A_1A_1$	$+ ¹⁄_{16}A_1A_2$	$+ ¹⁵⁄_{32}A_2A_2$	93.75
	.	.	.	
	.	.	.	
S_7[a]	$½A_1A_1$		$½A_2A_2$	~100

[a] Progenies are approaching homozygosity.

As the name implies, pedigree selection includes keeping detailed records of the selections during each stage of breeding activity. Pedigree selection can be initiated in any type of plant population, but it is often associated with selection in populations generated by hybridizing two desirable parental stocks that are pure lines. Usually, the parental stocks used for hybridizing are selected to complement the weaknesses (e.g., susceptibility to a pest in an otherwise desirable stock) of each other. Two parental stocks are crossed to form the F_1 hybrid. If the parents are homozygous and homogeneous, the F_1 also will be homogeneous but heterozygous; hence, there will be no genetic variation among the F_1 plants. The F_1 generation is usually selfed (or could be sib-mated) to produce the F_2 generation. The F_2 generation will be segregating for the genetic factors that were different in the two parents. Genetic variation, therefore, is present in the F_2 generation to permit selection.

Although selection can be practiced in the F_2 generation, selection would be restricted to phenotypic differences among plants. Pedigree selection is usually initiated by self-fertilizing individual F_2 plants, which can be done either naturally (self-fertilizing species) or by use of manual methods of self-fertilization (e.g., corn and sorghum). Seed harvested from the self-fertilized F_2 plants are S_1 progenies. S_1s are planted in progeny rows, and selection is initiated among progeny rows and for plants within the segregating progeny rows. S_1 progeny rows could be replicated with one (or more) replications in a pest nursery and one replication in a breeding nursery. Self-fertilization is effected in the S_1 progenies, and the selected individuals (S_2 progenies) are planted the next growing season in progeny rows. The process is repeated in successive generations with the breeder imposing selection for the desirable plant types (Table 2). During the course of inbreeding and selection, the breeder must maintain careful records of his selections to have a detailed diary (pedigree) of the parentage of his selections.

The mechanics of pedigree selection will vary among plant species, but they generally are repetitive from generation to generation. Though the pedigree methods are widely used and repetitive, the breeder, however, must make several important decisions in initiating and conducting pedigree selection. Some of the important decisions include: choice of parental material to make crosses; number of individuals to sample from the segregating population to permit the breeder an opportunity to select the desired genotype; extent of sampling within one or more segregating populations; intensity of selection in the different generations of inbreeding; heritability of the different traits and how effective will selection be for the different traits; and at what stage of inbreeding should testing for important performance traits be initiated. Specific guidelines for making choices are not available because of the wide range of plant species, specific traits under selection, complexity of traits under selection, amount of genetic variability available, and how effectively the breeder can select for the desirable plant type.

Choice of parental material to initiate pedigree selection may be relatively easy in some instances and difficult in others. If the breeder is experienced and knows his materials, the parental stocks needed for crossing for a specific breeding goal may be obvious. In other instances, a breeding program may be initiated for a plant species which has little genetic or breeding information available; choice of parental materials, therefore, may not be obvious. Proper choice of breeding stocks is very important to realize ultimate breeding goals. If the choice includes parental stocks that do not include the genes desired, all subsequent breeding efforts will be doomed. Before one chooses his parental material for initiating pedigree selection, it is necessary that all sources of information are explored to enhance the ultimate success of the program.

The number of individuals to sample and the number of populations to sample also are questions for which obvious answers are not always available. Guidelines can be developed for the number of individuals sampled, however, depending on the complexity of the trait under selection. If, for example, a trait is controlled by a single factor pair, the cross between two parental stocks that differ in their alleles would be $A_1A_1 \times A_2A_2$ to produce the F_1 generation, A_1A_2 (Table 2). The F_1 is self-fertilized to produce the F_2 of $(1/4)$ A_1A_1 + $(1/2)$ A_1A_2 + $(1/4)$ A_2A_2. The likelihood of a single plant being A_1A_1 or A_2A_2 is one fourth. The breeder needs to know, however, the minimum number of plants that are required to grow to have, say, 19:1 odds of obtaining one plant that is homozygous for A_1A_1 or A_2A_2. A formula was developed by Muller[5] to calculate this information. His basic formula was

$$N = -\log_e F (P - 0.5)$$

where N is the number of plants needed, $-\log_e F$ is the factor for a given probability, and P is the number of individuals which, on the average, will give one of the desired types. The numbers of plants needed to be 90, 95, and 99% sure of recovering at least one desirable homozygote in an F_2 are listed in Table 3. For example, if one factor was segregating in the F_2, it would be necessary to have 11 plants for a 95% chance and 16 plants for a 99% chance of recovering the desired genotype. If a triple recessive homozygote was desired from a trihybrid F_2 population, 190 and 292 plants would be grown to have a 95 and 99% chance, respectively, of recovering the triple recessive. It is clear that the fewer genetic factors involved, the easier the task becomes for recovering the desired genotype. The task of recovering genotypes that have more than three independent factors becomes increasingly difficult. It is not unreasonable to assume, for example, that a minimum of 20 independent factors influence yield of corn. Growing enough plants to have a 90% chance of recovering a selection with all 20 factors in the homozygous condition becomes nearly impossible. Even if one were able to produce this "idealized combination", there would still be the problem of recognizing and isolating it.

The other problem confronting a breeder is whether to extensively sample one population or sample several populations less extensively. This problem is generally one related to traits that are not simply inherited (e.g., one to three factors). It is realized that it is not generally possible to grow and sample populations for traits governed by 20 genetic factors and expect

Table 3
NUMBER OF PLANTS NEEDED TO RECOVER AT LEAST ONE OF THE REQUIRED GENOTYPES FOR THREE LEVELS OF PROBABILITY

Ratio of required genotype to total	Number of different pairs of factors	Level of probability (%) 90	95	99
1:4	1	8	11	16
1:16	2	36	46	71
1:64	3	146	190	292
1:256	4	588	765	1,175
1:1024	5	2,354	3,066	4,708
1:4096	6	9,420	12,270	18,839
$1:4^n$	n	$N = -\log_e F (4^n - 0.5)$[a]		

[a] For 9:1, 19:1, and 99:1 odds $-\log_e F$ is 2.3, 2.996, and 4.6, respectively.

to recover genotypes with all of the favorable factors. Hence, compromises have to be made for the extent of sampling for quantitatively inherited traits. Considering the numbers required for, say, six independent factor differences (12,270 for 19:1 odds, Table 3), breeders of most plant species have to be content with recovery of F_2 plants that, while not homozygous for the desired combination, contain variability that may provide more favorable combinations in later generations of selfing. Hence, breeders tend to select within more than one F_2 population. This is a conservative approach to protect the breeder in case of a poor choice of parents in producing the F_2 population. If a breeder restricted extensive sampling (e.g., 10,000) to only one population and the proper combination of factors was not present, progress from selection would be futile. Bauman,[6] from an extensive survey of U.S. corn breeders, reported that U.S. corn breeders preferred to sample 500 individuals from four to five populations.

Selection intensity is usually greater in the initial generations of pedigree selection. If the traits are controlled by one or two genetic factors, selection is intense to fix the traits in homozygous condition. Further selection is practiced after the traits are fixed for other morphological and agronomic traits. Selection pressure for traits controlled by several or many genes will vary during self-fertilization because of the distribution of genetic variance among and within progeny rows. Selection may be intense in the early generations (S_1 or S_2) for pest resistance, maturity, morphological traits, etc. Selection would be emphasized among progenies for some traits and within progenies for others. The distribution of the additive genetic variance among and within progenies is listed in Table 4. It can be seen that the variation among progenies increases and the variation within progenies decreases with self-fertilization. Selection within progenies will be more effective initially because the level of homozygosity increases at the rate of $[(2^m - 1)/2^m]^n$, where m is the segregating generation and n is the number of factor pairs involved. In contrast, the genetic variation among nearly homozygous progenies is double that among F_2 plants, assuming all lines of descent are maintained. In reality, compromises in selection are made based on the experience of the breeder and the traits under selection. The judgments are empirical with selection in each generation of self-fertilization to reduce the number of progenies included in final evaluation trials.

The stage of testing progenies extracted by pedigree selection depends on the traits under selection, their heritability, type of plant propagation, and the judgment of the breeder. There are two schools of opinion for stage of testing — early (S_0 or S_1) vs. late (S_5—S_7). Early testing proponents test early to discard progenies that do not have promise, and practice more intense selection and testing in later generations for a more elite group of progenies.

Table 4
DISTRIBUTION OF THE ADDITIVE GENETIC VARIANCE AMONG AND WITHIN PROGENIES DERIVED FROM A POPULATION BY SELF-FERTILIZATION,[a] ASSUMING $p = q = 0.5$ OR NO DOMINANCE EFFECTS

Generation	Level of inbreeding (%)	Additive genetic variance Among progenies	Within progenies
F_2—S_0	0	1	—
F_3—S_1	50.0	1	1/2
F_4—S_2	75.0	3/2	1/4
F_5—S_3	87.5	7/4	1/8
F_6—S_4	93.75	15/8	1/16
F_7—S_5	96.87	31/16	1/32
F_8—S_6	98.8435	63/32	1/64
F_9—S_7	99.4227	127/64	1/128
F_n—S_{n-2}	100	2	0

[a] See Table 2.

Supporters of late testing suggest that effective selection can be made among and within progenies prior to conducting the expensive performance trials. Consequently, many breeders compromise and test in the S_2 or S_3 generation after practicing selection for more highly heritable traits in the S_1 or S_2 generations. Stage of progeny testing also may be dictated more by the method of plant propagation. Plant species that are self-fertilizing (e.g., soybeans, oats, and peas) may not be tested until they are nearly homogeneous (e.g., S_7); this is their normal mode of commercial production and testing is delayed until homozygosity is reached for most traits. Stage of testing progenies developed by pedigree selection, therefore, varies among breeders and the different plant species.

Pedigree selection is the most popular breeding system used in plant improvement. It is one of the original breeding systems used with the rediscovery of Mendel's laws of inheritance and has played a prominent role in plant improvement for the past 70 years. Though pedigree selection may be critized for having a restricted genetic base (as commonly used by crossing two pure-line stocks) and a long generation interval (13.3 years in corn[7]), it will continue to be important in future plant improvement.

Backcrossing

The mechanics of the backcross method of selection are similar to those described for pedigree selection. Two primary distinctions between the two systems are (1) repeated crosses are made to one parent (recurrent parent) to recover its genotype except for the introduced trait; and (2) backcrossing is very useful if it is desired to transfer one or two traits from one parent (nonrecurrent parent) into an otherwise desirable parent (recurrent parent). Because of the segregation of alleles for the different traits transferred by use of backcrossing, the backcross method of breeding is used extensively to correct minor faults in otherwise acceptable cultivars. The method is particularly useful for the improvement of self-fertilizing cultivars or inbred lines of cross-fertilizing cultivars and is most useful for transferring a single highly heritable trait. In each generation plants carrying the desired alleles of the nonrecurrent parent are backcrossed to the recurrent parent. Genes of the nonrecurrent parent, therefore, are carried in the heterozygous condition throughout the backcrossing. After the backcrossing is completed, the traits being transferred are made homozygous by self-fertil-

Table 5
EXAMPLE OF BACKCROSSING FOR A CROSS OF TWO PURE-LINE PARENTS DIFFERING FOR ONE FACTOR PAIR

Generation	One factor pair Recurrent——Nonrecurrent	Homozygosity (%)
Parents	A_1A_1 x A_2A_2	100
F_1	A_1A_2	0
BC1	A_1A_1 x A_1A_2 = $\frac{1}{2}A_1A_1 + \frac{1}{2}A_1A_2$	50
BC2	A_1A_1 x $[\frac{1}{2}A_1A_1 + \frac{1}{2}A_1A_2]$ = $\frac{1}{2}A_1A_1 + \frac{1}{2}[\frac{1}{2}A_1A_1 + \frac{1}{2}A_1A_2]$ = $\frac{1}{7}A_1A_1 + \frac{1}{4}A_1A_2$	75
BC3	A_1A_1 x $[\frac{1}{7}A_1A_1 + \frac{1}{4}A_1A_2]$ = $\frac{1}{7}A_1A_1 + \frac{1}{4}[\frac{1}{2}A_1A_1 + \frac{1}{2}A_1A_2]$ = $\frac{7}{8}A_1A_1 + \frac{1}{8}A_1A_2$	87.5
BC4	A_1A_1 x $[\frac{7}{8}A_1A_1 + \frac{1}{8}A_1A_2]$ = $\frac{7}{8}A_1A_1 + \frac{1}{8}(\frac{1}{2}A_1A_1 + \frac{1}{2}A_1A_2)$ = $\frac{15}{16}A_1A_1 + \frac{1}{16}A_1A_2$	93.75
⋮		
BCn	A_1A_1	~100

Note: A_1A_1 is the recurrent parent.

ization. Little attempt is usually made to select plants carrying genes of the recurrent parent because they are automatically recovered by the backcrossing procedure.

Requirements for a successful backcrossing program include: (1) a satisfactory recurrent parent; (2) an acceptable donor parent (nonrecurrent parent) having the alleles required to improve recurrent parent; (3) a trait controlled by a few genes (i.e., highly heritable); and (4) a genotype of the recurrent parent that can be recovered with a reasonable number of backcrosses with populations of manageable size. As mentioned previously, the main objective of most backcrossing programs is to transfer traits controlled by one or two genes that can be identified. The genetics of backcrossing are similar to those for pedigree selection (Table 5): the approach to homozygosity (assuming no linkage) is the same whether backcrossing or selfing; i.e., $(2^m - 1)/2^m$ where m is the number of backcrosses. Comparisons of Tables 2 and 5 show that four generations of self-fertilizing (S_0 to S_3, Table 2) and backcrossing (Table 5) result in the same percentage of homozygosity (93.75%). The rate of approach to homozygosity is the same, the difference being that in backcrossing every homozygous genotype is of the desired type (A_1A_1, Table 5) whereas under selfing this is not the case (A_1A_1 and A_2A_2, Table 2). Each of the homozygous loci will be homozygous for the genotype of the recurrent parent, but at the heterozygous loci 50% of the alleles are from the recurrent parent and 50% are from the nonrecurrent parent. If five generations of backcrossing are completed [$(2^5 - 1)/2^5 = 96.87\%$], the recovered line will have 96.87% homozygous loci and 3.13% heterozygous. The germplasm of the recovered line will be 96.87 + 1.56 or 98.43%.

The ease in transfer of traits by backcrossing depends on whether the trait being transferred is dominant or recessive in its expression and the stage of plant development at which the trait can be identified. If the allele being transferred is dominant in its expression and can be classified before flowering (e.g., rust resistance in oats), all plants that do not carry the resistant allele can be eliminated and only those plants that carry the resistant allele are backcrossed. If, however, the expression of the allele does not occur until after flowering, all plants will have to be backcrossed to the recurrent parent and selection made at the time

Table 6
NUMBER OF PLANTS NEEDED TO RECOVER AT LEAST ONE OF THE REQUIRED GENOTYPE BY BACKCROSSING FOR THREE LEVELS OF PROBABILITY

Ratio of required genotype to total	Number of different pairs of factors	Level of probability (%) 90	95	99
1:2	1	3.4	4.5	6.9
1:4	2	8.0	10.5	16.1
1:8	3	17.3	22.5	34.5
1:16	4	35.6	46.4	71.3
1:32	5	72.5	94.4	144.9
1:64	6	146.1	190.2	292.1
$1:2^n$	n	$N = -\log_e F (2^n - 0.5)$[a]		

[a] For 9:1, 19:1, and 99:1 odds — $\log_e F$ is 2.3, 2.996, and 4.6, respectively.

of expression. If recessive alleles are being transferred, their presence is masked by the dominant allele in the backcrosses to the recurrent parent. Selfing after each backcross generation can be done to readily select those plants carrying the recessive alleles. Or, alternatively, larger populations and progeny testing can be used to recover the desirable genotype after each backcross.

The number of plants to grow to ensure recovery of the desired genotypes in the backcross populations is similar to that of pedigree selection. Consider the situation of crossing cultivar A *(rr)*, which is rust susceptible, to cultivar B *(RR)*, which is rust resistant. It is desired to transfer rust resistance from B to otherwise desirable cultivar A (recurrent parent). Half of the backcross progeny will be resistant *(Rr)* and half will be susceptible *(rr)*. If the breeder has one plant, the odds are 1:1 that the genotype of the plant is *Rr* or *rr*. Better odds are necessary to reduce failure in recovery of the desired genotype. Using methods similar to those described for pedigree selection, the breeder can improve his odds to 99 out of 100 if seven backcross plants are grown (Table 6). If the trait is conditioned by a recessive allele, a backcross population of seven plants also will ensure by odds of 99 out of 100 that at least one plant will carry the desired recessive allele. Suppose the inheritance of the trait being backcrossed is governed by two genes; the backcross population size needed for odds of 19:1 is 11 plants. Comparison of Tables 3 and 6 shows that fewer plants are needed for backcrossing than for pedigree selection because of the greater homozygosity of the desired types recovered by backcrossing.

It was discussed previously that backcrossing is usually used for transferring traits having high heritability; i.e., traits governed by one to three genes. Backcrossing has not been used extensively for transferring quantitative traits. The difficulty of using backcrossing for quantitative traits is the dilution of the traits being transferred. Richey[8] suggested use of convergent improvement for the transfer of desirable traits between two elite lines of corn. Convergent improvement is a special case of backcrossing, but it never has been used extensively. Penny and Dicke[9] reported on use of the backcross method for transferring European corn borer resistance, which is inherited quantitatively, into elite, high-combining corn lines. Backcrossing was generally not effective per se, unless one generation of intermating of resistant progenies was conducted between each backcross. Although used extensively, backcrossing is usually restricted to the transfer of traits governed by one or two genes in pure lines of self-fertilizing species or inbred lines (e.g., corn) of cross-fertilizing species.

Bulk-Population Breeding

Several types of breeding systems have been suggested and used to develop cultivars by use of bulk populations. This system of breeding has been used in small grain crops (e.g., barley, oats, soybeans, and wheat), beans, and corn, but it has not been suitable for, say, fruit and vegetable crops. Bulk populations can be formed in several ways and for many different types of plant materials. In the simplest case, bulk-population breeding may be initiated in the F_2 population formed from a cross of two pure-line cultivars, which is identical to the populations usually used in pedigree selection. Mechanical mixtures of more than two pure-line cultivars, or their crosses, and original landrace cultivars (e.g., corn) are other types of populations used in bulk breeding. Whatever type of population used, the general principles of bulk breeding apply.

Within the general category of bulk-population breeding, there are several methods that can be used for handling the populations. The selection methods used depend on the extent to which the breeder relies on either natural or artificial (or both) selection pressures; but, in most instances, breeders modify the methods used to accommodate selection for specific traits. All of the bulk-population breeding methods, however, have the common goal of developing superior cultivars. The advantages accredited to bulk-population breeding, in contrast to pedigree breeding, are simplicity, economy of resources, and taking advantage of natural selection for some traits.

Bulk

Natural selection forces are emphasized in bulk breeding. The method has been frequently used for self-fertilizing species. The bulk population of the desired cultivars is formed and, for example, 10,000 seeds are planted in one or more environments. The plants are harvested in bulk, and a sample of seed taken for planting the following season. Similar procedures are continued until the plants are reasonably homozygous. During the time that the populations have been handled as bulks, natural selection operates continuously to shift gene frequencies within the population. The role of natural selection in changing gene frequencies may be systematic for some traits (winter hardiness, maturity, survival, and competitive advantage), but oscillating for others (pest resistance, straw strength, and yield). Effectiveness of natural selection will vary among environments for different traits and the presence of natural forces (e.g., occurrence of pest).

The classic study of bulk breeding was reported by Harlan and Martini[10] in barley. A bulk mixture of 11 barley cultivars was grown for 4 to 12 years at ten sites within the U.S. Final census data showed large differences in frequency of the different cultivars for the ten sites. Conditions at each site, therefore, were sufficiently diverse to cause differences in survival of cultivars. The greater frequency of one cultivar than another would depend on (1) the number of seeds produced and (2) the proportion of seeds that attained maturity and produced viable offspring. Harlan and Martini's data clearly showed that natural selection was a significant force at all sites, and at some more than others.

Survival and competitive advantage seem to be significant factors in the final census of populations handled by bulk breeding. Experiments have shown, however, that those genotypes that survive best in mixtures may not be the best for use in commercial agriculture; i.e., performance in mixtures is not correlated with pure stands. Hence, one of the advantages cited for bulk breeding — survival of the fittest in mixed stands — may not be an advantage in modern agricultural production situations. Bulk breeding would select for those that survive a range of environments, either in time or space, but the forces of natural selection would not be consistent for some traits (e.g., pest resistance) and neutral for others (e.g., seed color), which may be important and could be emphasized by use of other breeding methods.

Evidence is not consistent for the utility of bulk selection in self-fertilizing plant species. Effectiveness of bulk breeding may be realized in the long-term, but the predictability of natural selection may not be sufficient for identifying cultivars for agricultural uses. In comparison with other breeding methods that emphasize artificial selection among and within progenies, bulk breeding may not have broad appeal. Bulk breeding is probably included in most breeding programs of self-fertilizing species, but the populations are carried in conjunction with other breeding systems.

Modified Bulk (Mass Selection)

Modified bulk breeding, or mass selection, includes both natural and artificial selection. Mass selection, therefore, is practiced to reduce fluctuations in gene frequency among environments for specific traits. For example, greater pressure for rust resistance can be imposed with a uniform application of inoculum in the bulk population. Selection pressure for maturity, seed color, kernel size, and plant height also may be imposed to direct the evolution of the bulk population. These modifications of bulk breeding are imposed to develop populations that have a greater potential of isolating cultivars for agricultural purposes. Mass selection is based on individual plants and would include the forces of natural selection, as well as the artificial selection that is practiced.

Types of populations used in the modifications of bulk breeding would be the same as those for bulk breeding. The mechanics of forming the populations and handling the material would be similar, except at harvest. Instead of bulk harvesting the populations, seed is harvested only from selected plants; seeds from the selected plants are bulked to form the population for the next generation of selection, preferably an equal quantity of seed from each plant. Mass selection would reduce the genetic variability of the traits under selection and selection pressure would be in the desired direction in each generation. Gene frequency changes, therefore, are under greater control by the breeder. Changes in gene frequency for the traits not under direct mass selection would behave in the same manner as for bulk breeding; i.e., there would be survival of the most fit and competitive genotypes. Because of the additional forces of artificial selection for desired traits, the duration of mass selection in bulk populations will not be as long as for bulk breeding. The approach to homozygosity, however, will be the same for both methods.

Several modifications of mass selection have been suggested for both self- and cross-fertilizing species. Modifications have been made to increase the efficiency of selection and the heritability of the traits under selection. These include mechanical means of selecting sound, plump seeds (fanned mass selection), clipping (plant height), and whip test (straw strength). Gardner[11] suggested the grid method of mass selection to reduce the effects of the microenvironment on selection in corn, a cross-fertilizing plant species. All of the suggested modifications were made to reduce the time required to improve the bulk population, control the direction of gene frequency changes, and improve the heritability of the traits under selection. The breeder, therefore, has greater control on the destiny of the population rather than leaving it to the forces of natural selection. Mass selection is particularly well suited for use in applying greater selection pressure for easily recognizable, desirable plants in the early generations of selection.

Single-Seed Descent

Although single-seed descent is not a new breeding system, it has received renewed interest in recent years in the breeding of self-fertilizing crop species, particularly in oats, soybeans, and wheat. In contrast to the other bulk-breeding methods, the forces of natural and artificial selection are minimized as much as possible. Hence, the primary objective of single-seed descent is to maintain during inbreeding individuals of each genotype in the original segregating population; each F_2 (or S_0) plant is theoretically represented at some future generation when artificial selection is imposed on progenies that are relatively homozygous. The single-seed descent method is included with the other bulk-breeding systems

Table 7
COMPARISON OF THREE METHODS OF SELECTION WITHIN BULK POPULATIONS OF SELF-FERTILIZING PLANT SPECIES

	Relative advantages		
Materials and methods	Bulk	Modified (mass) bulk	Single-seed descent
Type of population	Any type	Any type	Segregating
Duration of selection	Longest	Intermediate	Shortest
Genetic variability of F_7	Least	Intermediate	Greatest
Effects of natural selection	Greatest	Intermediate	Least
Effects of artificial selection	Some	Greatest	None
Rate homozygosity attained	Same	Same	Same
Stage for extraction of progenies	Latest	Intermediate	Earliest
Relative cost	Intermediate	Greatest	Least
Resources needed	Intermediate	Greatest	Least
Record keeping	None	None	None

because one (or more) seeds are saved from each plant in each generation of self-fertilization and bulked to form the next generation bulk. No pedigree records of lines of descent are kept with single-seed descent, which also is similar to the other bulk methods of breeding.

Populations used for single-seed descent are usually segregating F_2 populations formed from hybridization of pure lines. F_2 populations, however, are not a prerequisite for single-seed descent. Populations may be formed from use of complex hybrid populations, varietal mixtures, or landrace varieties. But the theoretical advantages of single-seed descent would be nullified if there was no segregation within the original population. If, for example, the original population was a mixture of ten pure-line cultivars, individuals recovered in some future generation (e.g., S_7) would represent only the ten original cultivars, unless mutations or some accidental hybridization had occurred among the ten cultivars. Segregating populations, therefore, are a prerequisite for effective use of single-seed descent.

Segregating populations are grown and the breeder harvests at least one seed from each plant. Seeds are bulked and planted for the next generation. This method is continued until the desired level of homozygosity (F_5—F_7) has been obtained. At this time, the breeder harvests all seeds from each plant to grow in a progeny row for seed increase. Evaluation trials are conducted to determine the relative performance of the single-seed-derived progenies for productivity, agronomic traits, pest resistance, etc. Until the evaluation trials, no intentional selection is imposed by the breeder and no effective natural selection is considered. Breeding programs that emphasize single-seed descent usually try to complete as many generations of self-fertilization per year as possible. Use of glasshouses and winter nurseries is common. Thousands of plants may be sown in dense stands in a very small area. The only requirement is that the plants have sufficient area and nutrients to produce one or two viable seeds; vegetative development, quality, productivity, agronomic traits, and pest resistance are not considered at this time. Effective use of glasshouses may permit obtaining F_5 or F_7 generation seed in 2 years. Bulk populations are then grown in the field, and selection for obvious traits can be made in the plant selections used for seed increase.

Single-seed descent has, it seems, several important advantages for modern breeding programs (Table 7). Populations either can be extensively sampled (say, 10,000 individuals) or 20 populations can be adequately sampled (say, 500 individuals) with use of relatively limited resources to obtain relatively homozygous individuals in a short time that represent the range of variability of the original segregating populations. Also, the variability among the homozygous individuals will be twice the variability of the original segregating population (Table 4). Because of the rapid "turn-around time" from F_2 to F_7, the disadvantages of no selection during self-fertilization are negated. A new series of segregating populations can be initiated at any stage and not require extensive resources. The method also requires no pedigree records until selection is imposed. Although single-seed descent breeding methods are more attractive to use in self-fertilizing species, they can be used for cross-fertilizing species except that the ease of use is reduced; artificial means of self-fertilization would be needed, which reduces the advantage of rapidly obtaining homozygous genotypes. Also, space requirements would be much greater for large plants, such as corn, so that it would not be possible to advance large populations in the glasshouse by single-seed descent.

Population Improvement

Breeding systems discussed so far have had one primary objective — the development of improved cultivars. Except for some of the bulk methods of selection, little or no consideration was given to long-term breeding objectives. Germplasm included for selection was in most instances crosses of elite pure lines (pedigree and single-seed descent), incorporation of a specific gene(s) in an otherwise elite pure line (backcrossing), and mixtures or multiple crosses of elite pure lines (bulk methods). The germplasm base was, in most instances, of a restricted nature, although not necessarily so. All of the breeding systems can be used in broad, genetically based populations, such as landrace cultivars of corn or introduced germplasm, but the breeding methods discussed are often used to improve otherwise elite pure lines for traits that have a relatively high heritability. For many of the important economic traits, where productivity is usually of greatest concern, it was questioned if adequate additive genetic variability was available and if the breeding methods used were adequate to expect continued genetic progress. Because most of the economically important traits are quantitatively inherited (controlled by many genes), it was suggested that other breeding systems would be more appropriate for these types of traits. Breeding systems, designated as recurrent selection, were suggested to select for quantitative traits.[12-15]

Recurrent selection methods were developed (1) to emphasize selection for quantitatively inherited traits, and (2) to conduct selection in a cyclical manner to affect a gradual increase in the relative frequency of the genes for the traits under selection. Also, recurrent selection usually is conducted for populations that have a broad genetic base. In comparison to the breeding systems discussed previously, recurrent selection methods are used to improve germplasm sources for future use. Several different schemes for conducting recurrent selection were developed to emphasize selection for different types of gene action considered important in the expression of the traits under selection. If the traits under selection are considered to be under control of primarily additive genetic effects, Hayes and Garber,[12] Jenkins,[13] and Gardner[11] suggested methods that emphasized selection for general combining ability. Hull,[14] however, believed that overdominant genetic effects were of greater importance and suggested selection should be emphasized for specific combining ability. Because the relative importance of additive and nonadditive genetic effects was not established, Comstock et al.[15] suggested reciprocal recurrent selection that capitalized on both general and specific combining ability. All of the different recurrent selection schemes, however, can be classified in two groups: those based on simple phenotypic (or mass) selection, and those based on selection with progeny testing. The specific scheme used for improvement of a population depends, to some extent, on the type of gene action considered predominant in a particular plant species, type of plant propagation (e.g., self- or cross-fertilizing by seed or vegetatively), and the primary form of the superior cultivars derived from the improved populations (pure lines or hybrids). All of the different recurrent selection methods can be used for most crop species, but the mechanics of conducting them will vary.

Except for simple mass selection, recurrent selection includes three phases that are conducted in a repetitive manner: (1) development of progenies for evaluation; (2) evaluation of progenies in replicated trials conducted in different environments; and (3) recombination of the superior progenies (based on information obtained from replicated trials) to synthesize the population for the next cycle of selection. The purpose of the repetitiveness of these three phases, therefore, is to select the best performing progenies from the population in each cycle of selection. The reasoning for this is to recombine those progenies that have the greatest frequency of desirable alleles for the trait(s) under selection. Hence, the recurrent selection methods were designed to cause a small, gradual increase in the frequency of desirable alleles within the population. An increase in gene frequency would enhance, theoretically, the chances of obtaining superior cultivars from the populations improved by recurrent selection.

Recurrent selection methods are further classified by whether they emphasize selection within one (intra) population or between two (inter) populations, and by the types of progenies evaluated. Use of one or more populations usually depends on the primary objective of selection and whether pure-line or hybrid cultivars are the primary product. But there are exceptions. More different types of recurrent selection have been conducted in corn than any other plant species. The ultimate breeding goal in most modern corn-breeding programs is selection of pure lines that have superior performance in hybrids; hence, parallel improvement of two populations that exhibit high heterosis in their cross seems to be a logical breeding approach. Methods that emphasize selection within one population, however, have been as successful as selection between two populations for developing lines that when crossed to elite lines produced superior hybrids. Recurrent selection methods that have been used for intra- and interpopulation improvement and some of the different types of progenies used to identify individuals for recombining are listed in Table 8. A more detailed account of these breeding systems was discussed by Hallauer and Miranda.[16]

Intrapopulation Recurrent Selection

Several different methods of conducting selection within a population have been suggested (Table 8). They range in complexity from simple mass selection of individuals to crosses of individuals to testers and developing inbred lines with different levels of homozygosity.

Mass

Mass selection was one of the original forms of plant selection and is frequently used today, particularly for traits that have relative high heritabilities. Mass selection can be conducted either with or without genetic recombination. In self-fertilizing plant species, mass selection usually is without recombination (see modified bulk breeding), but Matzinger et al.[17] has conducted mass selection with recombination within tobacco populations, a self-fertilizing plant species. Mass selection without recombination, however, does not fit the definition of recurrent selection because recombination is one of the essential features for the long-term improvement of populations. In cross-fertilizing species, mass selection has had mixed results. It has been successful in some populations for some traits and not in others. Mass selection as a breeding system was seemingly ineffective in the early part of the 20th century.[18] The ineffectiveness of the early studies on mass selection was attributed to poor experimental technique. Mass selection received greater interest after it became obvious that most corn populations had adequate additive genetic variance for mass selection to be effective. Renewed interest in mass selection occurred after Gardner[11] suggested several modifications to improve efficiency of selection, primarily for control of environmental effects. Although results of several mass-selection studies in corn have been reported during the past 20 years, results are not consistent, particularly for complexly inherited traits. The reasons are the same as reported previously: it seems some type of progeny tests are required to accurately determine the breeding values of individuals. Mass selection, however, will continue to be used in specific instances because of its simplicity and economic considerations.

Table 8
SOME OF THE TYPES OF PROGENIES TESTED IN RECURRENT SELECTION SCHEMES FOR INTRA- AND INTERPOPULATION IMPROVEMENT

Selection method	Years per cycle[a]	Parental control	σ_G^{2b} σ_A^2	σ_D^2
Intrapopulation				
Phenotypic (mass) with recombination				
One sex	1	1/2	1	1
Both sexes	1 or 2	1	1	1
Ear-to-row				
One sex	1	1/2	1/4[c]	0
Both sexes	2	1	1/4[c]	0
Half-sib				
Self recombined	3	2	1/4	0
Half-sibs recombined	2	1	1/4	0
Full-sib				
Self recombined	2	2	1/2	1/4
Full-sibs recombined	2	1	1/2	1/4
Inbred				
S_1	2	1	1[d]	1/4[e]
S_2	3	1	3/2[d]	3/16[e]
S_n	n + 1	1	s[e]	0
Interpopulation				
Reciprocal recurrent: half-sib				
Self recombined	3	2	1/4	0
Half-sibs recombined	2	1	1/4	0
Modified I[f]	3	1/4	1/4	0
Modified II[f]	1 or 2	1/2	1/4	0
Reciprocal recurrent: full-sib				
Self recombined	2	2	1/2	1/4
Full-sibs recombined	2	1	1/2	1/4
Modified I[g]	2	1	1/2	1/4
Modified II[g]	2	1	1/2	1/4

[a] Years per cycle based on temperate zone conditions of one growing season per year and use of a winter nursery. Other combinations of growing season were listed by Hallauer and Miranda, p. 257.[16]
[b] Variation among progenies (plants for mass selection) that are tested.
[c] If within-plot selection is practiced, 3/8 σ_A^2 should be added.[19,20]
[d] Not equal to σ_A^2 unless $p = q = 0.5$ or no dominance.
[e] Coefficient difficult to define unless $p = q = 0.5$.
[f] Modifications of reciprocal recurrent selection were suggested by Paterniani.[26]
[g] Modifications of reciprocal recurrent selection based on full-sib progenies were suggested by Marquez-Sanchez.[31]

Ear-To-Row

Ear-to-row selection also is a form of half-sib selection, but it was listed separately because of its historical significance. Hopkins[4] suggested ear-to-row selection to enhance selection for oil and protein content in the corn variety 'Burr's White'. Hopkins[4] used ear-to-row methods to evaluate progenies, but no systematic manner was used in the recombination of superior progenies. Hence, ear-to-row selection as a breeding system did not seem to be effective in the early part of the 20th century and was generally ignored until recently. The seemingly poor results from ear-to-row selection also could be attributed to poor parental control and experimental technique. As for mass selection, ear-to-row selection has received greater interest in recent years because of the modifications suggested by Lonnquist[19] and Compton and Comstock.[20] These modifications included more than one replication to obtain a better measure of the breeding values of individuals, an isolation planting to control parentage, and isolations for recombination that included only the superior progenies. Ear-to-row selection, as presently used, is more complex to use than the original form, but it has proven effective. Because of the suggested changes to enhance selection, ear-to-row selection is being used more extensively now than after the original suggestion by Hopkins.[4]

Half-Sib Progeny

Selection based on half-sib progeny tests has been used more frequently than any of the other forms of progeny tests. Half-sib selection also is often called topcross, testcross, or polycross tests. As the name implies, half of the parentage is common to all the progenies (hence, half-sib progenies), and this may be the only common parameter for the possible methods of half-sib progeny selection. Usually, a group of individuals is crossed to a common tester to produce the half-sib progenies. But the individuals may be at the S_0, S_1, S_2 ... S_n level, and the tester may be the original source population from which the individuals were sampled, an unrelated population, a bulk population of individuals being crossed, or an unrelated inbred line or hybrid. Additionally, the progenies recombined may be from remnant half-sib seed or S_1 seed if the individuals crossed to the tester were self-fertilized. Possible combinations for the following list are 16, not considering inbreeding of individuals crossed beyond S_1 and possible options in types of testers.

Individuals crossed	Tester	Progenies recombined
S_0	Broad genetically based	Half-sibs
S_1	Narrow genetically based	S_1s
	Unrelated (population, line, hybrid)	
	Related (population, line, hybrid)	

Choices of individuals to cross, types of testers to use, and types of progenies to recombine depend on how the plant species are propagated, plant materials and facilities available, previous breeding information, type of gene action considered of greatest importance, and the level of plant-breeding effort. Half-sib progeny selection, however, is restricted to crops that are naturally cross-fertilized (e.g., alfalfa, corn, rye, sunflower) or can be easily crossed (e.g., sorghum, cotton, tobacco). Half-sib progeny selection has been used extensively in corn, but the specific method suggested for use depended on what was considered the predominant type of gene action expressed in the crosses.[13-15] Selection based on half-sib progenies therefore anticipates that the elite genotypes will be candidates for use in hybrids, particularly if the tester is an inbred line. In these instances the population is improved per se. The improved population is projected to be the source population for the extraction of lines to be used in crosses with the tester.

Data reported on the use of half-sib progeny selection in corn were summarized by Sprague and Eberhart[21] and Hallauer and Miranda.[16] Although several different methods of half-sib progeny selection have been used, rate of progress by the different methods was similar — about 2 to 3% per cycle. Additive genetic effects seemed to be emphasized in all instances. Comparisons of rate of progress with other selection methods were not consistent, although theoretically other selection methods (e.g., S_1 or S_2 progeny selection) had greater predicted rates of gain. Other methods of intrapopulation selection may be more effective for emphasizing selection of additive genetic effects, but half-sib progeny selection will continue to be widely used. For those plant species that are committed to hybrid breeding programs, half-sib progeny selection will be used empirically for determining general combining ability of lines and progenies.

Full-Sib Progeny

Selection based on full-sib progenies for intrapopulation selection has not been used as extensively as other methods. In contrast to half-sib progeny selection, full-sib progenies have both parents common, but both parents are different for each full-sib progeny. Full-sib progenies are established within a population by crossing two individuals. If the crosses are made reciprocally between the two individuals, seed produced is bulked for testing, and remnant full-sib seed used for recombination. If multiple pollinations can be made for each individual (e.g., corn, sunflower, tobacco, tomato), both cross- and self-fertilized seed can be produced; for these instances, selfed seed can be used for recombination. Though full-sib progeny selection has not been used to the same extent as ear-to-row and half-sib progeny selection, full-sib progeny selection has two advantages over both of these selection methods. First, full-sib progeny selection has greater emphasis on selection for additive genetic effects ($1/2$ vs. $1/4$, Table 8). Secondly, if dominance effects are of importance, full-sib progeny selection will include one quarter of the variance due to dominance deviations. Full-sib progeny selection is more complex to conduct, but the number of years per cycle is no greater than for ear-to-row and half-sib progeny selection, and greater additive genetic variance is available among progenies as a basis of selection. Full-sib progeny selection may have greater use when populations have been improved to the extent that additive genetic effects are smaller and dominant effects have greater importance.

Inbred Progeny

Inbred progeny selection has been popular in applied breeding programs of self-fertilizing species for many years, but it has only recently been used in recurrent selection programs, both for self- and cross-fertilizing plant species. In contrast to the other progenies evaluated for intrapopulation improvement, no crosses are produced. Progenies evaluated in trials are established by either mechanical methods of self-fertilization (e.g., corn and sorghum) or by natural self-fertilization (e.g., soybeans and wheat). Level of inbreeding of the progenies depends on the effects of inbreeding depression, type of selection practiced within and among progenies, and the traits under selection for the different generations of inbreeding. Possible options with inbred progeny selection are many, and they have advantages and disadvantages for conducting recurrent selection. For example, greater levels of inbreeding increase the additive genetic variance among progenies, but this also increases the duration of each cycle of selection (Table 8). Additional generations of inbreeding permit selection for certain traits in each generation before conducting expensive yield trials, but unfavorable genetic correlations may reduce the effects of selection in the trials. In some plant species (e.g., oats and soybeans), single-seed descent may be used to obtain the inbred progenies, in which case minimum selection would be used before conducting the evaluation trials. Procedures used for conducting inbred progeny selection will have to be adapted for the plant species and traits under selection.

The primary advantage of inbred progeny selection is that it emphasizes selection for additive genetic effects. If additive genetic variance is the predominant type of genetic variance within a population, inbred progeny selection would have an advantage over the other methods of intrapopulation improvement. For self-fertilizing plant species, inbred progenies are easily obtained, but recombination may be a problem. Most of the evidence, however, indicates additive genetic variance is of greater importance in naturally self-fertilizing species.[22] If recombination is not a serious problem, recurrent selection based on inbred progeny selection would be a logical method for naturally self-fertilizing plant species, particularly if progenies are derived by efficient methods of single-seed descent to reduce the time between cycles of selection.[23]

Inbred progeny selection in naturally cross-fertilizing plant species is complicated by having to mechanically produce the progenies. But this is generally not a serious restriction in its use; opportunities for selection exist at each generation of selection and recombination among selected progenies can be made rather easily. In corn, for example, inbred progeny selection conducted at the S_2 generation permits selection among S_0 plants and S_1 progenies before conducting evaluation trials. Selection at the S_0 and S_1 generations is for traits that have a relatively high heritability (e.g., pest resistance), and replicated S_2 progeny trails are conducted for more complex traits (e.g., yield). Hence, multistage selection permits selection for several traits to improve the populations for several different traits. Though inbred progeny testing seems to have, for example, a theoretical advantage over half-sib progeny selection, the empirical results have not been consistent for the two distinctly different methods of selection.[16,24] It seems, however, that inbred progeny selection will continue to increase in its use in the future. Its use is amenable for most plant species.

Interpopulation Recurrent Selection

The only distinction between the methods used for interpopulation improvement and those for intrapopulation improvement is that selection is emphasized to improve the cross of two (or more) populations rather than for the improvement of the populations per se. Obviously, interpopulation improvement was designed for those plant species that have the potential of producing hybrids as the final product of selection. As a corollary, types of gene action involved in the expression of heterosis were instrumental in the development of interpopulation recurrent selection methods.[15] Because the main objective of interpopulation recurrent selection is the improvement of the cross between populations, the number of selection methods is restricted to those that include progenies produced by crosses. Hence, interpopulation recurrent selection methods include either half- or full-sib progenies. Most of the methods used for interpopulation improvement also tend to be more complex than those for intrapopulation improvement.

Reciprocal Recurrent

The original suggestion of interpopulation recurrent selection was made by Comstock et al.[15] to accommodate selection for both general (primarily additive genetic effects) and specific (nonadditive genetic effects) combining ability in corn. The objective of selection was to improve the gene frequency of two populations in a complementary fashion for the improvement of the cross between the two populations. Two sets of half-sib progenies are used in the evaluation trials, one for each of the two populations, say A and B. Plants in A are used as males to pollinate plants (designated as females) in B. Conversely, plants in B are used as males to pollinate plants in A. Each male is crossed to a different set of females. Equal quantities of seed on the female plants pollinated by each male are bulked to form the half-sib progenies for test. For example, if 100 males in A are crossed to female plants in B and 100 males in B are crossed to female plants in A, there will be 200 half-sib progenies for test; i.e., 100 testcrosses of A male plants with B (the common tester) and 100 testcrosses of B male plants with A (the common tester). The plants used as males are usually, but not always, selfed to produce S_1 seed on each male plant. The two sets of half-sib progenies are grown in replicated tests repeated over environments to determine the breeding values of the males sampled from each population. As with half-sib progeny selection within one population, either S_1 seed of the males or remnant half-sib seed of the superior half-sib progenies are recombined to reconstitute the population for the next cycle of selection. Because the males used in each population represent an unselected sample, it is not uncommon to self the S_0 plants in each population to form a series of S_1 progenies for the A and B populations. Plants used as males to produce the half-sib progenies are from selected S_1 progenies; selection among S_1 progenies include those traits that can be screened before pollination.

Reciprocal recurrent selection has had greatest use in improvement of corn populations. Although the method was designed to include selection for both additive and nonadditive genetic effects, most evidence suggests that response to selection was affected primarily by additive effects. Direct response to reciprocal recurrent selection has been positive in all instances in the population crosses, but correlated response of the populations per se has been inconsistent.[16] Crosses of lines extracted from populations improved by reciprocal recurrent selection were 8% better, on the average, than crosses of lines extracted from the original unselected populations.[25] General conclusions on the effectiveness of reciprocal recurrent selection are not available at this time. Only a limited number of studies have been reported; number of cycles of selection completed is few in most instances; and the confounding effects of inbreeding due to small effective population sizes have limited, in some instances, the interpretation of the results.

Modified Reciprocal Recurrent

Paterniani[26] suggested two modifications of reciprocal recurrent selection to reduce the amount of hand pollinations required to produce the half-sib progenies. Both modifications require isolation plantings of the two populations under selection. In the first modification, open-pollinated ears of B are planted ear-to-row in a detasseling block with the male rows including plants from A; in a second detasseling block, ears of A are planted ear-to-row with the male rows including plants from B. Half-sib families are obtained by open pollination, harvested in bulk, and evaluated in replicated trials. Remnant half-sib seed of the selected half-sib progenies is used for recombination. The second modification requires plants that produce seed on more than one flower, such as prolific corn. The same procedures are used in the two isolation blocks as described for the first modification. Additionally, bulk pollen of the male plants in each isolation block is collected to pollinate another flower of the male plants. The crossed half-sib progenies are evaluated in replicated trials and the hand-pollinated ears (half-sibs) are used for recombination.

Reciprocal Full-Sib

Reciprocal recurrent selection based on full-sib progeny evaluation was suggested by Hallauer and Eberhart[27] to emphasize selection for nonadditive genetic effects. The mechanical methods of conducting reciprocal full-sib selection are very similar to those for reciprocal recurrent selection. The major difference is that individual plants of two populations, say X and Y, are crossed to produce full-sib progenies rather than half-sib progenies. Full-sib progenies are produced in the same manner described for intrapopulation full-sib selection except plants are sampled from two populations rather than within one population. Individual plants in X are crossed on to individual plants in Y, and vice versa. Full-sib seed for the two plants included in the crosses is bulked to use in replicated trials repeated over environments.

The original suggestion of reciprocal full-sib selection was in corn and required multiple pollinations on each plant. It required plants that produced viable seed in sufficient quantities on two ears. Full-sib crosses of individual S_0 plants in X and Y are reciprocally crossed to produce full-sib seed and the other ear on each plant was selfed to produce S_1 progeny seed; hence, S_1 seed was produced on each plant included in the full-sib crosses, and full-sib seed was available by bulking the reciprocal crosses between each of the two plants. S_1 progeny seed was used for recombination to resynthesize populations X and Y for the next cycle of selection.

Reciprocal full-sib selection has had limited use because of the complexity of the system, multiple pollinations on the same plant, and also the fact that other methods are as effective if additive genetic effects are of greatest importance. Accurate record keeping is essential because of the different types of seed produced on each plant. Multiple pollinations on the same plants restrict its use to plant species with multiple flowers, such as prolific corn, that produce adequate numbers of seeds; the ease in making the cross- and self-fertilizations also will influence the use of reciprocal full-sib selection. Although reciprocal full-sib selection was designed to maximize selection for nonadditive effects in the development of single-cross hybrids,[28] Hoegemeyer and Hallauer[29] reported that the lines isolated from the original populations also had excellent general combining with other elite lines. Additional cycles of selection will be necessary to determine if reciprocal full-sib selection will effectively select for the seemingly minor importance of nonadditive genetic effects. Hallauer[30] also suggested use of S_1 progenies rather than S_0 plants in making the full-sib progenies. This suggestion would remove the restrictions of multiple flowering and simplify record keeping of the S_1 and full-sib progeny seed produced on different plants within an S_1 progeny row.

Modified Reciprocal Full-Sib

Two modifications of reciprocal full-sib selection were suggested by Marquez-Sanchez.[31] Both modifications remove the requirement of requiring multiflowered plants to produce the full-sib and S_1 progenies. In the original method of reciprocal full-sib selection, both the full-sib and S_1 progenies were produced on individual plants. Marquez-Sanchez suggested that progenies (S_1, full-sib, or half-sib) be developed in the populations before developing full-sib progenies. For the first modification, individuals plants chosen to produce the full-sib progenies between two populations also are used to pollinate one or more of their respective siblings. Hence, the same plant was used as a male to produce the two types of progenies, but the seed to propagate the plant used to produce the full-sib progeny was produced on a sibling rather than by self-fertilization. The second modification uses the same procedures except the pollen of the plants within the progeny row is bulked to produce the full-sib progenies. Hence, the full-sib progenies are produced from a bulk sample of pollen from the progenies rather than from crosses of individual plants. These modifications have not been tested, but their main advantage is the use of progenies rather than individual plants. Accurate record keeping, however, will be required to maintain pedigrees of the different progenies.

Genetic Gain

Response to any of the recurrent selection schemes can be of two forms: (1) predicted gain based on genetic information available and the specific method of selection used; and (2) realized gain measured from replicated trials. In the first instance, predicted gain provides guidelines, under certain assumptions, of the response expected in a population for a given selection method. Realized gain is a measure of the gain actually obtained after selection has been completed. Usually, predicted and realized gain are compared to determine how the realized gain compares with the predicted gain; i.e., how accurately the variables used to predict gain agreed with the actual response to selection. If information is available, predicted gain is a useful parameter to calculate the most efficient selection method to use for a given situation.[16]

Genetic gain from selection is a measure of the expected response of a population based on the heritability of the trait under selection. In its simplest form predicted genetic gain (Δ_G) is determined as $\Delta_G = H(\overline{X}_S - \overline{X})$, where H is the heritability of the trait under selection, $\overline{X}$ is the mean of all progenies tested, and $\overline{X}_S$ is the mean of the selected progenies used for recombination. The expression ($\overline{X}_S - \overline{X}$) is usually called the selection differential.

It is obvious that the greater the heritability and the selection differential, the greater the predicted response to selection.

There are several variables, however, that can influence predicted gain, particularly when comparing different methods of recurrent selection. A useful formula has been developed that includes the different variables for predicting genetic gain from different methods of recurrent selection:

$$\Delta_G = kc\sigma_{A'}^2/(y\sigma_P)$$

where k is the standardized selection differential (selection intensity) (see p. 166, Hallauer and Miranda[16]), c is a function of the parental control of the progenies used for recombination (see Table 8), $\sigma_{A'}^2$ is the additive genetic variance among progenies (see Table 8), y is the number of years per cycle of recurrent selection; and σ_P is the phenotypic standard deviation among progenies (see the above section titled "Breeding Systems"). This expression includes five variables, each of which can affect genetic gain. In making calculations to compare different methods of selection k can be constant, but the other variables will change. An increase for any of the three variables (k, c, and $\sigma_{A'}^2$ in the numerator (with the others held constant) will increase expected gain, whereas a decrease in either y or σ_P (with the others held constant) will increase genetic gain. Greater genetic response to selection will be expected by increasing the selection differential, greater parental control, greater additive genetic variance among progenies tested, fewer number of years to complete each cycle of selection, and small phenotypic variation of the progenies tested (i.e., relatively high heritability). Usually, trade-offs have to be made in comparing the relative efficiency of different selection methods. One cycle of mass selection can be made each year, but the heritability of the traits under selection may be low. S_2 recurrent selection may have a higher heritability, but it requires 3 years to complete one cycle of selection. The prediction formula, therefore, is a very useful relation to calculate predicted genetic gain to provide guidelines in choosing a selection method that would be more effective for a given situation.

Selection studies that have been conducted for a long term have permitted comparisons between predicted and realized response to selection. In most instances, realized response was less than that predicted. It has not been obvious in all instances why the realized response was less than the predicted. Some possible causes include inadequate sampling of the populations to include the range of variability available (σ_A^2 overestimated); too high selection differential, which reduced number of progenies recombined, resulting in inbreeding depression, genetic drift, or both; genotype by environment interactions were underestimated which inflated the estimate of heritability; and poor experimental technique. Although the predicted gain tends to be overestimated, valid comparisons can be made among selection methods to determine which one is most effective for the parameters available. Prediction formulas for some of the more common methods of recurrent selection are listed in Table 9 on a per-cycle basis.

GERMPLASM

All of the breeding systems were discussed with reference to some source population or germplasm. The specifics of the types of populations to use for the breeding systems were not discussed because all of the breeding systems can, in most instances, be imposed on any type of population. There are, however, certain exceptions. Simple phenotypic, or mass selection, has different connotations if imposed on a mixture of self-fertilizing pure lines without genetic recombination vs. mass selection in a naturally cross-fertilizing plant species with genetic recombination. But mass selection with genetic recombination has been conducted in both self- and cross-fertilizing species. The backcross method of breeding is usually

Table 9
EXPECTED GENETIC GAIN (Δ_G) FOR SOME OF THE DIFFERENT INTRA- AND INTERPOPULATION METHODS OF RECURRENT SELECTION

Selection method	Expected genetic gain (Δ_G)[a]
Mass	
One sex	$k(1/2)\sigma_A^2/(\sigma_W^2 + \sigma_{DE}^2 + \sigma_{AE}^2 + \sigma_D^2 + \sigma_A^2)^{1/2}$
Both sexes	$k\sigma_A^2/(\sigma_W^2 + \sigma_{DE}^2 + \sigma_{AE}^2 + \sigma_D^2 + \sigma_A^2)^{1/2}$
Modified ear-to-row[b]	
One sex	$k(1/8)\sigma_A^2/[\sigma^2/re + (1/4)\sigma_{AEe}^2 + (1/8)\sigma_A^2]^{1/2}$
Both sexes	$k(1/4)\sigma_A^2/[\sigma^2/re + (1/4)\sigma_{AEe}^2 + (1/4)\sigma_A^2]^{1/2}$
Half-sib	
Remnant half-sib seed	$k(1/4)\sigma_A^2/[\sigma^2/re + (1/4)\sigma_{AEe}^2 + (1/4)\sigma_A^2]^{1/2}$
Self-seed	$k(1/2)\sigma_A^2/[\sigma^2/re + (1/4)\sigma_{AEe}^2 + (1/4)\sigma_A^2]^{1/2}$
Full-sib	$k(1/2)\sigma_A^2/\{\sigma^2/re + [(1/2)\sigma_{AE}^2 + (1/4)\sigma_{DE}^2]/e + (1/4)\sigma_D^2 + (1/2)\sigma_A^2\}^{1/2}$
Inbred	
S_1	$k\sigma_A^2/\{\sigma^2/re + [\sigma_{AE}^2 + (1/4)\sigma_{DE}^2]/e + (1/4)\sigma_D^2 + \sigma_A^2\}^{1/2}$
S_2	$k(3/2)\sigma_A^2/\{\sigma^2/re + [(3/2)\sigma_{AE}^2 + (3/16)\sigma_{DE}^2]/e + (3/16)\sigma_D^2 + (3/2)\sigma_A^2\}^{1/2}$
S_7	$k2\sigma_A^2/(\sigma^2/re + 2\sigma_{AE}^2/e + \sigma_A^2)^{1/2}$
Reciprocal recurrent	$\dfrac{k_1(1/4)\sigma_{A12}^2}{(\sigma_{12}^2/re + (1/4)\sigma_{AE12}^2/e + (1/4)\sigma_{A12}^2)^{1/2}} + \dfrac{k_2(1/4)\sigma_{A21}^2}{(\sigma_{21}^2/re + (1/4)\sigma_{AE21}^2/e + (1/4)\sigma_{A21}^2)^{1/2}}$
Reciprocal full-sib	$k(1/2)\sigma_A^2/\{\sigma^2/re + [1/4\sigma_{DE}^2 + (1/2)_{AE}^2]/e + (1/4)\sigma_D^2 + (1/2)\sigma_A^2\}^{1/2}$

[a] k, r, and e refer to the selection differential, number of replications, and number of environments, respectively. The components of variance include the additive genetic variance (σ_A^2), deviations due to dominance (σ_D^2), interaction of additive effects with environments (σ_{AE}^2), interaction of dominance effects with environments (σ_{DE}^2), experimental error (σ^2), and within-plot variance (σ_W^2).

[b] If selection is also practiced among plants within plots, $3/8\sigma_A^2$ should be added.[19,20]

limited to the transfer of one or two traits having a high heritability, but backcrossing also has been used in an attempt to transfer more complex traits. With minor exceptions, however, the principles of the different breeding systems apply regardless of the parental material included in the populations under selection.

Source germplasm in which one can initiate selection can range from related weedy species having seemingly little usefulness to F_2 populations derived from a cross of two elite pure lines. The choice of germplasm, of course, usually depends on the breeding goals. A poorly adapted, agronomically poor, weedy species may be an excellent source of pest resistance not available in highly selected elite germplasm. An F_2 population formed from crossing two elite pure lines may have restricted genetic variability, but the breeder may have greater opportunities of recovering lines that include the desirable traits of both parental lines. Backcrossing would be the logical method to recover pest resistance from weedy species and pedigree selection for extraction of lines from the F_2 population. Backcrossing could be used in crosses of elite lines, however, and pedigree selection may be used in landrace varieties that have a broad genetic base.

The first requirement in choice of germplasm is adequate genetic variability for the trait(s) to be selected. If the sources of germplasm are similar for genetic variability, mean performance would be considered. If genetic variability is similar, the one with the highest mean performance would be more desirable. Genetic variability is the raw material that the breeders manipulate to attain their projected goals. Choice of germplasm, therefore, often determines the ultimate success or failure of a breeding program regardless of the breeding strategies that may be used.

The genetic base of many of the economically important crops tends to be restricted.[32] This has occurred because a few elite genotypes were identified that had wide acceptance by the growers. These elite genotypes possessed the genetic factors that contributed to consistently superior performance, often over large areas. Subsequent breeding efforts were directed either to improve these superior genotypes (by pedigree selection in crosses with other elite genotypes) or to incorporate other traits that had a high heritability (by backcross selection). Consequently, the genetic base of many modern breeding programs has become restricted.[33,34]

Although parental materials within specific breeding programs include only a small portion of the germplasm available for specific plant species, a wide range of germplasm is available for most plant species. How long this status will prevail, however, is debatable. Modern technology is eliminating many of the wild relatives of our cultivated plant species and the original landraces are fast disappearing, but these types of problems are usually beyond the scope of specific breeding programs. The main problem, therefore, is how one chooses his parental materials to initiate a breeding program. Unfortunately, breeding information is lacking for many of the extensive germplasm collections. Detailed physiological and morphological data have been recorded in many instances, but information on performance, stability, adaptability, inbreeding depression, heterosis, etc. is not available to provide guidelines for choice of germplasm from the extensive collections. Because this information is lacking, breeders tend to use germplasm that has had proven performance based on experience by other breeders. This is especially true for breeding programs that have short-term breeding objectives. Lack of breeding information on germplasm restricts, therefore, use of broad-genetic-base materials in modern breeding programs.

Choice of germplasm to include as breeding materials depends to a large extent on the past experience of breeders, specific breeding goals, and limited information, published or unpublished, that may be available. Before initiating an extensive breeding program, it is essential that the breeder gather all the information available for possible sources of germplasm. Breeding systems can be changed in handling of the germplasm to make selection more effective, but the wrong choice of germplasm may reduce the chances of success. The only recourse seems to be for breeders to initially use proven sources of parental materials for the short-term, and gradually obtain breeding information on other sources that may have promise for the future. Similar care and effort should be given to choice of germplasm as is given to choice of breeding systems that will effectively extract superior genotypes from the source population.

SUMMARY

Plant breeding is the art and the science of improving the genetic pattern of plants relative to their economic use. Several plant-breeding systems to manipulate the genetic variability of populations to meet this goal were discussed. Breeding systems for improvement of plant species, however, are not limited to those discussed: variations, either large or small, are made by plant breeders to adapt the breeding systems for their specific programs, plant species, and breeding goals. All breeding systems are based on use of Mendelian genetic principles, which vary from transfer of traits controlled by one gene to increasing the gene frequency of quantitative traits by cyclical selection methods. Individual breeding programs include the use of one or more modifications to most of the breeding systems discussed. There is a tendency, however, to categorize the breeding systems into those for short-term breeding goals (e.g., pedigree, backcrossing, single-seed descent) vs. those for population improvement by recurrent selection (e.g., mass, half-sib, reciprocal recurrent).

It is difficult to categorize breeding systems for specific breeding goals. Though recurrent selection by its nature is cyclical and long term, pedigree selection also tends to be cyclical because the parents used are often derived from previous pedigree-selection programs. The main distinction between pedigree selection and recurrent selection, for example, is the type of germplasm included and the intensity of inbreeding. Pedigree selection usually implies that inbreeding is conducted within a population having restricted genetic variability. Recurrent selection, on the other hand, usually implies that selection is conducted in a population having a broad genetic base that minimizes the effects of inbreeding. But this need not be the case. Recurrent selection may be conducted in an F_2 population formed by crossing two pure lines, and pedigree selection may be conducted in landrace varieties that have a greater range of genetic variability, which was the case for the original inbred lines derived from open-pollinated varieties of corn. Breeding goals may be long term or short term, but the differences among breeding systems are not as distinct. Duvick,[7] in a comparison of pedigree and recurrent selection, calculated it took 13.3 years to complete one cycle of pedigree selection vs. 1 to 3 years to complete one cycle of recurrent selection. Objectives of the two breeding systems, however, were very different.

It seems the different breeding systems are most effective and efficient if they are used to complement each other, and this is usually the situation. For example, pedigree selection would complement recurrent selection to derive genotypes from the populations improved by recurrent selection. Recurrent selection methods have proven to be useful for the improvement of most types of plant species.[16] For recurrent selection to be a useful breeding system, populations must be developed that contribute to the goals of the breeding program. Hence, other breeding systems are needed to extract genotypes from populations under recurrent selection; one of the most effective methods is pedigree selection. To ensure systematic genetic advance in plant breeding, all types of breeding systems will be needed. Improvement of populations by recurrent selection for the sake of population improvement will not make a significant contribution to a breeding program unless they are used as sources to extract superior genotypes. Integration of the different breeding systems will provide improved germplasm sources in which other methods can be imposed to effectively exploit the improved germplasm sources.

REFERENCES

1. **Briggs, F. N. and Knowles, P. F.,** *Introduction to Plant Breeding,* Reinhold, New York, 1961, 211.
1a. **Fehr, W. R. and Hadley, H. H.,** *Hybridization of Crop Plants,* American Society of Agronomy, Madison, 1980.
2. **Hanson, W. D.,** Heritability, in *Statistical Genetics and Plant Breeding,* Hanson, W. D. and Robinson, H. F., Eds., National Academy of Sciences and National Research Council, Washington, D.C., 1963, 125.
3. **Simmonds, N. W.,** *Principles of Crop Improvement,* Longman, New York, 1979.
4. **Hopkins, C. G.,** Improvement in the chemical composition of the corn kernel, *Ill. Agric. Exp. Stn. Bull.,* 55, 205, 1899.
4a. **Sinnott, E. W., Dunn, L. C., and Dobzhansky, Th.,** *Principles of Genetics,* McGraw-Hill, New York, 1950, 27.
5. **Muller, H. J.,** A simple formula giving the number of individuals required for obtaining one of a given frequency, *Am. Nat.,* 57, 66, 1923.
6. **Bauman, L. F.,** Review of methods used by breeders to develop superior corn inbreds, *Annu. Corn Sorghum Res. Conf.,* 36, 199, 1981.
7. **Duvick, D. N.,** Genetic rates of gain in hybrid yields during the past 40 years, *Maydica,* 22, 187, 1977.
8. **Richey, F. D.,** The convergent improvement of selfed lines of corn, *Am. Nat.,* 61, 430, 1927.
9. **Penny, L. H. and Dicke, F. F.,** Inheritance of resistance to leaf feeding of the European corn borer, *Agron. J.,* 48, 200, 1956.
10. **Harlan, H. V. and Martini, M. L.,** The effect of natural selection in a mixture of barley varieties, *J. Agric. Res.,* 57, 138, 1938.
11. **Gardner, C. O.,** An evaluation of effects of mass selection and seed irradiation with thermal neutions on yield of corn, *Crop Sci.,* 1, 241, 1961.

12. **Hayes, H. K. and Garber, R. J.,** Synthetic production of high protein corn in relation to breeding, *J. Am. Soc. Agron.,* 11, 308, 1919.
13. **Jenkins, M. T.,** The segregation of genes affecting yield of grain in maize, *J. Am. Soc. Agron.,* 32, 55, 1940.
14. **Hull, H. F.,** Recurrent selection for specific combining in corn, *J. Am. Soc. Agron.,* 37, 134, 1945.
15. **Comstock, R. E., Robinson, H. F., and Harvey, P. H.,** A breeding procedure designed to make use of both general and specific combining ability, *Agron. J.,* 41, 306, 1949.
16. **Hallauer, A. R. and Miranda, J. B., Fo,** *Quantitative Genetics in Maize Breeding,* Iowa State University Press, Ames, 1981, chap. 12.
17. **Matzinger, D. F., Cockerham, C. C., and Wernsman, E. A.,** Single character and index mass selection with random mating in a naturally self-fertilizing species, in *Proc. Int. Conf. Quant. Genet.,* Pollak, E. O., Kempthorne, O., and Bailey, T. B., Eds., Iowa State University Press, Ames, 1977, 503.
18. **Sprague, G. F.,** Corn breeding, in *Corn and Corn Improvement,* Sprague, G. F., Ed., Academic Press, New York, 1956, chap. 5.
19. **Lonnquist, J. H.,** Modification of the ear-to-row procedure for the improvement of maize populations, *Crop Sci.,* 4, 227, 1964.
20. **Compton, W. A. and Comstock, R. E.,** More on modified ear-to-row selection in corn, *Crop Sci.,* 16, 122, 1976.
21. **Sprague, G. F. and Eberhart, S. A.,** Corn breeding, in *Corn and Corn Improvement,* Sprague, G. F., Ed., American Society of Agronomy, Madison, 1977, chap. 6.
22. **Matzinger, D. F.,** Experimental estimates of genetic parameters and their applications in self-fertilizing species, in *Statistical Genetics and Plant Breeding,* Hanson, W. D. and Robinson, H. F., Eds., National Academy of Science and National Research Council, Washington, D.C., 1963, 253.
23. **Fehr, W. R. and Ortiz, L. B.,** Recurrent selection for yield in soybeans, *J. Agric. Univ. P. R.,* 9, 222, 1975.
24. **Horner, E. S., Lundy, H. W., Lutrick, M. C., and Chapman, W. H.,** Comparison of three methods of recurrent selection in maize, *Crop Sci.,* 13, 485, 1973.
25. **Moll, R. H., Bari, A., and Stuber, C. W.,** Frequency distributions of maize yield before and after reciprocal recurrent selection, *Crop Sci.,* 17, 794, 1977.
26. **Paterniani, E.,** Selection among and within families in a Brazilian population of maize (*Zea mays* L.), *Crop Sci.,* 7, 212, 1967.
27. **Hallauer, A. R. and Eberhart, S. A.,** Reciprocal full-sib selection, *Crop Sci.,* 10, 315, 1970.
28. **Hallauer, A. R.,** Development of single-cross hybrids from two-eared maize populations, *Crop Sci.,* 7, 192, 1967.
29. **Hoegemeyer, T. C. and Hallauer, A. A.,** Selection among and within full-sib families to develop single crosses of maize, *Crop Sci.,* 16, 76, 1976.
30. **Hallauer, A. R.,** Hybrid development and population improvement in maize by reciprocal full-sib selection, *Egypt. J. Genet. Cytol.,* 1, 84, 1973.
31. **Marquez-Sanchez, F.,** Modifications to cyclical hybridization in maize with single-eared plants, *Crop Sci.,* 22, 314, 1982.
32. **Anon.,** *Genetic Vulnerability of Major Crops,* National Academy of Sciences, Washington, D.C., 1972.
33. **Darrah, L. L. and Zuber, M. S.,** 1979 U.S. corn germplasm base, *Annu. Corn Sorghum Res. Conf.,* 35, 234, 1980.
34. **St. Martin, S.,** Effective population size for the soybean improvement program in maturity groups 00 to IV, *Crop Sci.,* 22, 151, 1982.

Botany of Crops

PLANT PROPAGATION

Calvin Chong

INTRODUCTION

Most agriculturally valuable plants originated from plants growing in the wild, in cultivated populations or from mutations and breeding programs.[1] The successful propagation of these plants is basic to agriculture. The fundamental objectives are to increase their numbers and to retain their desirable properties. This is achieved either sexually by seed or asexually by various methods[2] described in Table 1.

SEED PROPAGATION

Most plants are propagated by seed. During embryo formation, the genes controlling plant characteristics segregate and recombine in many different ways resulting in plants resembling either, neither, or both of the parents.[3] Thus, seeds are used as the primary means of propagating plants that come true or nearly true to type. Many of our most popular crops such as the cereals and other self-pollinated annuals and biennials are propagated by seed. Most commercial plantings of vegetables are propagated from seeds sown directly in the field, but for an earlier crop, transplants (seedlings) started in greenhouses, hotbeds, or cold frames are often used.

Seed propagation is usually the cheapest method of reproduction. Since most viruses are usually not seed transmitted, seed propagation makes it possible to start plants free of virus infection. Seeds of most species can be kept viable for many years under proper storage conditions.[2,3]

For many cross-pollinated plants such as forage crops, fruit and nut species, and other perennials, which do not reproduce true to type, propagation by seed may be the only possible, practical, or economical method of reproduction. Seeds are also used to produce understocks (rootstocks), upon which to graft or bud selected types of fruit and other woody species, as well as to produce new plant varieties (cultivars) through hybridization.[3]

When seed is used for propagation, some are difficult to germinate or must have special treatment to overcome seed dormancy. Common treatments include softening or removal of the seed coat (scarification), extended exposure to warm and(or) low temperature (stratification), exposure to light, and possibly chemical treatments.[3]

VEGETATIVE PROPAGATION

Asexual propagation involves regeneration of vegetative tissues or organs into self-supporting individuals with similar properties as the source plant. A group of plants originating from one individual by vegetative propagation is collectively referred to as a clone. Theoretically, the life of a clone is unlimited.

The methods used in vegetative propagation[2] are summarized in Table 1. Vegetative propagation is used primarily for reproducing plants that do not produce true from seeds. Most fruit and nut crops and other woody crop species are in this category. Organs or pieces (cuttings) of vegetative tissues originating from stem or root are used to regenerate new plants by inducing adventitious roots and(or) shoots, often with the aid of growth regulators.[3] Grafting or budding two pieces of tissue together is often practiced when cuttings root with difficulty or not at all, or if a particular cultivar or understock is required. The twig or bud (scion) forms the top part and the understock becomes the root system.[1,3]

Under favorable conditions, vegetative propagation allows more rapid production of plants. Problems associated with seed dormancy or difficulty in seed germination are eliminated. Asexual propagation is also necessary to regenerate plant species or cultivars that produce no viable seeds. Some plant species reproduce themselves vegetatively from apomictic seeds (Table 1).

With the development of modern reproductive techniques in tissue culture (micropropagation), this methodology is now routinely used commercially to clone crops such as strawberries, apple cultivars and understocks, and many herbaceous species.[2,4,5] Tissue culture will eventually become a standard commercial method for asexual propagation of many plant species.[4,5]

Tables 2 and 3, respectively, list crops by commodity groups propagated predominantly by seed or by vegetative methods. Crops grown predominantly for their ornamental value are not included in these tables. Other sources should be consulted for more specific information.[1,2,3,5-9]

Table 1
METHODS OF PROPAGATION

Method	Description	Remarks
Sexual		
Propagation by seed	Each seed is a complete living organism — a result of the normal sexual process of two parent plants	
Asexual (vegetative)		
Utilization of apomictic seeds	Embryo originates from the seed tissue of the source plant by a vegetative process resulting in the development of seeds without the normal sexual process	Limited use in a few species such as citrus, mango, Kentucky bluegrass; resulting clones are usually virus-free
Induction of adventitious roots or shoots		
Cutting	Regeneration from vegetative part detached from the plant	Treatment of cuttings with root promoting chemicals is often beneficial; leafy cuttings are rooted more successfully under intermittent mist
Layering	Regeneration from vegetative part while still attached to the plant	Commonly referred to as marcotting
Graftage	The combining of plant parts by means of tissue regeneration resulting in a single plant; the part of the combination to provide the root is called the understock or rootstock; the part to become the top is called the scion	Often seedlings of the same or related species are propagated for use as understocks
Budding	When the scion consists of a single bud (scion bud)	
Grafting	When the scion consists of a twig with two or more buds, or when both scion and rootstock are connected to a growing root system	"Cutting-grafts" combine the rooting of cuttings with the operation of grafting; useful for difficult-to-root plants
Utilization of specialized vegetative structures	Stem or root modifications usually containing large amounts of stored resources	The process is referred to as *separation* when the structures divide naturally, and as *division* when they must be cut
Above ground		
Crown	A compressed stem usually located just above the soil line	
Offshoots	Branches growing from crown with buds and leaves	Often referred to as offsets, suckers, slips, pips
Stolon	Horizontal stems running usually above ground	
Runners	Stolon with very long internodes	
Below ground		
Bulb	A compressed stem with thick, fleshy leaves	Miniature bulbs (bulblets) developing at the axils of leaf scales are called offsets when grown to full size
Corm	A compressed stem with rudimentary, scaly leaves	
Rhizomes	Horizontal underground stems; roots and shoots develop at the nodes	
Tubers	Greatly enlarged fleshy portions of underground stems	
Tuberous root	Fleshy, swollen roots	
Tissue culture	Adventitious shoot and root formation from meristems, shoot tips, or other organs under aseptic condition on artificial nutrient media	Usually referred to as micropropagation

Table 2
CROPS BY COMMODITY GROUPS PROPAGATED PREDOMINANTLY BY SEED[a]

Crop	Family	Seed	Transplant	Vegetative method	Remarks
Cereals and Grains					
Amaranth, grain *Amaranthus leucocarpus* S. Wats.	Amaranthaceae	•[a]			
Barley *Hordeum vulgare* L.	Gramineae	•			
Buckwheat *Fagopyrum esculentum* Moench	Polygonaceae	•			
Chickpea *Cicer arietinum* L.	Leguminosae	•			
Corn *Zea mays* L.	Gramineae	•			
Oats *Avena sativa* L.	Gramineae	•			
Pulses *Phaseolus* spp., *Vigna* spp., *Vicia* spp. *Lens esculenta* Moench	Leguminosae	•			
Rice *Oryza sativa* L.	Gramineae	•	X[b]	X Rooted tillers	Seeds of most rice of the subsp. *indica* and the photoperiodic rices may exhibit dormancy
Rye *Secale cereale* L.	Gramineae	•			
Sorghum *Sorghum bicolor* Moench (*S. vulgare* Pers.)	Gramineae	•			
Triticale ×*Triticosecale* Wittmack	Gramineae	•			
Wheat *Triticum* spp.	Gramineae	•			

Forages

Common name	Scientific name	Family		Propagation	Notes
Alfalfa	*Medicago sativa* L.	Leguminosae	●		
Bahia grass	*Paspalum notatum* Flügge	Gramineae	●		
Bent grass	*Agrostis* spp.	Gramineae	●		
Bermuda grass	*Cynodon dactylon* (L.) Pers.	Gramineae	X	Stolons (sprigs)	Poor seeding habit
Bird's-foot trefoil	*Lotus corniculatus* L.	Leguminosae	●		
Bluegrass, Kentucky	*Poa pratensis* L.	Gramineae	X	Apomictic seeds	Seedlings are about 90% apomictic
Bromegrass, smooth	*Bromus inermis* Leyss.	Gramineae	●		
Canary grass, reed	*Phalaris arundinacea* L.	Gramineae	X	Stem sections	
Carib grass	*Eriochloa polystachya* H.B.K.	Gramineae	●	Runners Division of root clumps	
Carpet grass	*Axonopus affinis* Chase.	Gramineae	●		
Clovers	*Trifolium* spp. and *Melilotus* spp.	Leguminosae	●		
Dallisgrass	*Paspalum dilatum* Poir.	Gramineae	●		
Fescue	*Festuca* spp.	Gramineae	●		
Guineagrass	*Panicum maximum* Jacq.	Gramineae	●	Root division	Seed viability is usually low
Kudzu	*Pueraria lobata* (Willd.) Ohwi	Leguminosae	●	X Crown division X Cuttings	
Lespedeza	*Lespedeza* spp.	Leguminosae	●		
Lupine	*Lupinus* spp.	Leguminosae	●		

Table 2 (continued)
CROPS BY COMMODITY GROUPS PROPAGATED PREDOMINANTLY BY SEED[a]

Crop	Family	Seed	Transplant	Vegetative method	Remarks
Forages (continued)					
Millet *Panicum* spp., *Setaria* spp., *Echinochlea* spp., *Pennisetum* spp., *Eleusine* spp.	Gramineae	•			
Napiergrass *Pennisetum purpureum* Schumach.	Gramineae	X		• Stem sections	
Orchardgrass *Dactylis glomerata* L.	Gramineae	•		• Division of clumps	
Pangolagrass *Digitaria decumbens* Stent.	Gramineae			• Runner	Seedstalks produce few viable seeds
Paragrass *Panicum purpuracens* Radi.	Gramineae			• Runners • Division of root clumps	Extremely low seed germination
Redtop *Agrostis alba* L.	Gramineae	•			
Roughpea *Lathyrus hirsutus* L.	Leguminosae	•			
Ryegrass *Lolium* spp.	Gramineae	•			
Saint Augustinegrass *Stenotaphrum secundatum* (Walt.) O. Kuntze	Gramineae			• Stolons	Does not produce seed
Sainfoin *Onobrychis viciifolia* Scop.	Leguminosae	•			
Timothy *Phleum pratense* L.	Gramineae	•			
Vetch *Vicia* spp.	Leguminosae	•			
Wheatgrass *Agropyron* spp.	Gramineae	•			

Oil Crops

Common name / Scientific name	Family	Seed	Other	Notes
Castor bean *Ricinus communis* L.	Euphorbiaceae	●		Some castor seeds may require scarification. Seeds are actually whole fruits or seed nuts
Coconut *Cocos nucifera* L.	Palmae	●		
Corn *Zea mays* L.	Gramineae	●		
Flax *Linum usitatissimum* L.	Linaceae	●		
Jojoba *Simmondsia chinensis* (Link) C. K. Schneid	Buxaceae	●		
Oil palm *Elaeis guineensis* Jacq.	Palmae	●		Heat treatment of dried, depulped seed is usually required for germination
Peanut (groundnut) *Arachis hypogaea* L.	Leguminosae	●		
Rapeseed *Brassica campestris* L. and *B. napus* L.	Cruciferae	●	Cuttings	
Safflower *Carthamus tinctorius* L.	Compositae	●		
Sesame *Sesamum indicum* L.	Pedaliaceae	●		
Soybean *Glycine max* (L.) Merril	Leguminosae	●		
Sunflower *Helianthus annuus* L.	Compositae	●		
Tung (mu-tree) *Aleurites montana* (Lour.) Wils.	Euphorbiaceae	X	● Budding (patch) X Grafting	Unselected seedling material is variable; contains about 50% unproductive male trees

Fiber Crops

Common name / Scientific name	Family		Other	
Bowstring hemp *Sansevieria* spp.	Agavaceae		● Suckers ● Rhizomes X Leaf cuttings	

Table 2 (continued)
CROPS BY COMMODITY GROUPS PROPAGATED PREDOMINANTLY BY SEED[a]

Crop	Family	Seed	Transplant	Vegetative method	Remarks
Fiber Crops (continued)					
Cotton *Gossypium hirsutum* L. and *G. barbadense* L.	Malvaceae	●			
Flax *Linum usitatissimum* L.	Linaceae	●			
Hemp *Cannabis sativa* L.	Cannabidaceae	●			
Jute *Corchorus capsularis* L. and *C. olitorius* L.	Tiliaceae	●			
Kapok *Ceiba petandra* (L.) Gaertn.	Bombaceae	●		X Cuttings	Easily propagated by cuttings 5—8 cm in diameter and 1.2—2.0 m long using 2- to 3-year-old wood
Kenaf *Hibiscus cannabinus* L.	Malvaceae	●			
Ramie (Rhea, China grass) *Boehmeria nivea* (L.) Gaud.-Beup.	Urticaceae	X		● Division of rhizome X Suckers X Stem cuttings	
Sann (Sunn) hemp *Crotalaria juncea* L.	Leguminosae	●			
Sisal *Agave sisalana* Perr.	Agavaceae	X		● Aerial bulbils X Suckers	Plants seldom set seed; bulbils are preferred; more are produced and plants from them are more robust and more uniform
Vegetables (Above-Ground Parts)					
Artichoke *Cynara scolymus* L.	Compositae			● Root pieces ● Off-shoots	
Asparagus *Asparagus officinalis* L.	Liliaceae	X	X	● Crown divisions	

Common name / Scientific name	Family			Notes
Beans *Phaseolus vulgaris* L. (snap or green) *P. limensis* Macf. (lima)	Leguminosae	•		
Brocolli *Brassica oleracea* L., Italica group	Cruciferae	•	•	
Brussels sprouts *Brassica oleracea* L., Gemmifera group	Cruciferae	•	•	
Cabbage *B. oleracea* L., Capitata group	Cruciferae	•	•	
Cauliflower *Brassica oleracea* L., Botrytis group	Cruciferae	•	•	
Celery *Apium graveolens* L. var. *dulce* (Mill.) Pers.	Umbelliferae	•	•	Seeds germinate slowly
Chard, Swiss *Beta vulgaris* L., Cicla group	Chenopodiaceae	•	•	
Chickpeas *Cicer arietinum* L.	Leguminosae	•	•	
Chive *Allium schoenoprasum* L.	Amaryllidaceae	•	• Bulb	
Cho-cho (Choyote) *Sechium edule* (Jacq.) Swartz.	Cucurbitaceae	•	•	Seeds germinate readily from whole, fleshy fruit
Collards and kale *Brassica oleracea* L., Acephala group	Cruciferae	•	•	
Cucumber *Cucumis sativus* L.	Cucurbitaceae	•	•	
Endive (escarole) *Cichorium endiva* L.	Compositae	•	•	
Eggplant (aubergine) *So.anum melongena* L.	Solanaceae	•	•	
Kohlrabi *Brassica oleracea* L., Gongylodes group	Cruciferae	•	•	

Table 2 (continued)
CROPS BY COMMODITY GROUPS PROPAGATED PREDOMINANTLY BY SEED[a]

Vegetables (Above-Ground Parts) (continued)

Crop	Family	Seed	Transplant	Vegetative method	Remarks
Leek *Allium ampeloprasum* L., Porrum group	Amaryllidaceae	●	●		
Lettuce *Lactuca sativa* L.	Compositae	●	●		
Muskmelon (Cantaloupe) *Cucumis melo* L., Reticulatus group	Cucurbitaceae	●	●		
Mustard *Brassica juncea* L. and *B. hirta* Moench	Cruciferae	●			
Okra *Abelmoschus esculentus* (L.) Moench (*Hibiscus esculentus* L.)	Malvaceae	●	●		Seeds slow to germinate
Parsley *Petroselinum crispum* (Mill.) Nyman ex A.W. Hill	Umbelliferae	●	●		
Peas *Pisum sativum* L.	Leguminosae	●			
Pepper *Capsicum annuum* L. (sweet); *C. frutescens* L. (hot)	Solanaceae	●	●		
Pigeon pea *Cajanus cajan* (L.) Millsp. (*C. indicus* Spreng.)	Leguminosae	●			
Pumpkins and squashes *Cucurbito* spp.	Cucurbitaceae	●			
Rhubarb *Rheum rhabarbarum* L.	Polygonaceae			● Crown	

Common name / Scientific name	Family	Seed	Cuttings
Spinach *Spinacia oleracea* L.	Chenopodiaceae	●	
Sweet corn *Zea mays* L. var. *rugosa* Bonaf.	Gramineae	●	
Tomato *Lycopersicon esculentum* Mill.	Solanaceae	●	
Watercress *Nasturtium officinale* R.Br.	Cruciferae	●	●
Watermelon *Citrullus lanatus* Thunb. (*C. vulgaris* Schrad.)	Cucurbitaceae	●	

[a] (●) Indicates commercial or most common and practical method of propagation.
[b] (X) Indicates less desirable or limitedly used method of propagation.

Table 3
CROPS BY COMMODITY GROUPS PROPAGATED PREDOMINANTLY BY VEGETATIVE METHODS

Roots and Tubers

Crop and family	Seed	Cuttings	Layering	Budding	Grafting	Commonly used understock	Specialized structures and organs	Remarks
Arrowroot (edible canna) *Canna edulis* Ker. (purple) *Maranta arundinacea* L. (West Indian) Cannaceae							[a] Rhizome tips (2—4 nodes) [a] Tubers [a] Suckers	
Beets *Beta vulgaris* L. Crassa group J. Helm. Chenopodiaceae	a							
Carrots *Daucus carota* L. Umbelliferae	a							Seeds germinate slowly
Cassava (manioc, tapioca) *Manihot esculenta* Crantz Euphorbiaceae		Herbaceous[a]						Mature stem cuttings 15—20 cm in length with 6—10 nodes; those without branched shoots are preferable
Celeriac (turnip root) *Apium graveolens* L. var. *rapaceum* Gaud-Beaup. Umbelliferae	a							
Garlic *Allium sativum* L. Amaryllidaceae							[a] Separation of clove (compound bulb)	Seed is never used
Horseradish *Armoracia rusticana* P. Gaertn., B. Mey., & Scherb. Cruciferae		Root[a]					[a] Crowns	Seed is never used
Onion *Allium cepa* L. Amaryllidaceae	a						[a] Small bulbs ("sets")	
Parsnips *Pastinaca sativa* L. Umbelliferae	a							Seeds germinate slowly
Potatoes *Solanum tuberosum* L.							[a] Small whole tubers	

Solanaceae			
Radish			
Raphanus sativus L.	a		
Cruciferae			
Rutabaga (Swede)			
Brassica napus L.; Napobrassica group	a		
Cruciferae			
Salsify			
Tragopogon porrifolius L.	a		
Compositae			
Sweet potato			
Ipomoea batatas Lam.	Herbaceous,[a] leaf[b]	[a] Tuberous root (slips)	Cuttings taken from vine are usually 30—45 cm long from the apical growth of mature plants
Convolvulaceae			
Taro (dasheen, cocoyam)		[a] Top "pieces" of corms, bearing roots and leaf stalks	
Colocasia esculenta (L.) Schott		[a] Division of corms	
Araceae		[b] Small whole tubers	
		[b] Suckers or side shoots	
Turnip			
Brassica rapa L., Rapifera group	a		
Cruciferae			
Yams		[a] Small tubers	Tubers should be in dormant state
Dioscorea spp.		[a] Crown ("head" end)	Top part of enlarged tubers from which shoots arise
Dioscoreaceae		[b] Aerial bulbils	

Fruit crops

Akee (ackee)			
Blighia sapida Koenig	a		
Sapindaceae			
Almond	T-bud[a]		Seedling, *P. persica*, Marianna 2624 plum clonal rootstock, almond-peach hybrids
Prunus dulcis (Mill.) D. A. Webb			
P. amygdalus Batsch			
Rosaceae			

Root and stem cuttings not too successful

Table 3 (continued)
CROPS BY COMMODITY GROUPS PROPAGATED PREDOMINANTLY BY VEGETATIVE METHODS

Fruit Crops (continued)

Crop and family	Seed	Cuttings	Layering	Budding	Grafting	Commonly used understock	Specialized structures and organs	Remarks
Apple *Malus sylvestris* Mill., (*M. pumila* Mill., *M. domestica* Borkh.) Rosaceae		Softwood[b]	Mound[b]	T-bud[a]	Cleft, whip[a]	Seedlings of common species, many clonal *Malus* stocks		This species and many clonal stocks are also propagated by tissue culture Mound layering is commonly used for propagating rootstocks
Apricot *Prunus armeniaca* L. Rosaceae				T-bud[a]		Seedling, *P. persica*, *P. cerasifera*, *P. besseyi*		
Avocado *Persea americana* Mill. Lauraceae	b	Softwood[b]		T-bud[a]	Cleft, whip, side-veneer[a]	Seedling, *P. drymifolia*, West Indian-Guatamalan hybrids		
Banana *Musa acuminata* Colla Musaceae							[a] Rhizomatous "sword" suckers [b] Large (3—4.5 kg) rhizome pieces ("heads")	Does not produce seeds
Blackberry *Rubus* spp. (upright type)		Root, softwood, leafbud[b]					[a] Suckers with attached root piece	
(trailing type) Rosaceae		Root, softwood, leafbud[b]	Tip[a]				[b] Rhizome	
Blueberry, lowbush *Vaccinium angustifolium* Ait. Ericaceae	b	Softwood[a]						
Blueberry, highbush *Vaccinium corymbosum* L. Ericaceae		Hardwood, softwood; [a] leaf-bud[b]		T-bud[b]		Seedling, *V. ashei*		Normally difficult to propagate by hardwood cuttings although good results can be obtained; also propagated by tissue culture
Blueberry, rabbiteye *Vaccinium ashei* Reade Ericaceae		Softwood[a]						Also propagated by tissue culture
Breadfruit *Artocarpus altilis* (Parkins) Fosb. (*A. communis* Forst.) Moraceae		Root;[a] other[b]	Other[b]				[a] Root suckers	Usually seedless; seeded types known as breadnuts, can be

Crop	Cutting	Layering	Budding	Grafting	Seedage	Comments
Cherimoya *Annona cherimola* Mill. Annonaceae		[a]	T-bud[a]	Cleft[a]	Seedling, *A. squamosa*, *A. reticulata*	propagated by seed; branch cuttings 30—40 cm long with 3—4 nodes can be used., air layering of roots also used limitedly
Cherry *Prunus avium* L. (sweet) *P. cerasus* (sour) Rosaceae		Trench[b]	T-bud[a]		*P. avium*, *P. mahaleb*, clonally propagated *P. cerasus*, 'Stockthon Morello'	
Citrus Orange (*Citrus sinensis* (L.) Osbeck) Grapefruit (*C. × paradisi* Macf.) Lemon (*C. limon* (L.) Burm. f.) Persian lime (*C. aurantifolia* Christm.) Pummelo (*C. grandis* L.) Tangerines (mandarins) (*C. reticulata* Blanco) Rutaceae	Root, semihardwood, softwood, leaf-bud[b]	Air[b]	T-bud[a]		*C. sinensis*, *C. aurantium*, *C. limon*, *C. paradisi*, *Poncirus trifoliata*, *C. reticulata*, *C. microphylla*, hybrid seedlings, nucellar seedlings	Some cultivars are seedless, propagation methods are the same for all species, excessive sap bleeding can be a problem in graft healing; "cuttinggrafts" are also used, also from apomictic seedlings (nucellar embryos) from polyembryonic seeds
Crabapple *Malus* spp. Rosaceae	Hardwood, softwood[b]		T-bud[a]		Seedlings of common species	
Cranberry *Vaccinium macrocarpon* Ait. Ericaceae	Other[a]				[a] Runners	Cuttings are from upright vines
Currant *Ribes rubrum* L. (red) *R. nigrum* L. (black) Saxifragaceae	Hardwood[a]	Mound[b]				
Custard apple *Annona reticulata* L. Annonaceae		[a]	T-bud[b]	Cleft[b]	Seedling	
Date palm *Phoenix dactylifera* L. Palmae (Arecaceae)					[a] Off-shoots	These arise from auxillary buds near the base of the trees. Each well-rooted offshoot weighs between 18 and 45 kg
Fig *Ficus carica* L. Moraceae	Hardwood[a]	Air[b]	T-bud, patch[a]		Seedling	

Table 3 (continued)
CROPS BY COMMODITY GROUPS PROPAGATED PREDOMINANTLY BY VEGETATIVE METHODS

Fruit Crops (continued)

Crop and family	Seed	Cuttings	Layering	Budding	Grafting	Commonly used understock	Specialized structures and organs	Remarks
Gooseberry *Ribes hirtellum* Michx. (American) *R. uva-crispa* L. (European) Saxifragaceae	b		Mound[a]					
Grape *Vitis vinifera* L. (European) *V. labrusca* L. (American) *V. rotundifolia* (Muscadine) Vitaceae		Hardwood;[a] softwood[b] Hardwood;[a] softwood[b] Softwood[a]	Simple, trench, mound,[a] compound[b]	Chip[b] Chip[b]	Cleft, whip, other[b] Cleft, whip, other[b]		Special disease resistant clonal rootstocks	Some grape cultures are seedless. T-budding is not practical due to the large size of buds. The muscadine type is more difficult to start by cuttings. Induced by cutting the roots 60—100 cm from the trunk
Guava *Psidium guajava* L. Myrtaceae	a	Softwood[b]	Simple, mound, air[b]	Patch;[a] chip[b]	Side-veneer, approach[b]	Seedling	[b] Suckers	Seeds lose their viability quickly and should be planted fresh
Jackfruit *Artocarpus heterophyllus* Lam. Moraceae	a				Approach[b]	Seedling, *A. champeden*, *A. hirsuta*		
Jujube (Chinese date) *Zizyphus jujuba* Mill. Rhamnaceae	a	Root, hardwood[b]		Other or unspecified[b]	Whip[a]	Seedling		
Kiwifruit (Chinese gooseberry) *Actinidia chinensis* Planch. Actinidiaceae	b	Root, softwood[b]	Simple[b]	T-bud[a]	Cleft, whip[a]	Seedling		
Litchi *Litchi chinensis* Sonn. Sapindaceae		Hardwood, softwood[b]	Air[a]					Large limbs air layer easier than small ones
Loquat *Eriobotrya japonica* (Thunb.) Lindl. Rosaceae	b		Air[b]	T-bud[a]	Cleft, side-veneer[a]	*Cydonia oblonga*		
Mamey (mammee apple) *Mammea americana* L. Guttiferae	a			Other[b]	Approach[b]	Seedling		
Mango *Mangifera indica* L.	a	Softwood[b]	Air[b]	T-bud, patch, chip[a]	Side-veneer, approach[a]	Seedling of common species, clonally		Also from apomictic seedlings (nucellular embryos) from po-

Common/Scientific name	Cutting type	Budding	Grafting	Other rootstocks	Notes
Anacardiaceae				propagated rootstocks, nucellular seedlings	
Mulberry *Morus alba* L. Moraceae	Hardwood[a]	Other[b]		Seedling	lyembryonic seeds.
Olive *Olea europaea* L. Oleaceae	Hardwood, semihardwood,[a] other[b]	T-bud, patch, chip[a]	Whip, side-veneer, bark, other[a]	Seedling or clonal rootstock	[a] Suckers from old trees
Papaw (paw paw, papaya, mamao) *Carica papaya* L. Caricaceae	Other[b]	a	Other[b]		Large branches 8—10 cm diameter cut into 30-cm pieces and buried below the soil can also be induced to form plantlets Vegetative methods usually are not economical
Passion fruit (Granadilla) *Passiflora edulis* Sims. Passifloraceae	Hardwood, herbaceous[b]	a	Other[b]	*P. flavicarpa*	
Peach and nectarine *Prunus persica* (L.) Batsch. Rosaceae	Hardwood, softwood[b]	T-bud[a]		Seedling, *P. armeniaca, P. amygdalus, P. besseyi, P. tomentosa*	
Pear *Pyrus communis* L. Rosaceae	Hardwood, softwood[b]	T-bud[a]	Trench[b] Whip[a]	Seedlings of common species, rooted cuttings of *Cydonia oblonga*	
Persimmon *Diospyros virginiana* L. (American) *D. kaki* L. (Oriental) Ebenaceae	Root[b]	T-bud[a] Other[b]	Whip[a] Whip[a]	Seedling Seedling, *D. lotus, D. virginiana*	
Pineapple *Ananas comosus* (L.) Merr. Bromeliaceae					[a] Slips [b] Suckers [b] Crown
Plaintain *Musa paradisica* L. Musaceae					[a] Rhizomatous "sword" suckers [b] Large (3—4.5 kg) rhizome pieces ("heads")
Plum and prune *Prunus domestica* L. (European) *P. americana* Marsh. (American) *P. salicina* Lindl. (Japanese) Rosaceae	Hardwood, softwood[b]	T-bud[a]		*P. cerasifera, P. persica, P. armeniaca, P. amygdalus, P. angustifolia, P. besseyi,* seedlings of common seedlings, various clonal *Prunus* stocks	Side-shoot from fruiting stem Above-ground types Located on top of the fruit Does not produce seeds

108 CRC Handbook of Plant Science in Agriculture

Table 3 (continued)
CROPS BY COMMODITY GROUPS PROPAGATED PREDOMINANTLY BY VEGETATIVE METHODS

Crop and family	Seed	Cuttings	Layering	Budding	Grafting	Commonly used understock	Specialized structures and organs	Remarks
Fruit Crops (continued)								
Pomegranate *Punica granatum* L. Punicaceae	b	Hardwood,[a] softwood[b]	Other[b]				[a] Suckers	
Quince *Cydonia oblonga* Mill. Rosaceae		Hardwood[a]	Mound[a]	T-bud[b]		Seedling		
Rambutan (hairy litchi) *Nephelium lappaceum* L. Sapindaceae	b			Other[a]		Seedling		
Raspberry, red *Rubus idaeus* L. (European) *R. idaeus* var. *strigosus* Michx. (American) Rosaceae		Root, softwood[b]					[a] Suckers	Also propagated by tissue culture
Raspberry, black *Rubus occidentalis* L. Rosaceae		Other[b]	Tip[a]					Most cultivars are difficult to propagate vegetively; root cuttings taken in the fall require a cold period to satisfy dormacy
Saskatoon serviceberry *Amelanchier alnifolia* Nutt. Rosaceae	a	Root,[a] softwood[b]	Other[b]	T-bud[b]	Whip[b]	*Cotoneaster acutifolius, C. lucidus, Sorbus* spp.	[b] Crown division (shrubby types) [b] Root sprouts [b] Root suckers	
Star apple *Chrysophyllum cainito* L. Sapotaceae	a	Other[b]	Other[b]		Cleft, other[b]	Seedling		
Strawberry *Fragaria* × *Ananassa* Duchesne Rosaceae							[a] Runners [b] Crown division (everbearing types)	Also propagated by tissue culture
Sweetsop (sugar apple) *Annona squamosa* L. Annonaceae				T-bud[a]	Cleft[a]	*A. cherimola, A. reticulata*		
Nut crops								
Butternut				Inverted T[a]	Bark[a]	*J. nigra*		Excessive sap bleeding can be a

Plant		Cutting type	Layering	Budding	Grafting	Rootstock	Remarks
Juglans cinerea L. Juglandaceae							problem in graft healing
Carob Ceratonia siliqua L. Leguminosae	a						Cuttings can be rooted but with difficulty
Cashew Anacardium occidentale L. Anacardiaceae	a	Softwood[b]	Air[b]	T-bud, patch[b]	Approach[b]	Seedling	Seedlings transplant with difficulty
Chestnut, Chinese Castanea mollissima Blume. Fagaceae	a			Inverted T[b]	Whip, bark, other[b]	Same species	Difficult to graft or bud; "buried inarch" technique also used
Coconut Cocos nucifera L. Palmae (Arecaceae)	a					Same species only	Can be marcotted (layered) by placing rooting medium in a box around the trunk; the trunk can be cut below the new root system and the top planted Graftage is rarely practical
Filbert (hazelnut) Corylus avellana L. (European) C. americana Marsh. (American) Betulaceae		Hardwood, semihardwood, softwood[b]	Simple[a]				
Hickory (shagbark hickory) Carya ovata Koch. Juglandaceae				Patch[a]	Bark[a]	C. ovata, C. illinoensis	
Macadamia (Australian nut) Macadamia integrifolia Maiden & Betche M. tetraphylla (L.) Johnson Proteaceae	a	Semihardwood[b]	Air[b]	Other[b]	Side-veneer;[a] cleft, whip[b]	Common species	
Pecan Carya illinoinensis Wangenh. Juglandaceae	b	Root, softwood[b]	Trench, air[b]	Inverted T, patch;[a] T-bud, other[b]	Bark;[a] whip[b]	Same species, C. aquatica, C. cordiformis, Juglans cineraria	[b] Root sprouts
Pistachio Pistachia vera L. Anacardiaceae				T-bud[a]		P. atlantica, P. terebinthus	
Walnut Juglans spp. Juglandaceae				Inverted T, patch;[a] T-bud[b]	Whip, bark[a]	Common species	Excessive sap bleeding can be a problem in graft healing

Table 3 (continued)
CROPS BY COMMODITY GROUPS PROPAGATED PREDOMINANTLY BY VEGETATIVE METHODS

Specialty Crops (continued)

Crop and family	Seed	Cuttings	Layering	Budding	Grafting	Commonly used understock	Specialized structures and organs	Remarks
Betel pepper *Piper betle* L. Piperaceae		Softwood[a]						Vine cuttings 30—45 cm long are usually taken from terminal orthotropic (upright) vegetative shoots; cuttings from lateral branches may produce infertile plants
Black pepper *Piper nigrum* L. Piperaceae	b	Softwood;[a] leaf-bud[b]	Air, other[b]				[b] Rhizome	Vine cuttings 65 cm long with 6—7 nodes are usually taken from terminal orthotropic (upright) vegetative shoots; cuttings from lateral branches may produce infertile plants
Cardamom *Elettaria cardamomum* Maton Zingiberaceae	b						[a] Division of rhizome [b] Pseudostem	Seed germination is often irregular
Cinnamon *Cinnamomum zeylanicum* Breyn. Lauraceae	a	Softwood[b]	Air[b]				[b] Division of old rootstock	Depulped seeds quickly lose viability
Clove *Eugenia caryophyllata* (Sprengel) Bullock & Harrison Myrtaceae	a		Other[b]		Approach[b]	*Psidium guajava*		This species is notoriously difficult to propagate vegetatively
Cocoa *Theobroma cacao* L. Sterculiaceae	a	Semihardwood, softwood;[a] leaf-bud[b]	Other[b]	T-bud, patch[b]		Seedling		Seeds are viable for only a short time; they die quickly when dehydrated, fermented, or subjected to temperature extremes
Coffee *Coffea arabica* L. Rubiaceae	a	Semihardwood, softwood[b]		T-bud, inverted T[b]	Cleft, whip, other[b]			Buds, grafts, and cuttings must be taken from orthotropic (upright) shoots; those taken from primary fruiting branches do not produce upright growth, and cuttings made from them sprawl along the ground

Common name / Species / Family	Seed	Cutting type	Grafting/Budding	Layering	Other vegetative	Remarks
Ginger / *Zingiber officinale* Rosc. / Zingiberaceae					[a] Rhizome portions, each with at least one good bud	
Ginseng / *Panax quinquefolius* L. / Araliaceae	a					Seeds require stratification before sowing
Hops / *Humulus lupulus* L. / Cannabinaceae					[a] Rhizome	
Khuskus (Vetiver) / *Vetiveria zizanoides* (L.) Nash. / Gramineae					[a] Root divisions	
Kola / *Cola nitida* (Vent.) Schott & Endl. / Sterculiaceae	a	Softwood[b]		Air[b]	Patch[b]	
Maple / *Acer saccharum* Marsh. / Aceraceae	a	Softwood[b]	Side-veneer[b]			Seedling
Mint / *Mentha* spp. / Labiatae	b	Herbaceous[b]				Seedling
Nutmeg (mace) / *Myristica fragrans* Houtt. / Myristicaceae	a	Hardwood, semihardwood[b]	Other[b]		Other[b]	Seeds lose their viability quickly; they are planted in the shell; seedling material will give about 50% of each sex; vegetative methods not very successful
Olive / *Olea europea* L. / Oleaceae		Hardwood, softwood[a]	Other[b]		Other[b]	
Pimento (allspice) / *Pimenta officinalis* (L.) Merrill / Myrtaceae	a	Other[b]	Other[b]		Other[b]	Use freshly harvested seeds or germination will be greatly reduced; vegetative methods not very successful
Pyrethrum / *Chrysanthemum cinerariaefolium* (Trev.) Bocc. / Compositae	a				[b] Division of mature plants into "splits"	Germination often poor due to presence of unfertilized or nonviable seeds
Quinine / *Cinchona* spp. / Rubiaceae	a	Softwood[b]	Side-veneer[b]		Patch[b]	Common species

Table 3 (continued)
CROPS BY COMMODITY GROUPS PROPAGATED PREDOMINANTLY BY VEGETATIVE METHODS

Specialty Crops (continued)

Crop and family	Seed	Cuttings	Layering	Budding	Grafting	Commonly used understock	Specialized structures and organs	Remarks
Rubber *Hevea brasiliensis* (Willd. ex Adr. de Jues.) Muell.-Arg. Euphorbiaceae	b			Patch[a]	Other[a]	Seedling		Seeds lose viability quickly; exuding latex can be a problem in graft healing
Sapodilla *Achras zapota* L. (*Manilkara zapota* (L.) Van Royem) Sapotaceae	a		Air[b]	T-bud, other[b]	Side-veneer[b]	Seedling		
Screw pine *Pandanus odoratissimus* L.f. Pandanaceae		Other[b]					[a] Offsets [a] Division of suckers	
Sorrel *Hibiscus sabdariffa* L. Malvaceae	a	Other[b]						
Sugar beets *Beta vulgaris* L. Crassa Group J. Helm. Chenopodiaceae	a							
Sugarcane *Saccharum officinarum* L. Gramineae		Herbaceous[a]						The herbaceous stalk cuttings usually contain 3—4 nodes; single bud "seed pieces" and young cane tops 40—50 cm in length are also used
Tea *Thea sinensis* L. (*Camellia sinensis* (L.) O. Kuntze) Theaceae	a	Leafbud[a]		Patch[b]	Cleft[b]	Seedling		Seeds lose viability quickly
Thyme *Thymus* spp. Labiatae	b	Herbaceous[b]						
Tobacco *Nicotiana tabacum* L. Solanaceae	a						[a] Division of rhizomes	

Vanilla	Herbaceous[a]	Cuttings taken from vines are usually between 30—100 cm long; longer cuttings flower earlier
Vanilla planifolia Andrews		
Orchidaceae		

Note: Extent of use of propagation methods is indicated by a or b: a[(a)] indicates commercial or most common and practical method of propagation; b[(b)] indicates less desirable or limitedly used method of propagation. Methods are as follows:

Cuttings: root, hardwood, semihardwood, softwood, herbaceous, leaf-bud, other
Layering: tip, simple, compounds, trench, mound, air, other
Budding: T-bud, inverted T, patch, chip, approach, other or unspecified
Grafting: cleft, whip, side-veneer, bark, approach, other or unspecified.

REFERENCES

1. **Hartmann, H. T., Flocker, W. J., and Kofranek, A. M.,** *Plant Science — Growth, Development, and Utilization of Cultivated Plants,* Prentice-Hall, Englewood Cliffs, N.J., 1981.
2. **Janick, J., Schery, R. W., Woods, F. W., and Ruttan, V. W.,** *Plant Science. An Introduction to World Crops,* 3rd ed., W. H. Freeman, San Francisco, 1981.
3. **Hartmann, H. T. and Kester, D. E.,** *Plant Propagation — Principles and Practices,* 4th ed., Prentice-Hall, Englewood Cliffs, N.J., 1983.
4. **Conger, B. V., Ed.,** *Cloning Agricultural Plants Via In Vitro Techniques,* CRC Press, Boca Raton, Fla., 1981.
5. **George, E. F. and Sherrington, P. D.,** *Plant Propagation by Tissue Culture. Handbook and Directory of Commercial Laboratories,* Exergetics, Ltd., Eversley, Basingstoke, England, 1984.
6. **Evans, D. A. Sharp, W. R., Ammirato, P. V., and Yamada, T.,** *Handbook of Plant Cell Culture,* Vol. 1, *Techniques for Propagation and Breeding,* Macmillan, New York, 1983.
7. **Purseglove, J. W.,** *Tropical Crops. Dicotyledons,* Vol. 1 and 2, Longman, Green, London, 1968.
8. **Purseglove, J. W.,** *Tropical Crops. Monocotyledons,* Vol. 1 and 2, Longman Group, London, 1972.
9. **Morensen, E. and Bullard, E. T.,** *Handbook of Tropical and Subtropical Horticulture,* U.S. Department of State, Agency for International Development, Washington, D.C., 1970.
10. **Heath, M. E., Metcalfe, D. S., and Barnes, R. C.,** *Forages — The Science of Grassland Agriculture,* 3rd ed., Iowa State University Press, Ames, 1973.

BOTANY OF CROP PLANTS

S. B. Helgason and A. K. Storgaard

LIFE CYCLES

This section presents information concerning the timing of developmental phases involved in the growth of crop plants leading to the formation of products useful to man. Some of these products are obtained from the vegetative parts of the plant — the roots, the stems, the leaves, or portions of these — and the essential developmental features concerned can be confined to very few phases as suggested by Milthorpe.[1] He recognized three stages as essential for describing development in root crops: (1) that of preemergence, (2) that in which leaf growth is predominant, and (3) overlapping the second stage, that of growth of the storage organs. Crops for which the end product consists of flowers, seeds, fruits, or extracts of these deserve a more detailed description of growth patterns, but practical considerations limit the number of stages described. Clearly, the complex stage descriptions exemplified by systems presented for application to studies on wheat,[2] soybeans,[3] and sorghum,[4] useful as they are for dealing with phenomena relating to nutrition, physiological reactions and the like, are not appropriate for the purposes of this chapter. In any case, such detailed descriptions of growth stages, particularly in terms of developmental time frames, have been developed for only a few species. As Monteith remarks in his excellent summary of climate in relation to crop ecology,[5] "The study of developmental timing in relation to the calendar, otherwise known as phenology, is returning to fashion after many years of neglect." Evidence of this neglect is to be found in some volumes devoted entirely to a single crop, but in which phenology receives scant allocation.

Developmental patterns in plants are influenced profoundly by environmental and genetic factors. The work of Aitken[6] provides a number of examples of the reactions of different cultivars of a number of species to growth conditions in widely divergent natural environments involving a broad range of latitudes, altitudes, and climatic zones. In addition to showing striking differences in the total growth period (sowing to ripening) in response to variations in environment, she demonstrates that wide differences in reaction of cultivars frequently occur, and that various developmental phases of the crop life cycle may react quite differently to a specific change in environment. She also provides examples of developmental timing differences brought about when crops adapted to cool temperate regions are transferred from their natural habitat to be grown in the winter season of subtropical climatic zones, or in high-altitude regions of the tropics. The importance of studies of this nature lies partly in the demonstration that certain principles deduced from precise studies in growth cabinets of environmental and genetic phenomena are shown to apply on a broad scale, and also in showing effects that would be very difficult or impossible to duplicate in controlled conditions. A major contribution to information in this field of study is the collection of reviews edited by Alvim and Kozlowski,[7] which presents information on crop ecophysiology based on both field and laboratory research.

Crops adapted to tropical and subtropical regions present some challenges in defining developmental patterns beyond those characteristic of temperate climate crops, principally because the influence of seasonal temperature differences is insufficient to affect the reproductive cycle in a major way. Seasonal rainfall variations characterize many tropical areas, and the effect of a protracted dry period often has an effect on the timing of flowering somewhat analogous to the effect of winter on crops grown in temperate climates. Though the timing of dry seasons in an area may be relatively constant, their relation to the calendar varies widely between regions to which a crop is adapted, making phenological generalizations impractical. In such instances, an attempt has been made to describe the environmental influences which bear on the reproduction time frame, on the basis that this will have some predictive value in relation to the known climatic patterns of a region. In the absence of such information, an example of reaction to the climate of a specific region is provided, on the assumption that this can be applied through study of climatic analogs.

The variations in timing of developmental stages in crop plants revealed by the references cited and reinforced by much of the literature reviewed makes obvious the fact that a detailed description of the phenology of the major crop species would be a monumental task requiring a voluminous report. The excellent volumes of Nuttonson[8-10] exemplify the complexity of compendia of the global phenology of individual species. The same can be said of the phenological charts developed by Splittstoesser[11] for the less broadly based environments of the U.S.

The objective of this presentation is relatively modest. Considering the large number of crop species involved, the complexities of the reactions of each to various ecological and genetic factors, as well as the interactions among them, and the multiplicity of behavior patterns brought about by the wide distribution of many of them, only those developmental phenomena most crucial to attainment of a useful end product will be described. An attempt is made to provide descriptions which apply wherever the crop is produced on a major scale. Actual calendar dates are used minimally, as the regions to which they apply must generally be specified. The assumption is made that descriptions of time frames based on seasons and ranges can be applied to regions widely divergent in latitude, altitude, and climatic classification. Climatic analogs would doubtless be useful in applying these descriptions. Information is presented in tabular form for groups of species which have similar life cycles, with expansion through a preamble or footnotes where deemed desirable. Descriptive terminology is chosen to enable direct comparisons to be made among species of a group which fit a somewhat similar ecological niche, as well as having a comparable developmental pattern. It is the authors hope that this presentation of some basic knowledge of developmental patterns of crop species may not only be a source of information, incomplete as it is acknowledged to be, but that it will stimulate interest in expanding and refining our understanding of the reactions and interactions involved in the phenology of crop species.

CEREAL CROPS

Among species of this group are the most widely produced and important of man's food crops.[12] It follows that a wide range of types must exist to fit the diverse ecological circumstances inevitably encountered by these crops, encompassing a great variety of temperature, daylength, and moisture regimes, as well as numerous edaphic and biotic variations. Indeed, much effort in plant breeding is concerned with improving adaptation to conditions of the environment in specific regions, partly by altering the timing of developmental phases. Cultivars of spring wheat, for example, have been placed in three maturity groups: early, 75 to 95 days from sowing to maturity; midseason, 95 to 115 days; and late, 115 to 130 days.[8] In North America, most spring wheat cultivars are of the midseason category, with winter wheat taking over in the long-season areas. At high latitudes, early maturing cultivars of the cereals are grown to reduce the frost hazard, but in other regions they may be grown to reduce the effects of drought and heat. One example of this is the use of early to very early maturing oat cultivars in the central U.S. to promote development before the hottest part of the summer,[12] whereas midseason oat cultivars are in general use in Canada. Late maturing cultivars of these cereals are sometimes grown at high altitudes in the tropics.

Day length has a major influence on development, and cultivars of many crops have been bred to cope with these effects. Barley, wheat, and oats are basically long-day species, as are rye and triticale, but daylength-insensitive cultivars have been bred to broaden adaptability of these crops. This is one of the characteristics bred into cultivars of wheat, triticale, and barley selected at CIMMYT in Mexico,[13] and a number of northern breeders of these species expose their material to short days during segregating generations to enable selection for daylength insensitivity.[14] A complementary adaptation to day length has been achieved in maize, grown from 40°S latitude to 58°N latitude.[15] Though it is basically a short-day species, hybrids have been bred which are essentially day neutral, and adapted to the short seasons of high latitudes. Low temperatures are more limiting than long days in this instance, having a delaying effect on maturity as well as reducing growth and yield.[12,16]

Despite genotypic alterations to improve adaptation, the phasic development of cereals is influenced profoundly by environment. Reference has been made to some examples of these influences,[6] and some additional examples are provided from a moderately extensive literature on the subject. Nuttonson provides information on responses of wheat[8] and barley[9] based on data from numerous widely dispersed trials. An example of response to latitude is shown by the following data on spring wheat:

	Days from emergence to heading	Days from heading to ripe	Days from emergence to ripe
Baltic region	40	55	95
Black Sea region	57	36	93

The data show that though the total growth periods are similar in the two regions, the developmental phases are altered substantially. Similar effects are shown by the North American data, but the specific effect may be altered greatly by other factors of the environment, such as altitude. In fact, date of planting in a specific location, a factor partly under man's control, can have a decided (and generally predictable) effect on development. In wheat, a delay of 30 days in date of sowing was shown to shorten the period from sowing to heading by 16 days, but affected the time from heading to maturity only slightly.[8] Shortening of the period from sowing to anthesis is a general consequence of delayed sowing of cereals at a specific location in temperate climates.[10,12,16] A reduction in vegetative growth is another consequence of this effect. Early sowing in temperate climates, on the other hand, tends to shorten the period from pollination to maturity,[8,10] presumably because higher temperatures prevail during this phase than is the case with delayed seeding.

Both altitude and latitude influence the timing of phases. The average sowing date for spring wheat grown in the southernmost areas where the crop is adapted in the U.S. is in mid-March, whereas in the northern regions of Canada it is in late May.[8] The spread in maturity between the two regions is usually not greater, and sometimes less, than between sowing dates, perhaps reflecting the benefit of longer days in the northern regions.[8,10] At Njoro, Kenya, a few kilometers from the equator, daylength-insensitive wheat cultivars take 120 to 140 days from sowing to maturity,[17] or about 20 days longer than prevails in the temperate zones. This may be a reflection of both short daylength and high altitude (2160 m). Nuttonson[10] concluded from phenological studies on barley that (1) with altitude constant, degree-days required from sowing to maturity decreased with increased latitude, and (2) with latitude constant, degree-days required from sowing to maturity decreased with increased altitude. A heat-unit system of defining developmental progress might provide a narrower range of fluctuation than the time criterion, but this has yet to be developed for a great many crops.

The process of growth and development is composed of a relatively complex sequence of events which can be defined with different levels of detail, depending on the purpose of classifying the stages. In maize, for example, Hanway[19] defined ten stages from emergence

Table 1
DEVELOPMENTAL PATTERN OF PRINCIPAL CEREAL CROPS

Timing of Developmental Stages in Days

Crop	Type[a]	Sowing to floral different	Floral differentiation to flower	Flower to maturity	Sowing to maturity Standard[b]	Extremes	Ref.
Barley	A	17—32	30—37	30—40	75—110	65—140	10,12, 27,28
Hordeum vulgare	WA	8—25[c]	28—35	30—45	65—105	60—130	10
Maize	A	24—54[d]	28—75	40—60	90—180	65—330	12,15, 16,19, 20,30
Zea mays							
Millet-pearl	A	22—55	25—60	35—60	100—160	90—190	32,33,36
Pennisetum typhoides							
Millet-foxtail	A	12—25	30—40	20—30	65—100	60—120	12,17,34
Setaria italica							
Millet-proso	A	15—30	30—40	18—30	63—100	60—110	12,17
Panicum miliaceum							
Oat	A	24—45	28—40	30—45	80—125	70—160	12,26,27
Avena sativa	WA	10—30	28—40	30—45	70—110	65—150	12
Rice	AU	35—60	28—42	45—75	110—180	80—200	12
Oryza sativa	AP	45—70	30—45	50—75	120—190	90—240	23—25,37
Rye	A	20—32	32—40	33—45	80—115	70—150	9
Secale cereale	WA	5—25	30—40	30—60	60—110	55—140	9,22
Sorghum	A	28—60	30—60	35—70	95—190	90—230	4,31,35
Sorghum bicolor							
Triticale	A	20—40	35—40	40—60	90—130	85—170	39,40
× *Triticosecale*	WA	10—30	35—45	35—45	85—120	80—180	39
Wheat	A	20—35	32—40	38—45	85—130	75—160	8,14, 28,29, 41
Triticum aestivum	WA	19—25	32—40	35—45	75—110	73—140	8,22,25,38

[a] A = annual; W = winter; U = upland; P = paddy.
[b] Standard applies to areas of major production of the crop.
[c] For winter annuals the starting base is beginning of spring growth instead of sowing date.
[d] Range encompasses male and female; the male preceded the female as a rule, by a matter of 2 to 6 days.

to physiological maturity, a period of 126 days for the Iowa material on which they were based. While average dates were recorded for the stages, the influence of genotype and environment in modifying phenology was emphasized. Shaw,[20] using Hanway's description as a base, outlined reproduction in maize in seven stages of development, and these were further modified to produce the basic information in Table 1. Starzycki[21] used 12 stages to describe growth and development in rye, and others[2-4] have used various levels of detail to present descriptions in other crops.

The information shown in Table 1 involves only the most crucial phases in the reproductive cycle of cereal crops: sowing to floral differentiation, floral differential to flowering, flowering to maturing, and the sum of these, sowing to maturity. The growth of annual cereals consistently takes the form of a sigmoid curve, somewhat changed in the case of paddy rice because of modification through transplanting. Growth of the reproductive parts is a component of this pattern, each forming its own distinctive growth curve, but their development is characterized by the major events dealt with in Table 1. In the winter-annual forms the growth curve is interrupted by a more or less lengthy state of slow or arrested growth during which vernalization occurs. In other respects, their developmental pattern resembles that of their annual counterparts. In some circumstances the timing of autumn planting of winter types influences the rate of development in the seed-production year. Delayed planting of winter wheat and rye has been shown to result in delay in heading the following year.[22]

Rice is unique among the cereals in tolerating flooding, and even benefitting from it. This leads to variations in management practices which influence development, as mentioned previously. Upland rice, grown without flooding but requiring abundant and consistent soil moisture, is grown much like other cereals, but forms a minor part of world rice production. Traditionally, paddy rice is sown in beds and transplanted, at 30 to 70 days of age, into fields with the surface stirred to a soupy mud receptive to the plant roots.[12,13] Tillers usually begin to form 5 to 10 days after transplanting and may continue to form for up to 40 days.[23] By contrast, most of the U.S. crop, which is seeded directly into paddies, begins to tiller about 30 days after emergence.[12,23] Yoshida[24] points out that the length of the basic vegetative stage (juvenility) accounts mainly for the difference in time required by different rice cultivars to reach maturity given comparable management, and that this is modified by cultivar response to photoperiod. Almost all cultivars grown in the tropics are weakly to strongly short day requiring, whereas those grown in temperate zones are insensitive to weakly sensitive.[24] Production as far north as Hokkaido island in Japan is dependent on the use of insensitive cultivars,[24] which mature in about 90 days from transplanting.[25] In high altitudes rice reacts favorably to sowing (or transplanting) as soon as temperatures are favorable in spring,[12] whereas in tropical and subtropical areas it must be adjusted to provide the most favorable day length for cultivars used.[24] By these means the timing of developmental events can be harmonized with the environment.

The information in Table 1 is presented in terms of range of time intervals for each development phase; this is inevitable in view of the many influences bearing on development which have been outlined. An attempt to relate developmental patterns to specific environments and cultivars would require presentation of massive amounts of data, and could be deceptive except for information derived from controlled environment experiments, so this is not attempted. The headings in Table 1 imply that each developmental event can be assigned precise time, but this is not actually so. For example, floral differentiation proceeds in a number of stages over a period of days. An illustration of this is an investigation describing in detail the formation of the oat panicle,[23] showing that the branch primordia became visible at 26 days after planting, but that stamen and pistil are first evident on the 53rd day. The experiment was under controlled conditions, and the schedule would change with environment, but would always involve a number of days. There is evidence that the differentiation of a spiked inflorescence proceeds in a shorter time than in a panicled one.[27,28] For the purposes of Table 1, floral differentiation is defined as the visible onset of the development of the inflorescence, and the term "floral initiation" in the literature was assumed to have the same meaning unless otherwise defined. The terms "flowering" and "heading" appear to be used interchangeably in reports, although it is well known, for example, that although this is approximately the true situation in barley, flowering in wheat and rice is delayed well beyond full heading.[23,24,39] Such variations fade into relative insignificance in comparison with the effects of changes in environment. The limitations of the information given are recognized, and it is hoped that the references will provide adequate detailed information when it is required.

PULSE CROPS

Pulses, or grain legumes, second only to the cereals in importance as food for men and animals, encompass a wide range of genera and species. Some developmental characteristics of species which contribute substantially to the nutrition of man are presented in Table 2.[42]

Table 2
DEVELOPMENT OF PULSE CROPS

Crop[a]	Sowing to flowering (days)	Flowering to maturity (days)	Sowing to maturity (days) Standard	Sowing to maturity (days) Extremes	Ref.
Bean — common dry *Phaseolus vulgaris*	45—60	50—70	100—140	90—160	12,42,50,51,58
Chickpea *Cicer arietinum*	35—55	55—90	90—130	85—180	52—54
Cowpea *Vigna unguiculata*	35—70	25—35	80—160	60—210	12,43,47,54
Faba bean *Vicia faba*	40—60	45—75	95—135	90—150	55,56
Lentil *Lens culinaris*	45—58	40—50	75—96	70—130	48,50,54
Pea *Pisum sativum*	32—50	25—65	60—100	50—160	42,48,57
Peanut (ground nut) *Arachis hypogaea*	25—70	55—75	110—150	100—170	12,43,45,46,48,54
Pigeon pea *Cajanus cajan*	60—110	50—100	120—180	90—240	44,48,54

[a] All but pigeon pea are herbaceous annuals. Pigeon pea is a woody shrub normally grown as an annual.[42]

Some of them are also used in the unripe stage as vegetables, and most have other uses, as for fodder and green manure.[12]

A characteristic these species have in common, which has an important bearing on development, is indeterminate flowering. The intensity of it varies among the species, and cultivars of peas have been developed in which all the pods ripen over a short period.[42] At the other extreme is the cowpea, which usually continues to flower until checked by adverse environmental conditions.[12] An individual flower in cowpea can produce ripe seeds in as little as 3 weeks after pollination.[43] Pigeon pea is also persistent in continuing to flower, sometimes until 75 to 80% of the pods are ripe.[44] The other species listed in Table 2 generally continue to flower over a period of weeks, up to as much as 2 months in peanuts, for example.[45]

Though *Phaseolus vulgaris* is the most important species of bean, it is by no means the only bean utilized by man.[12,42] Moreover, within *P. vulgaris* there are distinct types, each with a characteristic developmental timing. Among those used for dry beans, the earliest maturing group of cultivars is referred to as the navy bean, grown widely in the higher latitudes of the temperate climatic zones, and at the other maturity extreme, the red kidney, requiring at least 140 days frost-free season, and therefore grown in the longer season areas of the temperate zones.[12] A somewhat analogous situation holds with peanuts, in which the Spanish-type cultivars are early maturing whereas the Virginia ones take approximately a third or more longer from planting to maturity.[46]

Temperature influences development in various ways. Peas and faba beans are favored by moderate to low temperatures, and if grown in hot climates, are usually grown during the coolest part of the year, or at high altitudes. This has the effect of lengthening the developmental phases. The cowpea and pigeon pea are at the other end of the scale in temperature requirement, thriving only at relatively high temperatures.[12,47]

Photoperiodic response also may alter developmental pattern. However, pigeon pea is the only one of the crops listed in Table 2 which is basically short-day requiring, the others

Table 3
DEVELOPMENT OF ANNUAL OILSEED CROPS

Timing of Developmental Stages in Days

Crop	Type[a]	Sowing to flowering	Flowering to maturity	Sowing to maturity Standard[b]	Extremes	Ref.
Castor bean[c] *Ricinus communis*	A/P	80—110	70—100	160—180	140—	12,54,68
Linseed *Linum usitatissimum*	A	45—75	35—55	90—110	85—130	12,61,85,86
Rapeseed	WA	25—45[d]	40—70	90—115	80—125	62,63,87
Brassica napus	A	45—65	40—55	80—120	70—150	62,64,88,89
B. campestris	A	35—50	32—48	70—100	60—115	62,64,88,89
Safflower *Carthamus tinctorius*	A	65—140[e]	35—60	110—200	100—250	12,54,65,66
Sesame *Sesamum indicum*	A	70—110	40—80	130—160	120—180	67—69
Soybean *Glycine max*	A	35—60	40—100	100—160	75—195	3,70—79
Sunflower *Helianthus annuus*	A	65—90	30—60	95—130	70—170	80—84

[a] A = annual; P = perennial; W = winter.
[b] Standard applies to areas of major production of the crop.
[c] Castor bean, usually grown as a woody annual for oil seed, will persist and bear as perennial in suitable environment.
[d] For this winter annual, the starting base is beginning of growth in spring instead of sowing date.
[e] Fall seeded in regions of mild winters, the safflower is delayed in early development.

being day neutral, or having only a few cultivars requiring short days.[48,49] Peanuts have given variable reactions to photoperiod, and it has been suggested that the cotyledons of the ungerminated seed may be already induced if grown in a suitable environment.[12]

Chickpeas are an important crop in India and Pakistan, where they are usually sown toward the end of the rainy season to avoid problems of disease and seed setting associated with wet conditions.[53] Early growth corresponds with the coolest part of the year, tending to delay development, but the ultimate effect of this depends on the region; the temperature rises as the dry season advances and the extent of this rise influences rate of development of the crop.

OIL-SEED CROPS

The annual oilseed crops (Table 3) are, with the exception of castor and sesame, adapted to temperate climates. Castor is, in fact, a short-lived pernnial in areas where frosts do not occur, but commercial production is generally on an annual cropping regime. Cultivars vary greatly in onset of flowering, which is indeterminate.[54] In Oklahoma, where flowering usually starts in June, the blossoms in a female raceme are likely to open over a period of 20 days or more, and the male ones over 35 days.[60] High temperatures favor rapid development, and a growing season in excess of 140 days is generally required for production of an abundant crop.[54]

Flax grown in upper latitudes, in which it produces acceptable yields of a high-quality product, generally takes a week or more from first to full bloom, with flowering tapering off over several days more. In some conditions, a flax crop may produce a second flush of bloom after an interval, and in long-season areas often produces two such flushes.[61] Cultivars differ in this tendency. In areas where the lowest temperatures are only slightly below freezing, as in parts of Texas, flax is sown in early winter, taking a relatively long season to mature. Cold-tolerant cultivars have been bred to meet these conditions.[12]

Rapeseed produced in Europe is predominately of winter-annual type; with conditions that permit good winter survival, the yield is substantially greater than from spring-type cultivars.[62] The level of winter survival influences maturity in that thick stands have a delaying effect on development. Moderate nitrogen deficiency tends to hasten maturity, whereas moderate phosphorus deficiency may delay maturity somewhat.[63] Winter rape grown in Western Europe may begin flowering as early as mid April, and continues to flower over a period of 3 to 5 weeks.

Rapeseed is the most important oilseed crop in Canada, where the winters are too severe for the production of the winter-annual type. In the northerly areas of that country, *B. campestris* is grown because of the short season to which cultivars of that species are adapted. Floral development is somewhat indeterminate in all types, with flowering continuing over 2 to 4 weeks, depending on the environment.[64] Summer rape is grown in most parts of India, generally sown toward the end of the wet season (September in the Punjab) to develop and mature in the cool, dry "rabi" season, taking 90 to 100 days from sowing to maturity.[63]

Safflower planted as a spring crop in Montana or Nebraska takes about 110 days from sowing to maturity.[12] In Arizona and California planting may be timed for flowering to occur during dry conditions, or in frost-free areas may be grown in winter in rotation with a heat-adapted summer crop. As a winter crop it may require from 120 to 200 days to complete the growth cycle, depending on time of planting.[54,65] Winter types may remain in the rosette stage until spring, when the plants elongate and produce numerous flower heads.[66] Long days hasten development in safflower.[54]

The developmental timing of sesame is difficult to characterize because the plant flowers over a long period. At New Delhi, as a mean of five cultivars, flowering began 55 days after sowing, reached a peak at 76 to 83 days, and was completed at 111 days.[67,68] Dry weight of the pods increased rapidly in linear fashion from 76 to 104 days after sowing, then tapered off to maturity. Breeding of nonshattering sesame cultivars suited to mechanical harvesting has stimulated interest in production of the crop in the southwestern U.S., where maturity is reached in about 120 days.[69] As a soil temperature of 21°C is required for germination, planting the crop 2 to 3 weeks later than cotton is recommended.

Genetic and physiologic factors influencing soybean development have been studied extensively and the interactions among them shown to be quite complex.[70,71] Cultivars differ in tendency to indeterminate flowering, but many of those grown in the central U.S. bloom over a period of 20 to 30 days, and some for as much as 50 days. Under most conditions the bloom of the first 3 to 5 days does not set pods. Even though blooming in a field sown to a uniform cultivar may continue over an extended period, all pods generally mature within a week of one another. Soybean cultivars vary widely in daylength reaction from short day to day neutral.[72-75] Despite the usual short-day adaptation of cultivars adapted to the central and southern U.S., delaying the date of planting generally shortens the development period, suggesting interaction of temperature with day length in developmental timing.[75,77] At relatively high latitudes, delaying planting beyond the date when soil temperatures are suitable is generally not feasible because of the increased danger of fall frost damage.

Robinson,[80] in a detailed study of sunflower phenology involving six cultivars planted at seven dates, found that most of the difference among cultivars in days from planting to maturity occurred during the emergence to head-visible period, and to a lesser extent during the head-visible to first-anther period. Thus cultivars could effectively be ranked for maturity based on either day or growing degree-day summations to first anther. The flowering period of first to last anther was constant at about 8 days among cultivars over the various planting dates. The use of the growing degree-days criteria clearly differentiated maturity among cultivars, but greatly reduced differences among planting dates and years compared to the use of day summations. Latitude has a substantial effect on development in sunflower, each degree of high latitude increasing the period from planting to ray flower stage by nearly 2 days, although growing degree-day summations were little changed.[81,82] Time of planting in North America ranges from early May at the highest latitudes of sunflower production, to as early as mid March in southern U.S. Delay in planting beyond the optimum period in temperate climatic zones has an adverse effect on yield and has been shown to alter the quality of the oil.[82]

Four tree species predominate among many which produce oils of commercial importance. These are so divergent in their life cycles that a tabular presentation of the information would have little value. An outline of the developmental process and timing is presented for each of the species.

Coconut (*Cocos nucifera*)

This monocotyledonous tree is grown from seed; attempts are being made to develop practical means of vegetative propagation for large plantations.[37] Germination of well-matured nuts takes about 11 or 12 weeks. The seedling is slow growing and depends on nutrition from the nut for a year or more.[90] Seedling leaves are entire, becoming pinnate as the tree grows.[91] The seedlings are usually grown in nurseries where the most vigorous are selected for transfer to the permanent plantation as early as 18 weeks after onset of germination and up to the four-leaf stage.[37] The true stem develops at about 5 years of age. Individual leaves differentiate and grow in the sheath about 30 months before opening; a new leaf opens about once each month. On an established tree, about 30 to 40 leaves are active at one time, each persisting about 3 years.[90,91] Flowering begins in 5 to 7 years after planting, though on some dwarf types it begins at 3 years. An inflorescence composed of many branches forms at each leaf axil, each branch bearing one (rarely more) female flower at the base of numerous male flowers to the extremity. Before emerging from the subtending leaf axil, the inflorescence is enclosed in a double spathe which persists for some weeks before growth bursts it open. The male flowers shed their pollen before the females of the inflorescence are receptive, encouraging cross-pollination.[92] Although the female flowers are numerous, about 3 to 9 nuts to an inflorescence is an average set, and 10 to 12 is considered a good crop.[90,91] The coconut shell is fully grown in about 68 days, the meat takes almost 300 days to develop, and the entire fruit is mature in about 12 months; in 14 to 16 months it will fall from the tree.[37] However, the maximum liquid content of the nut, for drinking purposes, is achieved in about 7 months.[90]

As inflorescence development is coordinated with leaf development, each at about monthly intervals, production is generally continuous over the year. The amount of the crop may vary with the environment, with more floral abortion at some seasons resulting in variation in the quantity of nuts produced. In parts of India, for example, the heaviest crop is produced in November.[90] The influence of environment in this regard has been the subject of considerable research.[37]

Oil Palm (*Elaeis guineensis*)

The African oil palm is grown from seed, although research is under way to develop means of vegetative propagation.[93] The seed is dormant at maturity, and may remain so up to 6 months.[94] Germination having occurred, the seedling grows fairly rapidly, achieving the three-leaf stage at about 50 days. It utilizes the seed for nutrition for up to 160 days, but most of the oil is used up in 90 days or so. By the age of 120 days the seedlings are transferred to a nursery, from which they are moved to permanent plantations at the age of 12 to 14 months.[93,94] In the seedling stage the leaves are entire, but by stage of planting, a number of divided leaves will have been produced.[37] The tree flowers at the age of 2 to 5 years. The inflorescence primordia form at the base of each leaf at leaf initiation, and take about 2 years to reach the spear stage, during which the branches are enclosed within a double spathe. About 9 to 10 months later flowering takes place.[95] Each inflorescence has numerous flowers on branches, and is usually composed either of female or male flowers, but an occasional inflorescence is composed of a mixture of female and male branches. Both sexes are on one tree, but cross-pollination is the rule.[95] Average time from fertilization to full development of the fruit is 140 days, with a range of 120 to 200 days.[37,94] The fruit abscisses and drops about 10 to 15 days after maturity. The rate of leaf production (and consequently inflorescence formation) ranges from 20 to 35 per year, depending primarily on the environment.[37] Production of nuts is normally continuous unless adverse conditions intervene. In most areas there is an annual rhythm in production that peaks at different times. For example, production in the Congo and Malaya is abundant through the year, but peaks in April to July in the former and November to February in the latter.[37] In Honduras, production reaches a low ebb in the January to April period, but rises to a high level in September to November. Production in Nigeria follows a somewhat similar pattern, with less extreme peaks and valleys. The precipitation pattern in a region appears to be related to the production pattern.

The leaves arise from the trunk in a spiral pattern which is readily apparent from the frond bases which persist for several years after the leaves drop. The naked trunk begins to appear at the base at about 20 years of age. The leaf canopy increases in size up to about the age of 8 or 9 years, as does productivity, leveling off until the tree begins to deteriorate.[37] Though trees may live as long as 200 years, commercial plantations are usually replaced in about 35 years.

Olive (*Olea europaea*)

Usually propagated from cuttings, the trees of the olive may begin to bear at 3 to 5 years of age. For cultivars which do not root well from cuttings, grafting onto seedlings can be done, but the seed takes 4 or 5 months to germinate, and the seedling must be 2 years old to be ready for grafting, thus delaying development of the reproductive stage.[91] The plants are usually moved to a permanent plantation after 2 years in a nursery, provided the stem thickness is approaching 2 cm. By the 6th year in the orchard the trees should be bearing a fair crop.[91]

Though the olive is an evergreen tree, it tolerates light frosts; indeed, it requires some chilling for floral initiation. The flowers are borne in panicles which develop in the axils of leaves on the growth of the previous year. Floral initiation takes place about 8 weeks before bloom. In California, full bloom is reached some time during the month of May, and the fruits ripen in early October.[96] The fruits will cling to the tree for several weeks after they reach full size, and during the early phase of this period oil content increases while water content decreases.[96] Oil content usually is maximum 6 to 8 months after bloom.[91] Though the olive tree is capable of very long life, productivity usually declines after about 50 years, and this tends to be the average life of an olive orchard.[97]

Tung (*Aleurites fordii*)

A deciduous tree which may grow to a height of about 12 m, the tung is grown mainly in subtropical and warm temperate regions.[97] It requires either a chilling period of several days or a dry season of about 4 months for the deciduous phase.[54,97] Transplanted to the orchard 2 years after grafting on a seedling which develops to the proper stage in 8 or 9 months, the trees flower and begin to bear 2 to 3 years later. They approach full bearing in 7 to 10 years.[42] The flowers are produced in early spring, terminally on the growth of the previous year. A tree bears separate male and female flowers, but some trees have mainly female flowers and are referred to as "bearers".[54] The fruit is a drupe about 6 cm in diameter, spherical in shape, and divided into five locules with a simple seed in each. On the U.S. Gulf Coast the seed will usually be full size by early July, but the embryo and endosperm develop from microscopic size at this stage to fill the seed with embryo and endosperm by October.[98] The oil is mainly in the endosperm and most of it is formed from mid July to mid September in that region. The species *A. montana* is said to be better suited than *A. fordii* to growing in the tropics, where plantations are usually at high altitudes.[54] In Malawi, the trees of this species reach full bearing in about 6 years.

STARCH-PRODUCING ROOT CROPS

Important as suppliers of high-energy human food are four species of starchy root crops: potato (*Solanum tuberosum*), cassava (*Manihot esculenta*), sweet potato (*Ipomoea batatas*), and yam (*Dioscorea* spp.). Two other species, taro (*Colocasia esculenta*) and arrowroot (*Maranta arundinacea*), are of lesser significance. The *Solanum* potato is the only one of these adapted beyond the subtropics; the others require relatively high average temperatures in the 21- to 30°C-range to develop well, and are favored by short day length. Sweet potato is an exception among the tropical group in day-length requirement, with apparently day-neutral cultivars grown in parts of Japan and the southern U.S.[99] Propagated readily by vegetative means, these species do not normally produce seed, and must be provided specific conditions to enable production of seedling variants for breeding purposes.[100-102] Indeed, some cultivars are not known to flower under conditions so far provided.

Despite the somewhat limited ecological adaptation of the tropical root crops, the timing of phases of plant development can be influenced by the environment. The onset of root bulking and the relative proportion of growth of plant parts are important in terms of ultimate food production, and are influenced by such factors as temperature (including diurnal range) and moisture supply, as well as by day length.[99,102]

Arrowroot (*Maranta arundinacea*)

Tuberous rhizomes constitute the useful portion of the arrowroot, yielding a flour considered highly digestible and therefore desirable for use in foods for infants and invalids.[42] The well-budded proximal end of the tuber is cut off for regeneration when required. However, the crop is often grown continuously over a period of 5 to 6 years on a semiratoon system.[54] In St. Vincent the crop is planted in May, at the beginning of the rainy season. The plants flower in about 3 months, and the rhizomes are ready for harvest in 10 to 12 months. Maturity is indicated by leaf fall, signifying that the rhizomes have reached maximum starch content. The crop is harvested at once when it reaches this stage, as aging of the rhizomes leads to difficulty in extracting the starch.[54]

Cassava (*Manihot esculenta*)

A perennial woody shrub, the edible roots of cassava are usually harvested annually. Propagated from stem cuttings, the roots remain fibrous for 7 to 8 weeks while the plant develops a leaf canopy. Tuberization is dependent on short-day conditions to the extent that the crop is most successful within 15° of the equator. With favorable conditions, up to 15 thickened roots may develop over a period of 8 to 10 weeks.[54,99] These roots are likely to reach maximum thickness in 25 to 40 weeks.[99] Harvesting is sometimes begun as early as 8 months after planting, but is usually delayed until 14 to 16 months to develop a maximum yield.[42] Should a period of drought intervene, the roots tend to become woody and may be harvested prematurely for the sake of quality.[99]

Cassava has the potential to develop flower buds at each branch, but flower development depends on both cultivar and environment.[102] Flowers have been known to emerge at 6 weeks, but usually take about 6 months from emergence.[54] Plants are monoecious, but the female flowers are usually receptive 1 to 2 weeks ahead of anthesis of male flowers on the same branch. However, a plant may flower over a period of 2 months, providing ample opportunity for selfing within the plant, or crossing if that is desired. The fruit reaches full size 2 weeks after pollination, but takes 3 to 5 months to grow, ripen, and split to release the seeds.

Potato (*Solanum tuberosum*)

The most widely adapted of the root crops, the white potato is grown from the tropics to 70°N latitude. Numerous cultivars have been developed for adaptation to various daylength and temperature regimes, as well as to provide resistance against a number of diseases to which the potato is subject. Short-season types have been developed to mature in as little as 90 days, not only to provide for high-latitude climates, but also for early markets in temperate regions and short rainy seasons in the tropics and subtropics. Although cultivars have been selected for adaptation to the higher latitudes, there is evidence that some European cultivars react favorably to short days, based on the demonstration that delayed planting usually slows tuberization and reduces yield in Europe, and that transfer to a tropical environment from a temperate one reduced the length of time to maturity.[99] However, late-maturing European cultivars are more responsive to short days than are the early-maturing cultivars of that region.[104] Temperature is an important factor in both the timing and quantity of tuber formation and growth. Tuber growth is favored by air temperatures below 21°C and soil temperatures below 17°C.[12] Consequently, in tropical and subtropical regions, the potato is favored by high altitudes. In subtropical climates the crop is generally grown in winter in areas not subject to frosts, thereby taking advantage of low temperatures and short days. On the North American continent, planting time in Florida is generally from late September to early December, whereas in the southern U.S. subject to frosts, it will be delayed to late January or February. At high latitudes, planting is delayed proportionately, taking place in early to late May as a general rule in prairie Canada.[11,12]

The growth pattern of potato in a typical potato-growing area in the north temperate climatic region is as follows: from planting to emergence takes 10 to 15 days; top growth expands slowly to day 30, then accelerates rapidly to about day 70, declining gradually from about day 75; tuberization begins at 25 to 30 days, with tuber weight increasing exponentially for 15 to 20 days, then in linear fashion from about day 55 to day 95.[12,104] Total tuber weight may increase fivefold between blossoming and leaf senescence, and starch content increases markedly during the last few days of this period. Usually the potatoes are left in the ground for a few days after they reach maximum weight to allow suberization of the skins and lessen harvest damage.

Potatoes usually flower freely about 55 to 70 days after planting, but seed is rarely set under field conditions. Pollinations for breeding purposes are carried out in a controlled environment by special techniques described by Plaistad.[100] The tomato-like fruits, about 2.5 cm. in diameter, take 6 to 8 weeks to ripen after pollination.

Sweet Potato (*Ipomoea batatas*)

Though basically a tropical plant growing best at average temperatures in the 24 to 26°C range, the sweet potato is produced commercially from 40°N to 32°S latitudes.[54] In climatic zones free of frosts, it functions as a perennial herb, with a prostrate to semierect growth habit, but it is generally grown as an annual. Propagated from stem cuttings in the tropics, but usually from small tubers or tuber parts in temperate zones, the plants are normally started in nurseries or hotbeds, to be planted in the field at 25 to 40 days of age.[12,99] Tubers begin to form after 30 to 60 days in the field, depending on cultivar and environment. Early tuberization is promoted by short days, more so with tropical than with temperate cultivars.[99] The time required from planting to maturity ranges from 110 to 180 days, depending on cultivar and environment.[12,42,99] In some areas (e.g., Africa), the tubers are often removed as they are needed without reference to total maturity of the plant.[99]

The sweet potato characteristically is low in seed production, and flowering rarely occurs in most cultivars grown in the continental U.S. In some tropical areas, including the West Indies, flowering is general and moderate seed production expected.[54] For breeding purposes, special techniques have been developed, generally employing the controlled environments of greenhouses or growth cabinets, and by these means relatively reliable seed production is obtained.[101] Under such conditions flowering may continue over a prolonged period of weeks or even months, as happens in nature when conditions favor flowering. Seed matures in about 30 days after pollination.[101]

Taro (*Colocasia esculenta*)

Taro is grown most widely in the tropical islands of the Pacific where it is propagated by planting cormels or tops of corms into mud fields resembling rice paddies.[105] Growth commences quickly and vegetative development is rapid over the first 4 to 6 weeks. The leaf canopy develops to reach a maximum in 20 to 25 weeks after planting. The corms begin to expand in 12 to 20 weeks after planting. The leaves gradually diminish after about a year of growth, and most of the leaves may have fallen by the 15th month, indicating that the corms are ready for harvest.

Yam (*Dioscorea* spp.)

A number of yam species are listed by Wilson,[99] but the principal ones grown commercially are *D. alta, D. rotundata*, and *D. esculenta*.[12] Propagated from whole or cut tubers, the plant produced is a vine requiring support, and reaching maximum length in about 70 days from emergence.[106] The maximum leaf canopy is produced in about 13 to 20 weeks, and persists to 22 to 30 weeks after planting. The tubers begin growth at about 65 to 75 days and are often harvestable after 150 days, though in some conditions they may take over 200 days to reach maximum weight. The tubers remain dormant for 60 days, and sometimes as much as 150 days. Growth buds are not present at harvest, but develop during storage. This can be enhanced by burying the cut top of the tuber, which can be divided later to give rise to several plants.[106,107]

Some yam cultivars never flower, and in those that do, seed set requires a highly specific environment. Seeds that are produced remain dormant for 3 to 4 months. Production of seed is of importance only in breeding.[106] The cultivated species are dioecous in flowering habit.

SUGAR CROPS

Virtually all of the commercial supply of refined sugar in the world is the product of two species, sugarcane (*Saccharum officinarum*) and sugar beets (*Beta vulgaris*). About 65% of the refined sugar is supplied by sugarcane, and the remainder from sugar beets.[42] Other species provide sweetening mainly in the form of syrups, and the development of two of these is dealt with in the presentation which follows.

Sugarcane (*Saccharum officinaram*)

The species encompasses strains with a wide variety of characteristics, including those involved in developmental timing. This is partly a result of natural or induced hybridization with other species of *Saccharum,* some of it carried out to improve disease resistance or other agronomic characteristics of importance.[108]

Sugarcane is a crop of the humid tropical lowlands grown within 35° latitude of the equator. In plantation practice it is propagated entirely from cuttings of three-bud stalk pieces, planted in furrows 5 to 20 cm. deep. They root readily from the nodes.[12] In areas where winter frosts occur, fall planting is practiced on the basis that little growth will occur until spring. If planting is to be done in spring, the stalk pieces must be stored over winter, usually in bundles covered with earth or straw.

A stem develops from one or more of the buds on a planted cutting. It elongates rapidly after emergence, and stools develop from the basal nodes, giving rise to a clump of stems varying in number with plant spacing and the environment.[37] Sugarcane plants take from 9 to 24 months to reach the stage of harvesting for sugar. Most of the crop is grown in 14- to 18-month seasons, often followed by a ratoon crop which takes about 12 months.[12,109] Clones adapted to short seasons are used in climates where either low temperature or a prolonged dry period intervenes to arrest growth. In some southern regions of the U.S. (e.g., Louisiana), a favorable growing season of 9 to 10 months is terminated by nature's lowering temperatures, which induce increased sugar content by slowing down growth.[109] In climates with favorable temperatures throughout the year, as in Hawaii, the same effect is obtained by withholding irrigation. In such environments the crop can thus be managed, through timing of planting and control of moisture supply, to provide a continuous supply of cane for processing.[12,109] Chemicals are also available which have the effect of increasing sugar content and purity as harvest time approaches.[109] In tropical climates a planted crop is usually followed by two ratoon crops before a plantation is renewed.[12]

The cane crop is normally harvested before flowering occurs, and the plants are topped if the inflorescence appears before the planned time of harvest. Indeed, chemicals are available to inhibit or delay flowering.[37] Reproduction by seed is important only for breeding purposes. Improvement work is often carried out in greenhouses to provide a favorable environment for flowering and seed set, but may also be done in the field if such problems as lodging can be minimized. James describes a typical greenhouse schedule based on work at Canal Point, Fla.[108] Seed pieces started in February are ready for transplanting to large containers in April. High fertility is provided through July, then discontinued to assure low nutritional status during floral initiation in September. The inflorescence, a panicle generally referred to as an arrow or tassel, remains tightly wrapped in the flag leaf for a period of several weeks.[110] It breaks free through elongation of the peduncle, revealing a much-branched, plume-like panicle, tapered at the apex to a cone shape. The timing of flowering varies widely with genotype, usually taking over 12 months outdoors, but is capable of modification in greenhouse culture.[12,108] Anthesis may begin when the panicle emerges, or up to a few days later. It progresses from the top downward and the branch tips inwards, taking 3 to 14 days. The stigmas are receptive when the flowers open. Effective means of inducing male sterility have been developed to enable numerous crossed seeds to be obtained without emasculation. The seeds ripen in 22 to 36 days after pollination begins. The panicle is usually bagged before the first seeds ripen, as they shatter readily.[108]

Sugar Beet (*Beta vulgaris*)

The sugar beet is normally an herbaceous biennial, though in some environments it may behave as an annual. As the economic value of the crop is based on sugar produced by the root in the establishment season, the first year of development has been studied in detail.[111] Under favorable conditions the sugar beet seedling emerges in from 4 to 10 days. Growth is favored by moderate temperatures, and planting is usually done in early spring when soil temperatures reach 5 to 15°C. During early stages the root elongates rapidly, reaching a length of 28 to 32 cm by the time the first full leaf is grown. The root continues to elongate and the leaf canopy to expand until about 60 days after planting, when the root begins to thicken rapidly. The top continues to enlarge through 90 to 120 days, depending on environment. In contrast with such root crops as potato and sweet potato, the tops of which reach a maximum at a similar stage but then diminish toward maturity, the tops of sugar beets continue at about the same level through the remainder of the growing season of the first year.[1,99] In long-season areas the root will generally achieve maximum expansion in 180 days from planting.[111] During the final period of growth, sugar concentration depends in part on having low levels of soil nitrogen available to the plant. This can be accomplished through management, but in relatively short-season areas, the low temperatures which occur toward maturation are helpful in achieving increased sugar levels in the roots.[111]

Sugar beet seed is generally produced in areas with mild winters. The roots require a period of low temperature for floral induction, varying in length for different cultivars. Those which bolt easily require a relatively short cold period, and are satisfactory for use in areas where spring planting is practiced. In areas such as California, where the crop is planted in fall or winter, bolting-resistant cultivars are required, which require up to 90 days or more of chilling treatment.[112] In regions too cold for seed beets to winter over in the ground, the roots may be stored at temperatures of 5 to 7°C for 75 to 140 days as required; the storage limit is about 230 days. In areas with mild winters, the beets are planted in autumn for a seed crop; for example, in Oregon in early August to mid September, while in the milder climate of Arizona about a month later.[112] Bolting begins in early spring; in Arizona in March, but in Oregon in late April. The seed stalks grow rapidly to a height of 2 to 3 m, and blooming begins 35 to 45 days after the start of bolting. About 15 days later, full bloom is reached.[111] The flowering habit is indeterminate, resulting in flowering over a period of several weeks in most environments. Cross-pollination occurs at a high level, through mediation of both wind and insects, aided by the fact that the stigma is receptive about 6 days before anther dehiscence, and remains receptive for up to 12 days after anthesis.[112] The seed crop begins to mature about 6 weeks after full bloom.

Sugar Maple (*Acer saccharum*)

The sugar maple is a long-lived, hardwood, deciduous tree indigenous to the northeastern U.S. and the eastern half of Canada. It was a primary source of sweetening for the natives before settlement by Europeans, and much of the currently produced several million liters of maple syrup is produced from natural stands of the tree.[42] To determine the stage at which a tree is ready for tapping, the diameter is measured 150 cm above the ground.[113] Tapping may begin when a tree reaches a diameter of 25 cm, a stage reached in 30 to 40 years.[114] It is limited to one taphole until the diameter reaches 38 cm, to two tapholes from that size to 50 cm, to three tapholes up to a diameter of 60 cm, and to four tapholes at any diameter beyond that size. Overtapping leads to excessive staining of the sapwood. The rate of increase in girth of the trees is influenced by management in terms of judicious thinning of the bush. A bush which carries 170 to 220 tapholes per hectare is considered well managed.[113] Mature sugar bushes may tolerate up to 330 tapholes per hectare, but that situation indicates insufficient numbers of young trees in the population to sustain long-term productivity. The maple is long-lived; a healthy tree may live for over 200 years. They are known to continue productive for over 100 years if well managed. The effect of overexploitation on sap yield and tree survival have not been clearly demonstrated.[113]

Mature maples flower intermittently in spring, but 6 or 7 years may elapse between heavy seed crops.[113] Flowering takes place in mid March to May in various areas of adaptation, and the seeds generally mature in September to October.[115] They are released at the time of leaf fall.

The timing of annual tapping is a matter of judging when the sap begins to flow, and this can occur as early as February and as late as April, depending on location and the weather. It is said to begin when warm, sunny days follow cool, crisp nights, with the day temperature rising above 7°C.[42,113] The actual beginning of flow must be determined by trial trappings. The average duration of sap flow is about 34 days, but being highly dependent on weather conditions, is known to vary from as little as 9 days to as long as 57 days.[42]

Seedlings are sometimes transplanted from volunteer stands or nurseries, as part of sugarbush management. This may be done while the young trees are dormant, when the stems have reached a diameter of 1 to 2.5 cm and a height of 1.2 to 3 m.[113]

Sweet Sorghum (*Sorghum bicolor*)

This common name refers to a somewhat heterogeneous group of cultivars characterized by high sugar content and high moisture levels in the stalks. The term "sorgo" is also applied to the group. The developmental characteristics ascribed to sorghum in general (Table 1) apply also, in the main, to the sorgo type. In the U.S., however, sweet sorghum is primarily grown in the southeastern states and those bordering the Mississippi River, rather than in the Great Plains where grain sorghum is grown widely.[116] Few, if any, of the sweet sorghums mature as early as the earliest grain sorghums.

The timing of harvesting is important in relation to syrup quality. Most cultivars will produce satisfactory syrup if harvested when the grain is in the dough to ripe stages.[12] Earlier harvesting may lead to difficulties in clarification, but some cultivars, such as 'Wiley', can be harvested for syrup at any time from flowering to the dough stage without posing a problem in this regard.[116] It has been shown that the processing season can be prolonged without serious detriment to yield or quality of syrup by varying date of planting.[117] A slight delay in harvesting or short-term storage of the stalks (up to 6 days) has been shown to improve the clarity of the resulting syrup without serious adverse effects on other properties.[116,117] On the other hand, either practice will reduce the yield of sugar, as inversion of sucrose to other forms of sugar takes place.[117]

BEVERAGE AND DRUG CROPS

Diversity of both developmental pattern and utilization characteristics of these crops renders impractical the presentation of their phenology in tabular form. The primary reason for this is that in only three of them is the fruit the primary plant part utilized. The majority are crops grown in tropical and subtropical regions as well, leading to less regularity in developmental timing than is characteristic in temperate regions.

Cacao (*Theobroma cacao*)

A native of tropical rain forests, this species is primarily grown in areas where annual mean temperatures range from 22.4 to 26.7°C, within 20° latitude of the equator. The tree is evergreen characterized by periodic heavy leaf drop with flushes of both vegetative growth and flowering.[118] In mature plants, the periodicity of growth and development appears to be influenced primarily by exogenous factors. The main effect of temperature in these patterns is to influence the rapidity of growth and development.

Cacao begins to flower and bear fruit in 3 or 4 years with standard farm management. However, using good management of some recently developed hybrids, flowering in 18 months after planting in orchards has been achieved. Alvim reports that the period from flowering to harvest in Bahia, Brazil ranges from 140 to 175 days for pods developing during the warmer months, as compared with 167 to 205 days for those developed during the cooler months of the year.[118] Full bearing is usually achieved in 8 to 12 years. Trees may bear well up to an age of 50 or more years, but are usually replaced earlier because of reduced yield.[91]

The periodic flushes of nut production have been the subject of much research. In reviewing the findings, Alvim points out that in Bahia, Brazil, there are two major flushes of fruit bearing and a period of minimum fruit load between December and February.[118] This pattern is relatively consistent between years. The major flush of pod numbers occurred in May to June, whereas the maximum fruit load on a dry-weight basis was in August to September. He concluded that the main internal factor controlling flowering is competition between fruits and flowers, and the most influential external factor is alternation between dry and wet periods.

Coffee (*Coffea arabica*)

The coffee tree is an evergreen adapted to tropical areas characterized by moderate temperatures. Although growth and flowering may persist throughout the year, both are subject to a degree of periodicity in most environments.

Plants grown from seed will flower and begin to bear in 3 to 4 years.[37] Dormant coffee buds grafted on stock up to 2 years old usually grow in 3 or 4 months, though some viable buds stay dormant longer.[119] The stock is cut back to the union when the scion has grown a few centimeters. Bearing may occur in 2 to 3 years, but is of some economic value after the 4th year. Full bearing is achieved in 7 to 12 years.[91] Coffee trees are long lived, and are usually replaced because of the inroads of disease or to take advantage of improved cultivars rather than because of age. They are generally replaced in 25 years.[42]

Floral initiation in coffee may occur several weeks before flowering. A period of low temperature is implicated as the major factor in floral initiation, though photoperiod may play an accessory role.[119] There is characteristically a principal bearing season and one or two secondary seasons, corresponding to fluctuations in rainfall.[42] Breaking of flower bud dormancy is clearly related to rainfall, though not necessarily to soil-moisture availability. A flowering flush generally follows closely after irrigation or rainfall.[119,120] However, other environmental factors also have a bearing on the timing of floral initiation.[37]

Vegetative elongation takes place mainly during the rainy period in areas of seasonal rainfall distribution, such as Riuru, Kenya. Floral development during the early part of a rainy period is likely to be followed by a period of floral initiation preparatory to a later bearing flush.[119] However, in some environments, floral initiation occurs mainly after a crop has matured.[91] Enlargement of the fruit after fertilization follows a pattern which is basically similar in various environments and for different cultivars.[119] For 20 to 40 days from fertilization, growth is almost negligible, followed by a period of rapid swelling of the berry up to 85 to 100 days of age. There follows a long period when growth is suspended, generally lasting 90 to 120 days. Resumption of growth involves filling of the endosperm and ripening, which takes 30 to 60 days depending on cultivar and environment. Thus, the total time from fertilization to ripening is 230 to 270 days. This applies to continental environments in the tropics in which temperatures are moderated by high altitude, such as in the highlands of Kenya. A considerably shorter fruit development period of 170 to 230 days has been reported for Puerto Rico.[91] It is characteristic of coffee that only a small proportion of blossoms result in mature fruits, partly because of failure of fertilization, but also due to fruit shedding 8 to 12 weeks after flowering, when the berries are in the first month of rapid expansion.[37]

Cola (*Cola nitida*)

Environmental requirements of this species are similar to those of cacao. Though the trees can be propagated from cuttings, they take 6 to 10 weeks to root, and require rather exacting conditions. Grown from seed, the trees begin to bear in 5 or 6 years, and may produce reasonably well for 50 years or so. Fruiting tends to be in two flushes annually.[42] Flowering to maturity takes 120 to 150 days in Nigeria, where the main crop matures in October to December.[92]

Hop (*Humulus lupulus*)

The hop plant is an herbaceous perennial vine, grown on supports to provide for several meters of annual growth. The cone-like strobiles, which are the inflorescence of the female plants of this dioecious species, are used to flavor malt beverages.[12] Propagation for plantations is by asexual means to maintain genotypic constancy. The crop survives mild winters;

indeed requiring 4 to 5 weeks of near-freezing temperatures to produce normally. Aboveground parts die off each winter, and spring growth begins from below the soil surface. In the northern hemisphere, plants reach the stage of training for support by early May. Floral initiation occurs in late May to early June, with full bloom attained by late June to mid-July, depending on the cultivar.[121] Flowering often continues over a 2- to 3-week period. Males forming 1 or 2% of the stand supply adequate pollen in a crop. Some sterility is not objectionable as sterile hops are high in quality, but some yield loss results when fertilization does not occur.[12] Hop seeds usually mature 45 to 50 days after pollination. Thus the season required for the crop to develop and mature is 130 to 165 days, depending on cultivar used and weather encountered.[121] Hop plantings are known to persist for 30 to 40 years, but in practical culture to achieve maximum productivity, plantations are often replaced in about 7 years.[42]

Quinine (*Chinchona* spp.)

Much of the commercial production of quinine is based on the species *C. ledgeriana*, grafted on hardy rootstock of *C. succirubra*.[42] Seed germination usually takes 20 days or more, and seedlings require shading. Budding is on 1-year-old stock, and the plants are moved to the plantation 8 to 10 months later. Cuttings are being used to an increasing extent, but are exacting in their propagation requirements. The plants are about 1.5 m tall when planted, about 1.3 m apart in row. After 6 years, harvesting begins with the uprooting of every second plant in the rows. The process is continued annually until the plants are all harvested. The bark is removed from the entire tree, roots, trunk and branches, and quickly dried to preserve the quinine alkaloid crystals yielded by the bark.[42]

Tea (*Camellia sinensis*)

The tea plant develops into a small- to medium-sized evergreen tree when allowed to grow naturally. Commercial plantations are usually grown from seed, but vegetative propagation from selected clones is increasing in popularity because of the realization of improved adaptation and quality available in the species.[42,122] Seedlings are started in nurseries, and selected for vigor before planting in the permanent location. The plantation is pruned annually to produce flat-topped bushes that are convenient for plucking. Plucking begins when the bushes are 4 or 5 years old, the terminal two leaves (three or four in lower-quality teas) being taken at each plucking.[123] Intervals between pluckings vary from 8 to 14 days, depending partly on the rapidity of growth and partly on the economics of high yield vs. top quality.[37] Plantations are sometimes revitalized by cutting the bushes to the ground and allowing suckers to replace the old plants.[123] The tea tree is long lived, but the maximum life of a plantation is about 40 years.[42] Should a stand be kept that long, many diseased or otherwise unproductive plants will have been replaced.[123]

Current seed requirements annually utilize seed from about 700 ha of mature trees. The fruit, made up of three cells, each containing one seed, takes 9 to 12 months from flowering to maturity, depending on the strain and the environment.[123]

Tobacco (*Nicotiana tabacum*)

Tobacco is propagated by seed planted in coldframes or hotbeds. Seedlings take 6 to 8 weeks under glass, or 8 to 10 weeks under cloth, to reach the proper stage for transplanting.[12,125] Timing of planting in the north temperate zone (U.S.) is varied to suit the type of tobacco; e.g., flue- and air-cured burley are planted about a month earlier than cigar-filler and shade-grown wrapper. Transplanting time in Florida and Georgia begins in mid-April, in Kentucky and Tennessee in May, and in southern Wisconsin and Ontario in late May to early June. In Puerto Rico, the planting season is in October to December, as it is in some areas of southern Africa, as a consequence of rainfall distribution.[125] In areas where the mean growing season temperature is 24 to 27°C, as in the southern U.S., harvesting begins in 70 to 80 days, whereas at the higher latitudes of production, where mean temperatures are in the 20- to 30°C-range, a period of 100 to 120 days is required.[12] The plant is topped 2 to 3 weeks before leaf-harvesting begins, to remove the inflorescence in tobacco production plantations. The plants at this stage have 16 to 24 leaves.

The seed required for propagation is produced in long-season areas to avoid the chance of damage by autumn frosts. Most strains are photoperiod insensitive. The inflorescence is normally terminal, but small inflorescences can develop from axillary buds in the leaf nodes, especially after "topping" has removed apical dominance.[124] With secondary and tertiary flowers, the potential number of capsules is large. To improve size, the seed head is often trimmed to 25 to 30 primary capsules. Each capsule contains up to 3500 seeds. The seed matures in 4 to 5 weeks after pollination.[124]

Pyrethrum (*Chrysanthemum coccineum, C. roseum*)

Pyrethrum is an herbaceous perennial adapted to tropical and subtropical climates. The flowers, which are the source of the insecticide, are produced most abundantly in sunny locations where the soil is well drained.[42] In subtropical areas the crop is started from seed in the late fall or early spring, usually in dense stands in seed beds. Vigorous seedlings are selected for field planting, spacing the plants about 60 cm apart. With favorable conditions an abundance of flowers is produced the following spring and summer. In tropical regions, where seasonal temperature fluctuations are minimal, timing of operations and development is governed primarily by rainfall distribution over the year. As a moderate chilling is required to initiate flowering, production in tropical regions is at high altitudes; i.e., about 2800 m. In Kenya, a major producer of the crop, the first picking begins about 4 months after planting, and continues at intervals of 2 to 3 weeks over the flowering period, which extends over 9 to 10 months. The flowers are picked when fully expanded. Delay in picking leads to seed set and consequent reduction in further flowering.[54]

FIBER CROPS

Cotton (*Gossypium hirsutum*)

Cotton is an herbaceous dicotyledonous plant, perennial in regions where the temperature average for the coolest month is not below 17°C, but usually grown in plantations as an annual.[126] In latitude, it is best adapted between 37°N and 32°S, requiring a frost-free season of no less than 180 days.[12,126]

A large proportion of the cotton crop is grown in the subtropics, where it is planted in mid-spring, and in March through May in the U.S.[12] Emergence is rapid — 4 to 7 days in favorable temperatures (22°C +) and moisture conditions.[127,128] In an early maturing cultivar, floral initiation occurs in 30 to 35 days, at the leaf base of the sixth to tenth node. Influence of cultivar, planting date, day length, and temperature can extend the period by 20 to 25 days and the first floral leaf base by several nodes. Similarly, the first flowers appear in 45 to 65 days, with full bloom at 60 to 85 days after emergence. It is not unusual for more than half the flowers which open to be shed by 10 days after pollination. Boll formation is apparent in 75 to 100 days, with bolls beginning to open at 105 to 118 days. As flowering is indeterminate, with additional flowers appearing on a branch at about 6-day intervals, it continues as long as growing conditions are favorable.[128-130] The boll grows rapidly, reaching full size in 20 to 25 days after pollination. The fibers begin growth at about the time the flower opens, and reach full length in 21 to 24 days, with full thickness developed in an additional 30 to 40 days.[128,130] The fibers thicken through increments on the inside of the cell wall. As they dry, they flatten and become convoluted, after the bolls open.

The Sea Island and American-Pima types, which produce extra-long, fine fibers, on average require about 30 days longer to mature than the Upland and Asiatic cultivars, which produce shorter and coarser fibers. Growing seasons as short as 125 days from planting to harvesting do occur with a low frequency, but the general range is between 145 and 190 days.[12]

Hemp (*Cannabis sativa*)

Hemp is widely adapted, but the best development of fiber is in mild humid climates of the temperate regions, whereas the greatest yield of narcotic resin is in hotter tropical conditions.[131] The plant grows from 1 to 5 m tall, producing a fiber valued for length, a component of strength in cordage. The plant is an herbaceous annual, usually dioecious, but sometimes monoecious. The crop is grown from seed, which emerges readily even at low temperatures (down to 10°C). Flowering begins in 110 to 130 days, and the crop is harvested for fiber at that stage.[131] Male plants produce the best fiber, and are sometimes harvested separately. If left to ripen, the females produce mature seed in 20 to 40 days after pollination. The seed contains 30 to 35% drying oil, which is sometimes extracted for use in paints.

Plants harvested for fiber are subjected to retting, which may take a few days or several weeks depending on method, followed by breaking and scutching to extract the fibers.

Jute (*Corchorus capsularis, C. olitorius*)

Jute, the most widely used fiber for sacking in which to store and ship produce, is derived from an annual plant grown from seed. It grows to a height of up to 5 m. The time from seeding to harvest for fiber ranges from 90 to 140 days, depending on cultivar and environment.[42,132] The plants are cut at ground level when the pods are beginning to form, and placed in small stacks for drying. Most of the dry foliage drops off on handling in preparation for retting. In India, a major producer of jute, retting is carried out in slow running streams at 30 to 35°C, taking 10 or 12 days.[132] In Brazil, two crops per year are grown in some regions, one planted in spring and the other in autumn, using early-maturing cultivars.

Jute produces seed readily in short-day conditions of 11 to 12 hr of daylight. Seed setting is favored by sunny days at pollination. As already noted, there is a wide range of maturity times among cultivars, the beginning of flowering ranging from 85 to 130 days. Bloom to seed maturity takes about 35 to 45 days.[132]

Kapok (*Ceiba pentandra*)

This evergreen tree may be started from seed, but selected strains are usually budded on 1-year-old stock or reproduced from cuttings. A year after budding the plants are transferred to their permanent location, where they may begin to bear after 2 to 4 years.[42,54] Kapok is the floss produced on the inner walls of the banana-shaped reproduction capsules, up to 25 cm in length, in which the seeds lie loose at maturity. Adaptation is tropical; the fruit will not set at night temperatures below 20°C. The trees reach full bearing at 7 to 10 years of age, may reach a height of 30 m, and continue productive for 60 years or more.[54] Flowering occurs once a year, as a rule induced by a dry period which may also lead to some leaf fall. The flowers open soon after sunset, and remain open into the next day when the petals fall off. They are capable of self-pollination, but it is reported that bats are a prominent cross-pollinating agent, and insects may cross-pollinate flowers in the early morning. The seed capsules are usually harvested before they dehisce, then dried, the floss removed, and the seeds separated from it.[54]

Linen (*Linum usitatissimum*)

Cultivars of this species selected for fiber production are taller and less branching than those used for oil production. Size and quantity of seed have been sacrificed to obtain longer and finer fibers. These characteristics are enhanced by growing more plants per unit area, and producing the crop in regions with a cool climate and adequate rainfall.[12] The net effect of these factors is to lengthen the time frame of the developmental stages. The stage of full flowering is reached in 55 to 80 days. Early cultivars are ready for harvest in 110 days after planting, but a growing period of 120 to 150 days is more general, in part because environments which extend the period of vegetative growth tend to produce the highest quality of fiber. The crop is harvested when the seeds have reached full size, while the stem is still yellow and about two thirds of the leaves have been shed; later harvesting leads to a coarser grade of fiber.[42] The plants are pulled to avoid cutting the fibers, as cutting results in truncated ends which make the "feel" of the linen cloth less pleasant. The process of "retting" follows harvest. Dew retting is seldom practiced because it takes an uncertain length of time, usually 14 to 21 days but sometimes longer, and because the fiber quality may be reduced. Tank retting leads to uniform results if the temperature is kept at 27°C, and the process is completed in 6 to 8 days.[12,42]

Sisal (*Agave sisalana*)

Plantation sisal for fiber production is propagated from suckers, or from bulbils that form on the flowering stem or "pole".[92] Given the most favorable conditions, harvesting may begin in 3 years, but often is delayed to an age of 5 to 8 years.[42] The lower leaves are ready for harvest when they approach a horizontal position to the ground, and can be removed progressively until the plant bolts. This has been known to happen in 4 years, but rarely occurs before 7 years of age.[92,133] Plants seldom go beyond 10 years without forming a flowering pole, but may go to 15 years.[42] A plant normally yields from 185 to 300 leaves during this period. After going to flower, the plant dies.

Seed is seldom produced on poles in fiber plantations.[92] It can be produced for breeding purposes fairly readily through proper fertilizer application in short-day environments (11 to 14 hr).[133] The seed is generally dormant for up to 5 months. Seedling plants are usually grown in a nursery up to 3 years of age, when they will be about 60 cm tall. They are then transplanted to a field with spacing at about three per meter. The onset of completion of the life cycle is brought about through emergence of the flowering pole from a rosette of small leaves around the growing apex. The pole is a large panicle, 6 to 7 m in height, the branches about 30 cm long with secondary and tertiary branches at their ends on which the hermaphroditic flowers are formed. The pole takes 60 to 70 days to complete growth. It bears 25 to 40 branches, and each carries a cluster of about 40 flowers. Flowering begins at the base of the pole, taking a week to complete on each branch. The anthers dehisce 36 to 48 hr after extrusion. The style begins to elongate at anther extrusion but does not become receptive until 72 to 96 hr later; thus individual flowers do not self pollinate. Though fertilization may take place through pollen falling on stigmas below shedding flowers, most of the pollination is considered to be through insects and bats. It takes about 6 months from fertilization for the seed capsules to mature.[133]

FRUIT CROPS

The principal developmental features of most fruit crops are amenable to general description through tabular presentation (Tables 4 and 5). The tropical fruits are less regular in behavior than the temperate ones, and some expansion of the tabular information is required. As the information on banana and pineapple does not fit the tabular format, separate descriptions of their life cycles are provided.

Woody perennial fruit species, with few exceptions, take at least 3 years from orchard planting to begin bearing. The period may extend to 7 or 8 years, depending on the species, cultivar, and growing conditions. Once bearing is established, flowering and fruit set follow an annual rhythm imposed by the environment and environmental-genotypic interactions.

Table 4
DEVELOPMENT OF TREE FRUITS

Crop	Type[a]	Time to bearing (years)	Flower initiation (season)[b]	Flowering (season)	Bloom to maturity (days)	Reach full production (years)	Production life (years)	Ref.
Apple								
Malus domestica	D	5—8	ESu	MSp	80—150	10—12	20—40	136,139,140,142
M. bacata	D	4—6	ESu	MSp	65—110	8—18	30—50	136
Apricot, *Prunus armeniaca*	D	4—6	LSu	ESp	60—120	8—10	15—20	136,146,147
Avocado, *Persea americana*	E	5—7	Su	W[c]	140—400	7—10	30—40	54,91
Cherry								
Prunus avium (sweet)	D	5—7	Su[d]	VE Sp	35—65	9—12	25—30	136,139,148
P. cerasus (sour)	D	3—6	Su	E Sp	50—80	8—10	20—25	136,139,142
Date, *Phoenix dactylifera*	E	5—6	LSu	E Sp[e]	160—240	8—10	50+	91,95,149,150
Fig, *Ficus carica*	D	3—5	LSu,Sp[f]	Sp; LSu	70—130	5—7	—	42,92,141
Grapefruit, *Citrus paradisi*	E	4—7	Ir	In[g]	190—380	7—12	30—50	37,91,92,138
Kiwifruit, *Actinidia chinensis*	DV	4—5	VESp	Sp	160—200	7—8	—	151,152
Lemon, *C. limon*	E	4—7	In	In	220—280	7—12	30—50	37,92,138,153
Lime, *C. aurantifolia*	E	4—7	In	In	140—230	6—10	30—50	37,91,92,138
Mandarin, *C. reticulata*	E	4—8	W	ESp	170—330	6—10	20—40	92,138
Mango, *Mangifera indica*	E	3—7	LA	VESp[h]	70—180	8—12	35—40	131,154,155
Orange, *C. sinensis*	E	4—7	In	In	170—370	6—12	20—50	37,91,92,138
Papaya, *Carica papaya*	EH	1	Ap	MSu[i]	80—220	—	4—8	37,54,156,157
Peach, *Prunus persica*	D	3—4	MSu	ESp	80—150	6—10	30—160	136,139,140,158
Pear, *Pyrus communis*	D	6—8	LSp,Su	MSp	75—160	8—12	30—58	136,139,142
Plum								
Prunus domestica	D	4—6	LSu	Esp	75—150	7—9	20—30	136,139,140
P. salicina	D	4—6	LSu	VESp	50—120	7—9	20—30	136,139,140

[a] D = deciduous; E = evergreen; V = vine; H = herbaceous.
[b] E = early; M = mid; L = late; Sp = spring; Su = summer; A = autumn; W = winter; In = intermittent.
[c] Purseglove describes three avocado types with widely different fruit development periods, all of which tend to flower from November to May in the northern hemisphere, but the flowering period may last up to 6 months.[54]
[d] Sweet cherries initiate flowering after harvest of the preceding crop.

Table 5
DEVELOPMENT OF SMALL FRUITS

	Type[a]	Begin to bear (years)	Floral initiation (season)[b]	Time of bloom (season)[b]	Bloom to maturity (days)	Fully productive (years)	Longevity (years)	Ref.
Blueberry, *Vaccinium* spp.								
Highbush, *V. corymbosum*	DS	4—5	Au	ESp	70—90	5—8	20—30	136,137,142
Lowbush, *V. angustifolium*	DS	1—4	Au	ESp	60—80	2—3	—	136,137
Cranberry, *V. macrocarpon*	DV	3—4	LSu	MSu	75—100	5—6	40—50	136,137,159
Currant, *Ribes sativum*	DS	3—4	MSu	ESp	45—80	3—4	12—15	136,137
Gooseberry, *R. hirtellum*	DS	2—4	LSu	ESp	45—75	4—5	10—12	136,137
Grape: *Vitis* spp.								
V. vinifera	DV	3—4	MSu	LSp	80—190	5—6	30—50	135—137,142
V. labrusca	DV	3—4	MSu	LSp	80—180	5—6	30—50	136,137,142
V. rotundifolia	DV	3—4	MSu	LSp	100—120	5—6	30—50	136,137,142
Raspberry: *Rubus* spp.								
Red: *R. idaeus*	DC	2—3	LSu	MSp	40—80	3—4	8—10	136,137,160
Black: *R. occidentalis*	DC	2—3	Au	MSp	60—85	3—4	8—10	136,137
Strawberry, *Fragaria vesca*	HP	1—2	Au	VESp	20—45	1—2	4—6	136,137,142,161

[a] D = deciduous; S = shrub; V = vine; C = cane; H = herbaceous; P = perennial.
[b] Au = autumn; E = early; M = mid; L = late; Sp = spring; Su = summer; V = very.

[e] Tolerant to only mild frosts, the date is a dry climate plant which flowers in February to May in Arizona and California, depending on cultivar, weather, and age.
[f] The fig may initiate flowers during the growing season, resulting in a second crop; in some environments the latter is the main crop.
[g] Citrus may flower mainly in spring grown in a climate with a cool winter, and may produce some flowers continuously in a tropical climate; characteristically there is a season (or two or three seasons) when a flush of flowering occurs in response to a dry period or other inducement.
[h] Mango may depart from the pattern shown, depending on timing of dry seasons.
[i] Flowering in papaya is seasonal only in year one; thereafter fruits develop in leaf axils, about one per week at maximum.

Species adapted to the temperate zones generally go through floral induction and bud formation during the latter part of the spring season and into late summer. In the early maturing cherry, floral initiation generally begins after ripening of the fruit, but in several species, such as apple, peach, pear, and plum, initiation often begins in June in the northern hemisphere, and may continue into late September with conditions being favorable.[134] Indeed, the development of a bud from calyx, through anther and pistil to ovary has been shown to take up to 75 days in apple, for example.[135] The bud goes into a dormant phase, the breaking of which requires a period of chilling at temperatures below 7°C. The chilling period varies between species, modified by cultivar, but in general terms ranges from 3 to 5 weeks at the low end for blueberries to 7 to 9 weeks for apple, plum, peach, and sweet cherry at the other extreme.[134,136] Some types of grapes require only a few days, as is the case for some subtropical species such as olive, fig, and almond.[134] The cranberry requires an exceptionally long chilling period of about 15 weeks.[137] Citrus species do not require a chilling period, and given a favorable environment may flower and bear throughout the year.[138] More commonly, flushes of bloom occur in response to periodic changes in temperature and moisture. In some locations in the Caribbean area, for example, the main flowering period is in August to September, following a dry period from April to July. A minor flush occurs in March following a brief dry period.[138] In subtropical areas with definite seasonal temperature variations, such as Florida, California, and Spain, one major crop results from spring flowering.[91,138]

Growth of the majority of fruits following fertilization takes the form of a sigmoid curve, but in the stone fruits a double sigmoid curve best describes growth.[136] The double sigmoid curve is explained on the basis that rapid growth prevails during expansion of the shell of the pit, followed by little increase in fruit diameter while the endosperm and embryo grow to fill the shell, with subsequent rapid growth to full fruit size. However, the double sigmoid curve is also characteristic of the grape and fig.[136] A study of fruit development in the peach (Elberta cv.) showed that the pit reached near maximum size in 30 to 40 days, though the fruit was about 1/3 of the final diameter at that stage.[139] The embryo and endosperm grew to fill the shell in the following 30 days, while the fleshy portion grew little during that period. The fruit achieved the remaining diameter increase in the next 40 to 50 days.

Species differ in how easily they break winter dormancy and begin to bloom. Cultivars within species also differ in this respect, but spring frost injury due to a tendency for early blooming is somewhat more prevalent in apricot, peach, and sweet cherry than in the other species listed in Tables 4 and 5.

The phenology of fruit production in species which are widely adapted is strongly influenced by latitude and modified by other factors, particularly altitude. On the North American continent, the timing of blossoming and maturity is influenced similarly by latitude in the apple, pear, and peach.[136,140] Flowering is general in February in southern states such as Georgia and Alabama, with ripening of most cultivars in June and July. In the northern states and Canada, flowering is in May, and ripening in September and early October.[140] Sweet cherries are the earliest tree fruit to mature in their area of adaptation, which does not include high-temperature zones.

Most fruits undergo changes if left on the tree after full growth in dry matter has been reached, and these changes are likely to be different from those which occur in air-conditioned storage. Many fruits will endure only a short period on the tree after maturity either because they drop or deteriorate.[136,139] Most citrus species may be left on the tree for a prolonged period, up to a few months, without deterioration, and may even improve in quality from a moderate delay in harvesting after they mature.[54,92] Grapefruit are frequently left on the tree for some time after full growth. At the other extreme among citrus is the mandarin orange, which deteriorates or drops if left on the tree much after maturity. The avocado can also be left on the tree for some time after maturity.[92]

The fig, which is dioecious, as are the date, kiwi, and papaya among the fruits dealt with here, has a somewhat unique pattern of development. Some cultivars produce fruit parthenocarpically, but others must be pollinated by a wasp which inhabits the inedible caprifig, which is the donor of pollen.[92] The details of the process were reexamined and described by Valdeyron and Lloyd.[141] In southern France, a parthenocarpic crop may mature in July, no pollen nor wasps being present to fertilize the flower. The main crop matures in September and October from fertilizations effected in May. In some environments the caprifig may produce up to three flushes of crops and wasps during the year, providing pollination for the female flowers within the syconium, a globose pyriform receptacle within which the edible fruit is formed.[42]

Fruit growth on the highbush blueberry conforms to a double sigmoid curve pattern. Though individual fruits may mature in as little as 60 days after pollination, Shoemaker[137] states that a growing season of 160 days is desirable, partly because indeterminate fruiting prolongs the harvest over a 2-month period, and also to permit hardening of the flower buds. He suggests that the superior hardiness of the lowbush blueberry may be attributable to more effective protection from snow cover. The productive life of the lowbush blueberry is difficult to assess because of continuous replacement of plants through adventitious growth from rhizomes.[137] In addition to the two species listed in Table 5, the rabbiteye blueberry (*V. ashei*) is grown in the southeastern U.S. It is taller than the highbush (3 m), and fruits somewhat later where the two are grown in similar environments.[137] It is also more tolerant of heat, drought, and canker.

Grape species vary somewhat in phenology (Table 5) and encompass a range of cultivar maturities that extend over almost 3 months.[136] Development can also be modified by training, pruning, and other management practices.[137] Gibberellins are sometimes used to alter tightness of the fruit cluster, but two dippings, one 10 days before bloom and the other 10 days after, are reported to advance maturity by 2 to 3 weeks.[142]

The canes of bramble fruits are biennial, vegetative the first year and bearing in the second year, after which they should be removed. Currants produce fruit on 1-year-old canes, but also bear on spurs of 2- and 3-year-old canes. As the fruit on wood older than 3 years is inferior, these canes are usually pruned out.

Banana (*Musa acuminata*)

The banana plant is propagated vegetatively, either from suckers or from corm parts with buds. Seeds are rarely produced in commercial plantings, though some cultivars produce an occasional seed which can be used in breeding.[37] To establish a new plantation, suckers are usually grown in a nursery for 6 to 8 months before transplanting to an orchard. Growth proceeds through formation of successive single leaves arising from the pseudostem in ascending rotation. New leaves are produced at a rate of from one per 7 to 20 days, largely dependent on temperature. The number of leaves produced before fruiting may vary from 23 to 50, but the number of functional leaves is about 10 to 20 at any time because of death of the older leaves. Time of flowering is dependent on development of a critical leaf area, and in most tropical areas takes place 7 to 9 months after transplanting.[37] In California, suckers left in place as a form of ratoon crop flowered in 12 to 18 months,[143] but in the tropics they may flower in 6 to 8 months.[91]

The inflorescence is formed as an extension of the true stem, hidden within the pseudostem prior to flowering. Hence the inflorescence is determinate and the plant dies after maturing the fruit.[91] The inflorescence is a spike, the basal nodes of which, 5 to 15 in number, bear female flowers, whereas the terminal nodes bear male flowers.[95] The fruit matures in about 85 to 100 days after pollination.

The timing of crops to provide a continuous supply of fruit can be achieved in tropical areas in which there is a relatively consistent water supply. In such situations the initial crop may be timed through planting dates, and maintained as ratoon crops over several years. In subtropical regions, this is not always feasible, as planting is usually done in spring, maintaining vegetative growth the first year with near dormancy during the winter months. Bearing occurs in the second summer and autumn, but the ratoon crop does not maintain the same pattern.[37]

Pineapple (*Ananas comosa*)

The pineapple is a tropical crop grown primarily within the areas on either side of the equator bounded by the 16°C isotherm for the coldest month of the year.[37] Propagation is vegetative, from terminal shoots, side shoots, suckers, or stem sections. Well-developed suckers, as compared with the other materials, have been shown to shorten the time from planting to mature fruit by as much as 3 to 7 months in Hawaii,[37] and by a somewhat greater difference in Florida.[144] The difference in development time is utilized in providing for a continuous supply of market fruit. For each plantation only one type of propagating stock is used.[37] The vegetative phase, which lasts from 10 to 24 months, consists of production of 60 to 90 leaves on a central stem which bears the apical meristem, and which increases to a length of 20 to 25 cm. The leaves bear axillary buds, some of which develop into side shoots which may be utilized as a ratoon crop.[37] The initial inflorescence forms at the apex of the stem. In a typical Hawaiian crop, the period from planting to floral initiation was reported to be 427 days. The time from initiation to the first open flower was in the range of 42 to 80 days.[37,145] The period of flowering was about 26 days, and from cessation of flowering to ripe fruit, 109 days. Thus the total time from planting to mature fruit in Hawaii was 642 days. Cultivars adapted to Florida take 420 to 540 days from planting to maturity for the first crop, and about 360 days for a ratoon crop.[144] Two ratoon crops are usually produced, but the number varies with different cultivars. Obtaining uniform flowering is a problem, and this can be enhanced by the application of a synthetic hormone, such as napthaleneacetic acid, 6 months before the time of natural fruit setting. The treatment also has the effect of delaying flowering by several days,[97] and influences fruit size in various ways, depending on the timing of the treatment.[37]

DEVELOPMENTAL PATTERN OF TREE NUTS

Seven species of nut trees account for over 95% of world production.[162] The following descriptions outline the life cycles of these seven species, recognizing that others are important in the economies of certain regions of the world. Among the nut species presented are some that are adapted to temperate climates, others which are confined to the tropics or subtropics, and some which are widely adapted. As a consequence, the group represents a complex of phenological patterns which are not readily presented in tabular form. Because of this, and the additional desirability of describing some flowering peculiarities which influence development, each of the seven species will be described in a brief essay.

Almond (*Prunus amygdalus*)

The almond tree is capable of surviving several degrees of frost, but is an important crop only in subtropical regions where frost rarely occurs. It characteristically blooms very early in spring, placing strict limits on adaptation. It is an important crop in the warmer areas of California, Spain, and Italy, where early cultivars generally bloom by February 10, and later ones by the end of March.[162] Propagated by budding on peach, plum, or almond stock, the almond tree usually is bearing well by the seventh year.[42] Floral initiation takes place in late summer.[135] The nut, which is encased in a thin, fleshy fruit that splits on drying, matures in 180 to 220 days, depending on cultivar and weather conditions.[142] Orchard trees may continue productive for up to 40 years, and occasionally longer.

Cashew (*Anacardium occidentale*)

The cashew nut is produced on an evergreen tree which is propagated from cuttings or by grafting and begins to bear in 3 to 4 years. Timing of flowering is primarily governed by rainfall pattern in this tropical crop.[163] A flush of leaf growth occurs toward the end of the wet season, extending into the dry season. At this stage, induction of flowering has taken place. With persistent dry weather and sunshine, inhibitors of floral development are reduced and flowering proceeds. The flowers, produced in large clusters, are bisexual or male. Cross-pollination is favored by flower structure, but the extent of self-compatibility is not known with certainty. The nut reaches full size about 40 days after fertilization, and the apple associated with the nut takes 10 to 20 days additional time to mature. This total development time of 60 days is extended to about 90 days at the highest latitudes where the species is grown.[163]

Nambiar[163] lists the following peak flowering periods of cashew at various locations: Tanzania — August and September; at 13°N in India — January; at 17°N in India — late January to early February; Phillipines — March; Mozambique — October. These mark the onset of the principal annual reproductive season in these regions, but one or two additional growth flushes followed by less intense flowering have been reported.

The usual productive life of a tree is 15 to 20 years, although some have been known to produce for up to 45 years.[162]

Brazil Nuts (*Bertholletia excelsa*)

Commercial production of Brazil nuts is confined to the Amazon basin of South America. Even in this area, plantations are of relatively recent origin, and much of the production is from wild trees in Brazil and adjacent tropical countries. Ecological requirements of the species are such that plantings outside this area, to which it is native, have had limited success.[42,91,162]

The nut is slow to germinate, taking at least a month, and usually 3 or more months to sprout.[162] The tree grows rapidly and may flower in 5 or 6 years, but produces little fruit until the age of 12 to 15 years. Flowering occurs primarily from October to March. It usually takes a year or more for the fruits to ripen, resulting in blossoms and fruits often being present on a tree simultaneously. The nuts are borne within a spherical capsule, resembling in arrangement the sections of an orange, and held in place by fibrous material.[91,162] The trees are tall (up to 50 m. in height) and difficult to scale, so the pods are gathered after they fall to the ground. The harvest period lasts from November to June, and may be sporadic from trees in the wild in response to price incentives and weather conditions. The pods contain from 12 to 25 nuts and weigh from about 900 to 1800 g; hence gathering is dangerous to the worker during windy weather.[162] In any case, protective padding for the head is desirable for pickers. The pods are collected at camps where they are cut open with a knife made for the purpose, called a tercado. The nuts are cleaned and delivered to a central trading post.[162]

The long life cycle of the Brazil nut tree is a deterrent to plantation production in competition with nuts from wild trees. It has been estimated that such a plantation would require 30 years to reach a profitable level of productivity. Little is yet known about the productive longevity of the trees in plantations, but jungle trees are long lived and seem not be subject to major disease problems.[162]

Chestnut (*Castanea* spp.)

The future of the chestnut as a plantation tree is in some doubt due to the inroads of chestnut blight, which has essentially eliminated the American chestnut from North America. The Chinese chestnut *C. mollissima* is reputed to be the least susceptible of the genus to blight. The Japanese chestnut *C. cremata*, though somewhat less resistant, produces the largest nut among members of the genus and is used to a considerable extent in Europe.[162]

Many chestnut plantations have been grown from seed, but superior clones can be grafted readily. Trees begin bearing in 3 to 6 years, and usually produce substantial crops at 10 to 12 years of age, but yield per acre continues to rise until the age of 15 years at least. A moderate cold shock is required for floral development. Flowering occurs in late spring; Chinese chestnuts flower in early June in the Maryland area of the U.S.[162] The nuts develop in burs, and are shed in autumn when they are mature. At this stage they contain about 50% moisture, and spoilage is a problem unless the nuts are picked promptly and refrigerated. If dried, this must be done slowly to preserve quality. It has been suggested that mixing the chestnuts with dry peat to absorb moisture is an ideal way of curing the nuts. Chestnuts produced by seedling trees are subject to substantial spoilage, but some selected cultivars are much less prone to postharvest deterioration.[162]

Filberts (*Corylus avellana*); Syn. Hazelnut

This species is grown in plantations in the Mediterranean region and on the west coast of North America. Selected cultivars are propagated by grafting or budding on rootstocks grown from seed usually planted in autumn. They begin to bear in 4 to 7 years.[42] Floral induction of the male inflorescence takes place in spring, but for the female in late summer.[162] A moderate chilling period is required for the flowers to develop; this having been satisfied, both sexes may bloom during a brief period of weather above freezing. In plantation areas of the west coast of North America this may occur any time from mid-January to March, resulting in fertilization of cross-pollinators. The open female flower tolerates temperatures of -7 to 8°C.[136] The fertilized bud remains dormant, however, until the warm weather of spring brings on bud break and leaf growth. Growth of the pistil becomes evident in late spring as the shell of the nut develops rapidly to full size. The endosperm and embryo then develop to fill the nut cavity. The nuts fall from the tree when ripe, usually in September and October, and some inducement may be added by shaking the tree.[162] For mechanized harvesting the ground in the plantation is smoothed and packed. Various machines have been developed for picking up the nuts and separating out leaves and trash. The nuts are normally dried after cleaning at temperatures of 32 to 38°C to an in-shell moisture of 7 to 8%, and sealed in plastic containers to stabilize moisture and prevent absorption of objectionable flavors from the environment.[162]

Pecans (*Carya illinoensis*)

This species is grown intensively in the south central and southeastern regions of the U.S.[162] Selected cultivars are grafted on 1- or 2-year-old stock seedlings, and normally begin to bear in 4 to 7 years.[42,142] Initiation of the male flowering buds takes place in early summer, whereas the female ones are initiated in early spring (February to April in the U.S.). Both sexes come into flower in late spring.[162] The nuts reach maturity in 150 to 200 days after flowering, and are usually harvested mechanically off the ground after falling naturally or being induced to fall by mechanical tree shakers.[142,162] The nuts are "cured" for a few weeks, and then shelled mechanically. They may also be polished before marketing.

Walnut (*Juglans regia*)

Propagated from grafted cultivars, the walnut begins to bear nuts in 3 to 5 years, and produces a commercial crop by the age of 8 or 9 years.[162] As with the pecan, the male flowers are initiated in early summer, whereas the pistillate inflorescence is initiated in late winter to early spring. The plant is monoecious, and both sexes flower in mid to late spring, depending on cultivar.[136] Flowering sometimes commences early enough for frost damage to occur at the higher latitudes of production (e.g., Oregon and Washington), and preference is given to late-blooming types in these areas. There is a spread of about 45 days in flowering time between the earliest and latest cultivars grown in North America.[136] Each cultivar flowers over a period of 10 to 15 days. The nuts require a long season, with the normal harvest period in North America being through October and November.

VEGETABLE CROPS

Vegetables comprise a highly divergent group of species. They supply a broad range of food products based on utilization of various plant parts, ranging from fruits and seeds to vegetative parts; i.e., the roots, stems, and leaves. Tables 6 and 7 are designed to provide phenological information applicable to three broadly defined growing season lengths in the northern hemisphere. Included are ranges of average growing periods (planting to harvest stage) which apply in a general way in the area of adaptation of each species. The information was condensed from the excellent handbook by Splittstoesser,[11] modified slightly based on data obtained from sources in the references listed in the tables. This information may be utilized more widely through application of climatic analogs. Data from western Australia suggest that some species require longer growing periods than those indicated,[164] but this may be in part a result of utilizing the coolest season of the year for the production of crops adapted to moderate temperatures.

The division of species between Tables 6 and 7 is somewhat arbitrary, as some of those in Table 6 could conceivably be transplanted in certain circumstances and most, if not all, of the species in Table 7 are grown from field-planted seed in environments where transplanting offers little or no advantage. In short-season areas, the latter species are started under shelter to prolong the production period. In longer-season areas this is an added expense, and some of the advantage gained by early starting of the crop is reduced by the developmental delay caused by transplanting shock. However, it is reported that in New Jersey, where direct seeding of tomatoes is quite successful, about 30 percent of the crop is transplanted as it matures 10 to 14 days earlier than the field seeded crop, thereby meeting the demand for an early local tomato crop.[165] Although indeterminate cultivars of tomato are used in home gardens and for greenhouse production, commercial plantations are made with determinate types to provide a uniform crop for mechanical harvesting.

The seed production aspect of the vegetable cycle is not dealt with here. Over 20 years ago Hawthorn[166] stated, "Men who grow vegetable seeds for sale engage in a specialized and highly competitive farming operation." The specialized knowledge required in seed production operations has greatly increased since that time, and much of the involved methodology has been developed by private concerns for their own use, and is thus not in the published literature. For a recent treatment of some developments in seed production practices, see Hebblethwaite.[167]

Asparagus (*Asparagus officinalis*)

This species does not conform sufficiently in life cycle to be included in the tables. It is a perennial which takes 4 or 5 years to reach full productivity and generally continues productive for 15 to 20 years from planting.[168] Asparagus is usually established from seed, though vegetative propagation is possible. In the northern hemisphere, seeds are usually sown indoors in February, the plants pricked out and transferred to cold frames in April, to be transplanted to a nursery in June. They are planted in a permanent location the following spring, left to develop for two seasons, with harvesting beginning in the spring of the fourth year.[168] Alternatively, seeds can be started in a field nursery by planting in late spring after soaking the seed overnight in warm water.[169]

The annual harvest of asparagus begins when spring growth of most perennials is well started. In the northern hemisphere this means late April to late May, depending on latitude and other environmental factors. Harvesting usually continues for from 4 to 7 weeks, depending on environmental conditions.[168,170]

Table 6
PRODUCTION CYCLE OF FIELD-SEEDED VEGETABLES

Field-Seeding Dates by Frost-Free Period[a]

Crop	Type[b]	Long season 2-28 to 12-10[c] First	Long season 2-28 to 12-10[c] Last	Medium season 4-10 to 10-10 First	Medium season 4-10 to 10-10 Last	Short season 5-20 to 9-20 First	Short season 5-20 to 9-20 Last	Ready for use (in days)	Ref.[d]
Beans, snap or green, *Phaseolus vulgaris*	A	4—10	9—1	4—25	7—20	5—15	6—1	48—60	168,170
Beans, lima, *P. limensis*	A	4—15	8—20	5—1	6—1	—	—	65—90	168
Beans, broad, *Vicia faba*	A	—	—	3—15	5—10	4—20	5—5	70—100	170
Beets, *Beta vulgaris*	B	3—1	9—1	3—20	6—15	5—10	6—1	50—80	168,170
Carrots, *Daucus carota*	B	3—1	9—10	4—10	6—15	5—1	6—10	70—90	168,170
Corn, sweet, *Zea mays*	A	4—1	8—1	4—25	6—5	5—10	6—1	70—100	168—170
Cucumber, *Cucumis sativus*	A	4—1	8—15	5—10	6—10	5—25	6—10	50—70	168—170
Endive, *Cichorium endivia*	A,B	3—1	9—1	4—1	7—1	5—15	6—10	60—90	
Kale, *Brassica oleracea*[e]	B	2—15	9—1	4—1	7—1	5—1	6—1	50—80	
Kohlrabi, *B. oleracea*[e]	B	2—20	9—1	4—10	7—1	5—1	6—1	55—80	
Leek, *Allium ampeloprasum*	B	2—15	8—15	4—1	—	—	—	100—150	
Lettuce, *Lactuca sativa*	A	3—1	9—15	3—15	8—1	4—5	7—10	40—50	168,170
Muskmelon, *Cucumis melo*	A	4—1	7—1	4—25	6—1	5—5	5—15	85—110	170
Onions, cooking, *Allium cepa*	B	2—1	9—15	3—20	4—15	4—15	4—25	80—125	168,170
Onions, bunching, *A. cepa*	B	2—1	9—20	4—15	7—9	5—1	7—15	35—75	168,170
Okra, *Hibiscus esculentus*	A	3—25	8—1	6—1	—	—	—	55—65	
Parsley, *Petroselinum crispum*	B	2—15	9—1	4—10	7—1	5—15	6—10	70—90	170
Parsnip, *Pastinaca sativa*	B	2—15	9—1	4—1	5—20	5—1	5—15	120—170	168
Peas, garden, *Pisum sativum*	A	3—1	10—1	4—1	6—5	4—25	6—1	55—90	168,170
Potato, *Solanum tuberosum*	A	3—1	7—20	4—20	5—20	5—1	6—1	75—100	99,168
Radish, *Raphanus sativus*	A	3—1	8—1	3—20	8—15	4—25	8—1	25—40	168,170
Rutabaga, *Brassica napobrassica*	B	1—20	10—1	6—1	—	6—1	—	80—100	170
Spinach, *Spinacea oleracea*	A	2—15	10—1	4—1	8—10	4—25	8—1	40—60	168,170
Squash, summer, *Cucurbita* spp.[f]	A	4—1	6—15	5—1	6—1	5—25	6—1	45—60	170
Squash, winter, *Cucurbita* spp.	A	7—15	8—1	5—20	—	5—15	—	85—120	168,170
Sweet potato, *Ipomoea batatas*	A	4—15	6—1	—	—	—	—	110—130	99

Table 6 (continued)
PRODUCTION CYCLE OF FIELD-SEEDED VEGETABLES

Field-Seeding Dates by Frost-Free Period[a]

Crop	Type[b]	Long season 2-28 to 12-10[c] First	Last	Medium season 4-10 to 10-10 First	Last	Short season 5-20 to 9-20 First	Last	Ready for use (in days)	Ref.[d]
Turnip, *Brassica rapa*	B	2—15	10—1	3—20	6—1	5—25	6—1	40—80	170
Watermelon, *Citrulus lunatus*	A	3—15	7—1	5—15	6—10	5—25	—	80—110	170

[a] Range in dates allows for a succession of plantings where appropriate.
[b] A = annual; B = biennial (in terms of seed production; produce harvest is in year of planting).
[c] Average dates of last spring frost and first fall frost.
[d] Splittstoesser[11] was referred to for all species.
[e] Cole crops, *Brassica oleracea;* kale = var. *acephala;* kohlrabi = var. *botrytis,* subvar. *gongylodes.*
[f] Squash: *Cucurbita* spp; Summer - *C. pepo;* Winter - *C. moschata;* Pumpkin - *C. mixta.*

Rhubarb (*Rheum rhaponticum*)

Rhubarb is one of few vegetables from which the petiole is the only part utilized for food. New plantings are almost invariably established from root divisions to ensure retention of the characteristics of improved cultivars. Seeds can be readily produced for breeding by encouraging bolting. Autumn planting is successful in many environments, but spring planting is preferrable in climates with harsh winters.[168,169] Plants need at least a year to establish before they are harvested, and in the first harvest year should be plucked for only 3 to 4 weeks to permit recovery for winter. Subsequently harvesting can be carried on for up to 8 weeks each summer. Except on nonbolting cultivars, such as 'Valentine', it is necessary to remove the seed stalks when they show.

SPICE CROPS

The tropical adaptation of perennial tree and vine spice crops (Table 8) leads to difficulties in specifying flowering time, as the initiation of flowering is controlled by factors other than seasonal temperature changes. Cloves grown in Zanzibar flower in two periods: July to September and November to January.[171] Nutmeg usually flowers continuously, though flushes may occur in response to environment.[171] Pepper grown in Sarawak flowers over a period of about 5 months, whereas in parts of India two crops are produced annually. Vanilla usually flowers once a year over a period of about 2 months.[172] In Mexico this period is April and May, but in other plantation areas, such as Reunion and the Comoro island, it is November to January.

Vegetatively propagated plants generally come into bearing earlier than those grown from seed of the same species, and have the advantage of securing a known cultivar and the proper sex ratio in dioecious plantations. Some species, such as nutmeg, are difficult to propagate vegetatively on a plantation scale; Purseglove et al. have reviewed recently developed procedures.[172]

The pepper vine is often pruned to delay reproduction until the plants are well established and shaped to provide for ultimate maximum productivity.[171] The basic information presented in essay form is derived from Rosengarten,[173] with modifications as noted from Purseglove et al.[171,172]

Table 7
PRODUCTION CYCLE OF TRANSPLANTED VEGETABLES

Crop	Type[a]	Sowing days before field planting	Long (>200 days)	Medium (130—200 days)	Short (<130 days)	Ready for use (in days)[c]	Ref.[d]
Cole, *Brassica oleracea*							
Broccoli, Var. *botrytis*, s. var. *cymosa*	B	40—50	2/15—7/15	3/15—7/1	5/25—6/15	60—90	168,170
Brussels sprouts, Var. *botrytis*, s. var. *gemmifera*	B	40—55	2/15—7/1	3/15—6/20	5/20—6/1	90—140	168,170
Cabbage, Var. *botrytis*, s.var. *capitata*	B	40—50	2/1—7/10	3/15—6/20	5/20—6/1	60—96	168,170
Cauliflower, Var. *botrytis*, s.var. *cauliflora*	B	40—55	2/20—7/10	3/25—6/15	5/25—6/5	90—96	168,170
Collards, Var. *acephala*	B	40—50	2/20—7/1	3/25—6/10	—	50—80	168,170
	B	55—70	2/20—7/1	3/10—6/1	—	120—135	168,170
Celery, *Apium graveolens*	A	40—55	4/10—6/20	5/15—6/10	6/1—6/10	65—90	168,170
Eggplant, *Solanum melongena*	B	60—70	4/20—6/1	5/1—6/10	—	75—90	168
Leek, *Allium ampeloprasum*							
Lettuce (head), *Lactuca sativa*	A	35—45	2/15—3/10	4/1—5/15	5/10—6/1	65—80	168
Onion (Spanish), *Allium cepa*	B	50—65	2/1—3/1	4/1—5/1	5/1—5/15	90—140	170
Pepper, *Capsicum annuum*	A	40—55	4/2—6/20	5/10—6/15	5/15—6/10	60—90	170
Tomato, *Lycopersicon esculentum*	A	40—50	4/2—5/15	5/10—6/1	5/24—6/10	65—90	166,168

[a] A = annual; B = biennial.
[b] Length of season based on frost-free period.
[c] Period from field transplanting to harvesting for edible use.
[d] Splittstoesser[11] was referred to for all species.

Table 8
DEVELOPMENTAL PATTERN OF PERENNIAL SPICE CROPS

	Type[a]	Time to bearing (years)	Time to full production (years)	Flowering time[b]	Bloom to harvest (days)	Bearing life (years)	Ref.
Allspice (pimento), *Pimenta dioica*	ETD	6—8	13—16	sesnl[c]	85—125	50—100	171,173
Cloves, *Syzigium aromaticum*	ET	7—9	15—25	sesnl	170—190[d]	50—100	171,173
Nutmeg,[e] *Myristica fragrans*	ETD	5—8	15—17	cont	170—250	30—50	171,173
Pepper, *Piper nigrum*	EWV	3—4	7—8	sesnl[f]	180—240	12—20	171,173
Vanilla, *Vanilla planifolia*	HV	3—4	7—8	sesnl	120—270	10—12	172,173

[a] E = evergreen, T = tree, D = dioecious, W = woody, H = herbaceous, V = vine.
[b] cont = continuous; sesnl = seasonal.
[c] In Jamaica, pimento flowers in March to June.
[d] The unopen bud is the commercial product.
[e] Mace is a companion product of nutmeg, forming part of the same fruit.
[f] A dry period checks vegetative growth and brings on floral initiation; subsequent moisture stimulates flowering which usually lasts about 2 months.

PHENOLOGY OF MISCELLANEOUS SPICE CROPS

Anise (*Pimpinella anisum*)

Spring seeded directly in the field in regions with a warm temperate climate, this herbaceous annual flowers in 80 to 95 days. The seed matures in 110 to 125 days (and must be dried at moderate temperature before storage).[173]

Basil (*Ocimum basilicum*)

The leaves are the commercial product of this herbaceous annual species, which is adapted to warm temperate climates. Seeds are planted directly in the field, and emerge in 10 to 14 days. The crop is cut just before flowering, the first cutting being taken 90 to 130 days after seeding. Cuts are made well above ground level to permit regrowth, and up to four cuttings can be taken annually. (Drying of the leaves is done under cover at temperatures less than 110°C to preserve color).[173]

Capsicum Peppers (*Capsicum annuum, C. frutescens*)

These species are the source of paprika, cayenne, and red pepper, as well as an ingredient in chili powder, and require a relatively long season of 110 to 210 days to mature. Selected cultivars, adapted to various ecological areas, are grown from seed, planted as early as danger of frost is past. Flowering commences in 85 to 100 days and may continue for up to 90 days, being indeterminate. Correspondingly, harvest continues over a period of 3 to 4 months. The seeds and placental sections are removed before drying for grinding.[173]

Caraway Seed (*Carum carvi*)

An herbaceous biennial type, caraway is adapted to temperate climates with rather mild winters, as exemplified by the Netherlands, a major producer of the seed, but often survives more severe winters. Usually seeded in early spring, sometimes with an early-maturing annual companion crop, caraway produces flowering stalks early in the spring of the second year. Flowering occurs in late spring, May, or early June in the northern hemisphere. The seeds are ready to harvest 60 to 70 days after flowering. (The plants are cut as soon as the earliest seeds mature to keep shattering to a minimum, and are left in small stacks for further maturing, over a period of about 10 days before threshing.)[173]

Cardamom (*Elettaria cardamomum*)

Indigenous to south India and Sri Lanka, cardamom is an herbaceous perennial evergreen adapted to moderately high altitudes of the tropics. Sections of rhizomes are used to establish new plantations, which are expected to bear seed in the fourth year. Alternatively, plants are started from seed.[172] Flowering shoots are produced at the base of the plants, usually over a period of 8 or 9 months of the year, depending on rainfall distribution and quantity. Buds develop into flowers about 31 days after initiation and capsules mature irregularly 3 to 4 months later.[172] A plantation area is harvested 5 or 6 times during this period to obtain pods which are filled but not overly ripe. The best quality is obtained by picking firm, well filled pods before they are quite ripe. They are then dried, either in the sun or in heated dryers, and are marketed either in the pod, as decorticated seed, or as ground cardamom. A plantation usually remains productive for 10 to 15 years.[173]

Cinnamon (*Cinnamomum zeylanicum*)

The spice is derived from the inner bark of this evergreen perennial bush. Grown in tropical regions (e.g., Sri Lanka), the crop is propagated from cuttings or division of root stocks.[171] The plants are pruned in 2 to 3 years to encourage branching. Branches are harvested when 180 to 240 cm long, in about 2 years after pruning. Shoots arise from the base which usually are ready for cutting in 2 to 3 years. Plants may be coppiced up to 10 years of age, after which serious yield declines occur.[171,173]

C. cassia is similar to cinnamon in some respects, but is a coarser spice with a more intense aroma and flavor. It is usually propagated from seed. The entire bark is used, rather than only the inner bark used in cinnamon. Cassia is cut to form a coppiced bush, but usually takes 5 to 10 or more years for a harvestable cutting to develop. It is considered a less expensive, lower quality substitute for the genuine cinnamon.[173]

Coriander (*Coriandrum sativum*)

An annual herb, coriander is indigenous to the Mediterranean region, and is widely adapted in warm-temperature regions where moisture is adequate. Sown in early spring, it requires from 100 to 125 days to mature. The crop must be left to ripen fully, as indicated by the seeds turning yellowish brown, to assure that an unpleasant odor, characteristic of the plant and unripe fruit, have disappeared. To avoid shattering, the plants are generally cut and stacked while damp with dew. They are threshed after several days of drying, which must be thorough to produce the best quality of product.[173]

Dill (*Anethum graveolens*)

A widely adapted annual plant, dill is produced as an herb and a source of oil if harvested before flowering, or for the mature-crop seed. Grown from seed planted directly in the field in early spring, the dill flowers in about 70 days, and produces seed in 110 to 130 days.[173]

Ginger (*Zingiber officinale*)

The thick, white, tuberous rhizomes of ginger are the source of the commercial spice. It is an erect, perennial herb, which thrives best in warm, moist climates with abundant sunshine. Seldom producing viable seed, it is propagated by dividing selected rhizomes. After about 9 months of growth, the stalks begin to wither, indicating that the rhizomes are ready for harvest. The clumps of rhizomes, referred to as "hands", are cleaned, scraped, boiled and peeled before drying in the sun about 8 days. For use as preserved ginger, the rhizomes are harvested at 7 months to provide a tender, high-quality product.[172,173]

Horseradish (*Armoracia rusticana*)

This is a hardy, perennial plant of the mustard family. The vigorous, deep roots, the cylindrical, fleshy portion of which forms the useful product, may be difficult to eradicate in favorable environments. It is found growing wild in moist, semishaded locations where winters are not severe. Though the plants produce flowers and pods, they often fail to mature viable seed. Propagated from seed in the first year, plantings are usually made by selecting roots about 8 mm in diameter and 25 to 35 cm in length in autumn for refrigerated storage until they are planted in early spring. These will grow to a diameter suited for the marketable product by late autumn. As low temperatures are considered to improve flavor, harvesting is delayed until after some frost, usually October or November in North America.[173]

Mint (*Mentha piperita, M. spicata*)

Hardy, perennial herbs, peppermint and spearmint are widely used flavorings in condiments and foods in general; hence they are cultivated extensively in temperate zones. Propagated from rooted cuttings or runners, planted in March in the northern hemisphere, full bloom will usually be achieved in late July to August, signifying that the crop is ready for harvest. The cut plants are partially dried in the field, then processed to remove the essential oil which is then used as flavoring of various products. Alternatively, the leaves may be dried and fragmented for use as an herbal spice. The plants usually continue productive for 5 or 6 years.[173]

Mustard (*Brassica hirta, B. nigra*)

These two species (known as yellow and brown mustard, respectively) are the most prominent of several plants yielding related products. In type, they are herbaceous annuals adapted to cool temperate climates. Seeded moderately early in spring, they flower in 50 to 65 days, with the pods reaching harvest stage 40 to 50 days later. As shattering is a problem, harvesting is best done before the latest pods are mature.[173]

Pimento (*Pimenta dioica*)

This is a small evergreen tree, the dried unripe fruits of which provide the pimento spice used commercially (also referred to as allspice). Usually propagated from selected seed, a plantation begins to bear in 5 to 8 years. The trees are functionally dioecious, and the proportion of males and females is likely to be about equal, requiring overplanting and thinning for a desired proportion of ten females to one male. To provide a better sex balance as well as to use improved clones, approach grafting has been practiced on a substantial scale in Jamaica in recent years. Grafted plants are expected to produce in 3 years. Trees produce heavily by the age of 20 to 25 years, and continue to produce for up to 100 years. Flowering occurs at different times in various regions, but in Jamaica usually is concentrated from March to June. The fruit is picked when full grown but still green, about 90 to 120 days after pollination.[171,173]

Poppy Seed (*Papaver somniferum*)

The condiment poppy is an herbaceous annual grown from seed. Seed sown in March in the north temperate zone normally reaches full bloom in July, and is ready to harvest in September, thus taking 140 to 160 days from seeding to maturity. The heads, each containing numerous seeds, are fairly resistant to shattering. The plants are usually windrowed at harvest to allow the heads to dry before threshing.[173]

Sage (*Salvia officinalis*)

Grown for spice, the leaves of this small perennial, deciduous shrub are the basic crop. The flowers are removed by pruning to enhance the vegetative growth. Two harvests are generally taken per season by removing the leaves from 15 to 25 cm of the terminal growth of the branches. The leaves are dried in shade or under controlled conditions in porous bags, before crushing and screening to produce the spice of commerce. In the Mediterranean region and California, flowers usually appear in July and August of the second year. The normal productive life of a plantation is 4 years.[173]

Summer Savory (*Satureja hortensis*)

An herbaceous annual, the tender tips and leaves of savory are used as spice, either fresh or dried. Grown in California and the Mediterranean region, the crop is usually sown in April. The crop may be harvested at intervals after reaching a height of 15 cm. When the plants reach the bud stage — in 75 to 120 days from sowing depending on the environment — the whole plant is generally harvested.[11,173]

Tarragon (*Artemesia dracunculus*)

The leaves of this shrubby perennial are the source of the spice. The French tarragon, primarily used in commercial planting, seldom produces viable seeds and is therefore propagated from crown splits. Harvesting begins as soon as plants are well established, with the young shoots being cut and dried for production of crushed or powdered leaves. This may be repeated up to 3 or 4 times per season. Stands are generally renewed every 4 years.[11]

Thyme (*Thymus vulgaris*)

This species is commercially most important of several used for production of the spice. It is an herbaceous perennial. The plants are harvested when in bud, and the material shade-dried and ground to provide the spice of commerce. In long-season areas, a second crop may be harvested, with the proviso that only the top third of the plant be taken, that the remainder provide for improved winter survival. Plantations are renewed in 3 or 4 years.[11,173]

Turmeric (*Curcuma domestica*)

The rhizomes are the source of the spice derived from turmeric, an erect perennial herb which is grown as an annual. In India, seed rhizomes are usually planted in May or June, and take 2 to 4 weeks to emerge. The plants flower in about 150 days, and begin to produce rhizomes at that stage. After 90 to 120 days of growth, the rhizomes are ready for harvest; dying of the basal leaves is a sign of maturity being reached. The harvestable roots consists of a central, ovate bulb, and lateral branches and secondary rhizomes referred to as "fingers".[11,172]

Turmeric rarely produces viable seeds, but a variability among clones has been exploited to derive cultivars which vary in productivity and quality characteristics.[172]

PERENNIAL FORAGE CROPS

Grasses

The life cycles of perennial forage grass species have a number of basic features in common. They differ, however, in many details, the most striking of which appear to reflect adaptation to the environments in which they are important. Yet there are significant differences among species which are suited to a more or less specific ecological niche. Of the several species adapted to the temperate zones, for example, a few important ones have a juvenile stage through which they must pass before becoming sensitive to conditions of temperature and daylength required to induce them to readiness for flowering.[174,175] A number of widely used species which have no juvenile stage respond to temperature and daylength variations in terms of both quantity and time of inflorescence production.

Typical of temperate-zone grasses is the life cycle of smooth bromegrass (*Bromus inermis*), outlined in detail by Lamp.[176] This species does not go through a juvenile stage, but characteristically requires the short days and cool temperatures of autumn for induction to normal flowering,[177,178] although sparse flowering may occur under some conditions in the year of seeding.[179] In the year after establishment and in subsequent years, new tillers begin to form shortly after flowering — mid-June in northern Illinois.[176] These tillers, which continue to emerge until early autumn if conditions are favorable, are the basis of the reproduction of the following year. Lamp's observations showed that growth ceased in early December, to be renewed in mid-March. During the period March 24 to April 28, tillers grew and floral initiation proceeded. Heads began to emerge after mid-May and heading peaked from May 26 to June 1. Anthesis took place in the period June 9 to 30, and the seed was ready for harvest July 11 to 15. The period from beginning of spring growth to ripening was, therefore, approximately 115 days.

Environmental factors profoundly influence the phenology of perennial grasses. Aitken has shown the effects on the developmental patterns of a limited number of grass species of growing them at widely different latitudes and altitudes.[6] The effects are attributed primarily to differences in temperature and daylength at the various latitudes where the species were grown, with secondary effects arising from variations in altitude and proximity to bodies of water. A profound effect of latitude on flowering dates in timothy was reported by Evans,[180] showing progressive delays in flowering from May 31 in Georgia to July 11 in Alaska. The change in date of bloom was not directly proportional to change in latitude. A shift of about 7° latitude between the Georgia and Ohio locations resulted in a flowering delay of 26 days, whereas the shift of 18° between North Dakota and Alaska locations led to a delay of only 9 days in flowering. These effects suggest a complex interaction of daylength and temperature influencing development.

A number of studies carried out in controlled environments have shown that temperature and daylength influence the rate of development in various ways during different stages of the life cycle. It has been established that a number of temperate zone species require exposure to autumn conditions for induction, but which of the factors of low temperature, short days, or the combination of the two is essential has been shown in only a limited number of species.[174,175,178,179,181] In some cases the induction environment has not influenced the rate of development,[174,183] but in others the temperature level during induction was shown to have a considerable effect on time of heading.[182,197,214] On the other hand, both temperature levels and daylength have a profound influence on initiation of the inflorescence and the rate of subsequent development.[175,184-186] Excellent reviews of the interaction of factors involved in floral development have been published.[186,187] Somewhat different insights concerning the role of temperature, in particular, are contained in a report by Davies,[189] and a review by Evans et al.[190]

The majority of grass species adapted to the tropics and subtropics are either indifferent to daylength or are short-day types.[191] Exceptions to this rule have been demonstrated in Bahia grass and carpet grass, which are favored by an intermediate daylength of near 14 hr.[192] Tropical species do not require low-temperature exposure for induction, but time of heading has been shown to be advanced in some of them by such treatment.[193] However, the most prevalent effect of low temperatures is growth suppression, and few of the species survive more than a few degrees of frost.[191,194] Burton lists a number of perennial forage grasses which are adapted to subtropical conditions,[194] but Boonman stated that only half a dozen species dominate the market for tropical pasture species, and that only Rhodes grass and setaria are important in Kenya.[195] Certain grass species adapted to the milder portions of the temperate zone are also highly useful in the subtropics and some areas of the tropics. The grama grasses require no chilling period for reproduction and are insensitive to daylength (or slightly accelerated by short days).[175] Tall fescue and orchard grass thrive well in high temperatures, though seed production may be restricted in most strains due to a need for low-temperature induction.[194,196] The developmental pattern of such species is influenced greatly by environmental differences over the wide range of latitudes in which they are useful.

The phenology of tropical grasses is strongly influenced by rainfall patterns, making difficult the establishment of any widely applicable developmental systems. As an example, gamba grass in Nigeria is reported to begin panicle formation just before the rainy season begins in June, taking about 5 months to reach maturity; whereas in Brazil, in an area with more constant rainfall, three seed crops per year may be harvested.[191] Boonman[195] states that much of the problem of low seed yields in tropical climates arises from the spread over time in heading, with the consequence that seed ripens over a period of several weeks or even months. The flowering of individual panicles also is spread over a considerable period. In consequence, it is difficult to pick a harvest date on which more than a small fraction of the seed is ripe and not shattered. A clear definition of the time from pollination to maturity is difficult in such situations, and seems to have been established for only a few species.

Genotypic differences in length of developmental phases are encountered in most species. Some of these are strain differences which may have developed through natural selection in response to shifts in daylength or other factors.[182,197,198,200] Others have arisen through varietal development by breeders.[6,189,194,199] The variations include many instances of change in period from onset of floral initiation to maturity, which may be a reflection of alterations in responses to temperature and daylength which lead to shifts in the lengths of individual phases in development and to differential reactions to latitude and altitude.

Some tropical grass species rarely produce enough seed for practical use, and are therefore primarily reproduced vegetatively. Examples of this are angola grass (*Brachiaria mutica*), pangola grass (*Digitaria decumbens*), elephant grass (*Pennisetum purpureum*), and the Coastal cv. of Bermuda grass. Angola grass produces a moderate amount of seed which must be harvested a few days after anthesis to avoid shattering, but a sparse stand established from cutting fills in rapidly through stolons.[191] Elephant grass produces very little seed, and is propagated much like sugarcane, which it resembles in size and growth habit.[191]

The main features of the life cycles of a number of perennial forage grasses are shown in Table 9. A number of annual and winter annual grasses are also important forage species, but their life cycles were presented previously. For the temperate-zone grasses, the timing of events is taken from the starting point of the beginning of spring growth. The range of periods to maturity is somewhat truncated at the upper end, as in some coastal areas (e.g., Oregon and Wales), growth may be interrupted for only a short period, if at all. For tropical species, the beginning of growth for seed production phenology was taken from a time representing mowing or grazing of the sward, wherever such information was available.

Legumes

The most important dicotyledonous forage species belong to the legume family. The ones in general use are perennials or biennials, although annual forms of most genera are available, and are useful in certain circumstances. Annual legumes, such as soybeans, peas, and cowpeas, are also important forage species, but their life cycles are dealt with elsewhere.

Alfalfa has been described as "the world's leading hay crop"[12] and "Queen of Forages".[225] It seems appropriate to use the developmental pattern of this species as an example of a long-lived perennial forage legume. The primary source of this description is Smith,[207] expanded through information from other publications.[226-228] Alfalfa seed usually germinates readily and emergence occurs within a few days of planting. A soil temperature of 25°C is ideal, and growth is minimal below 10°C or above 30°C. A diurnal temperature spread (30/15° C) has been shown to be more favorable than either constant temperature for seedlings beyond 8 weeks old. Long days in excess of 16 hr favor floral development, and early spring plantings usually flower the first year at high latitudes. Seed production is likely to

Table 9
DEVELOPMENT OF PERENNIAL FORAGE GRASSES

Crop	Induction[a]	First growth to heading (days)	Heading to bloom (days)	First growth to maturity (days)	Ref.
Agropyron cristatum Crested wheatgrass (Fairway)	V-SD	48—57	15—20	85—105	196,203,204,206
A. desertorum Crested wheatgrass (Standard)	V-SD	45—55	15—20	80—100	196,206
A. intermedium Intermediate wheatgrass	V-SD	60—70	15—25	95—120	186,196,201,206
A. smithii Western wheatgrass	SD	60—70	13—25	95—120	183,196,201
A. trachycaulum Slender wheatgrass	V-SD	60—70	15—25	95—115	196,201
Agrostis alba Redtop	SD	65—75	15—20	95—120	183,196,205
Andropogon gayanus Gamba grass	—	—	—	100—150	191, 196
Axonopus affinis Carpet grass	—	—	—	80—140[b]	194,196
Bouteloua gracilis Blue grama	—	45—60	15—25	85—110[c]	198,199,201
Bromus inermis Smooth bromegrass	V-SD	50—70	15—25	90—120	176—178,184,186
Buchloe dactyloides Buffalo grass	—	45—65	10—20	90—130	196,201
Chloris gayana Rhodes grass	—	120—150[d]	10—25	150—180[b]	191,196
Cynodon dactylon	—	—	—	95—130[b]	191—196
Dactylis glomerata Orchard grass	V-SD	50—65	8—15	80—110	184,189,202,206,207
Elymus canadensis Canadian wild rye	V-SD	57—75	15—30	95—125	196,201
Elymus junceus Russian wild rye	V-SD	55—70	15—25	85—120	201,217
Eragrostis curvula Weeping lovegrass	—	80—110	—	120—160[b]	191,196,201
Eragrostis trichoides Sand lovegrass	—	150—200	—	170—220[b]	191,196,201
Festuca arundinacea Tall fescue	V-SD	60—75	15—20	95—130	201,206,207,220
Festuca elatior Meadow fescue	V-SD	55—70	20—25	90—120	202,208,209,212,216
Festuca rubra Creeping red fescue	V+-SD	55—70	12—20	80—110	202,210,212,213
Lolium perenne Perennial ryegrass	OV-SD[e]	60—80	20—30	95—135	182,189,214—216
Panicum virgatum Switchgrass	O	60—85	15—35	100—140	196,202,218
Paspalum dilatatum Dallis grass	O	75—80	15—20	105—120	191,194,196,217
Phalaris arundinacea Reed canary grass	V-SD	55—70	15—20	80—105	184,207,219,221
Phleum pratense Timothy	V-SD[f]	65—75	20—30	95—140	184,189,205,222—224

Table 9 (continued)
DEVELOPMENT OF PERENNIAL FORAGE GRASSES

Crop	Induction[a]	First growth to heading (days)	Heading to bloom (days)	First growth to maturity (days)	Ref.
Poa pratensis Kentucky bluegrass	V-SD	30—50	15—25	65—85	184,205,206,219
Setaria anceps Setaria grass	—	60—80	—	100—140[b]	191,195

[a] V = vernalization advantageous or required at least in most cultivars. SD = short day generally required for induction. O = no.
[b] These tropical species usually head and bloom over a long period.
[c] More than one seed crop per year may be harvested in long-season areas.
[d] From emergence of crop grown from seed.
[e] Cultivars differ in requirements for induction.
[f] Most cultivars are induced more fully at low than high temperatures, but some require neither cold nor short days.

be sparse in the establishment year, but can be greatly increased by transplanting seedlings started in a greenhouse.[228] At Madison, Wisc., alfalfa begins to go into a state of dormancy in late October which is complete by early December. Spring growth is usually evident by mid-April, and rapid vegetative growth proceeds through May and early June. Floral initiation takes place after mid-June, with buds developing at the 10th to the 14th node by the third week of that month. The plant, which is indeterminate in flowering habit, reaches full bloom about July 10, and will continue to bloom over a number of days, extending into two or three weeks in some environments. The seed also matures over a considerable period of time, requiring judgment in timing harvest to secure most of the potential seed crop. The interval between full bloom and harvest is about a month. By taking a hay cut at early-bloom stage on June 3, regrowth from the crowns made possible another such harvest on July 16, and a third on August 24. For seed production in some environments it is advantageous to take a cutting before growing the seed crop, but this is desirable only when it improves timing in relation to pollinators or environmental factors. Substantial interactions have been shown to take place between temperature and daylength effects, and clipping delayed until midsummer in Arizona have been shown to delay flowering greatly in comparison with spring clipping.[226]

Birdsfoot trefoil establishes rather slowly from seed, but is persistent once started. It starts growth early in spring, but blooms for the rest of the season if conditions are favorable, resulting in a very unevenly ripening seed crop which is difficult to harvest.[207] From fertilization to maturity of individual florets takes from 24 to 47 days.[229] Birdsfoot trefoil has the rare trait of producing stem buds from the roots if the crown is damaged. It has been shown to suffer greater delays in flowering as a result of temperature extremes than alfalfa, red clover, and alsike.[207]

The two sweet clover species widely grown in continental temperate climates are among the most cold tolerant of legumes. Their growth pattern is typical of biennials; in the first year a moderate top growth and development of a deep and thick tap root, followed by extensive top growth and reproduction in the second year. Top growth in the first year consists of a single stem supporting numerous branches, reaching their maximum growth in early autumn. At this time the roots develop rapidly in weight and carbohydrate reserves. Crown rhizomes also develop, which give rise to the growth of the second year.[207,230]

Red clover is a species which encompasses a wide range of types with distinct developmental patterns. They are usually categorized into two groups: the early maturing medium, or double-cut type, and the later mammoth or single-cut type. Hawkins[231] clearly differentiated these two types, in terms of time of flowering and node number below the flowering node, working with typical red clover grown in the U.K. Clones of 'English Broad Red', a double-cut type, on average flowered June 6 with seven nodes, whereas English 'single-cut' flowered June 25 with 11 nodes. He also identified a late red-clover class which seemed distinct, flowering on average July 7 with 14 nodes. An extensive examination of continental red clovers showed the double-cut type grown almost exclusively in some countries, and predominating in general. Most of the red clover grown in North America is of the medium or two-cut type.[196,207] However, the mammoth type has proved more hardy in severe winter conditions, and some cultivar development in Canada is based on this premise.[232] It has been shown that early floral development is favored by moderate temperatures and long days, some strains requiring at least 15 hr of daylight, whereas some early maturing types need only 12 hr.[233,234] Indeed, some cultivars show improved flowering at daylengths up to 18 hr.[237] In seed production, the practice is generally followed of taking a hay crop at early bloom and using the regrowth as the seed crop.[233] Red clover is determinate in flowering habit, and the greater branching of the second growth is advantageous in some circumstances.[207] The delay in flowering resulting from this practice usually amounts to 3 or 4 weeks.[196,207]

The other true clovers in Table 10 are indeterminate in flowering habit, with the exception of crimson clover. Despite adaptation to lower latitudes, crimson clover requires vernalization and is favored by daylengths in excess of 12 hr.[235] It is a winter annual which is an important forage in the southeastern U.S.[12]

White clover is of two types, the common low-growing type which persists widely in pastures and other turf, and the taller-growing ladino type used extensively in pasture mixtures.[196] Although the two are of one species, ladino is recognized as a separate category for pedigreed seed purposes in the U.S. because of distinctive vigor and value in mixtures with grasses.[196] White clover plants are moderately long lived, but a stand may persist almost indefinitely through spreading by stolons and self-reseeding. It is rather slow to establish from seed, and requires moderately high temperatures before beginning growth in spring. It flowers early, and continues to flower in flushes over the summer. In seed production of ladino in Oregon, the crop is clipped or grazed up to about mid May to promote even flowering of the seed crop.[196,207] White clover does not require vernalization, usually flowering in the year of seeding, but cold treatment leads to earlier flowering.[236]

Self-sterility is characteristic of most of the clovers listed in Table 10, and in those that are self-fertile, as is the case in crimson and strawberry clovers, cross-pollination is prevalent.[236]

OTHER CROPS

Rubber (*Hevea brasiliensis*)

A tropical evergreen native to the Brazilian tropics, the rubber tree is now widely grown in tropical plantations, especially in Southeast Asia.[37] Selected strains are generally grafted on stocks grown from seed. Increase in girth of trees is almost linear for the first 10 to 15 years, with annual increments gradually declining with advancing age.[37] Tapping begins when the tree has reached a girth of approximately 50 cm, a stage likely to be reached in 5 to 8 years. The "jebong" method of tapping is favored in well-managed plantations.[42] With a special knife made for the purpose, a trained worker opens a cut at a 30° angle at about head height, usually halfway round the tree. At intervals of 2 to 4 days, the tapper cuts a thin slice (approximately 1 mm) off the lower side of the opening, severing the sealed ends of the latex vessels and renewing the latex flow. During a year of tapping, 15 to 20 cm of bark will be removed, and the panel will extend to the base of the tree in 6 to 10

Table 10
DEVELOPMENTAL PATTERN OF FORAGE LEGUMES

Crop	First growth to bloom[a] (days)	Fertilization to mature seed[b] (days)	First growth to harvest (days)	Ref.
Lespedeza cuneata Sericea lespedeza	70—100	28—42	120—130	12,92,196,246
Lotus corniculatus Birdsfoot trefoil	40—60	24—47	90—120	196,229,238,239
Medicago sativa Alfalfa	45—75[c]	28—40	110—140	196,207,226—228
Melilotus alba White sweet clover	68—80	25—40	105—120	196,204,230,240,241
M. officinalis Yellow sweet clover	60—70	18—30	95—110	106,207,230,240,241
Onobrychis viciaefolia Sainfoin	50—60	42—49	95—110	12,242—244
Trifolium hybridum Alsike clover	45—70[c]	21—25	90—120	12,48,196,207
T. incarnatum Crimson clover	50—60	26—30	90—115	12,196,235
T. pratense Red clover	50—75[c]	21—30	95—130	231—234,237
T. repens White clover	25—40[c]	20—28	80—110	12,196,207,236
T. subterraneum Subterranean clover	50—90	28—39	95—210[d]	6,92,196,245

[a] From first visible spring growth to initial stage of profuse blooms; many species bloom profusely over a long period.
[b] This refers to developmental time of individual florets, in view of the indeterminate flowering of several of the species.
[c] The shorter period applies to regrowth after a clipping or grazing usually at bud or early bloom.
[d] The long growth period refers to early winter planting in an environment such as Melbourne.[6]

years. Subsequently, the other half of the trunk surface can be worked in a similar manner. By the time this is completed, if tapping has been done with care not to impare the cambium, the first panel can be reworked. The timing of these events is a function of the frequency of cutting and the skill of the tapper in removing thin slices, as well as of environmental conditions. Full-spiral panels can be used in place of the half-spiral ones described, leading to greater yield initially, at the expense of reduced girth increments. It is seldom used except toward the end of the life of a tree. A number of chemicals, among them 2:4-D, have been found useful to increase the flow from sluggish latex vessels. The productive life of a tree in a well-managed plantation may extend to 40 years or more, but replacement may occur sooner because the labor of exploiting trunks above the 2-m level is too costly.

Reproduction through seed is important as a source of root stocks and for improvement through breeding. The tree is monoecious, with male and female flowers in the same panicle. Self-fertility varies with genotype, but cross-fertilization is favored, even for root stocks, to improve vigor. Flowers may be produced throughout the year, but Iearing is usually seasonal, generally with two flushes annually. In Malaysia, for example, the heaviest seed crop is produced in August to September, with a somewhat lighter crop in January to February. Wide variations occur in response to seasonal environment, and genotype-environment interactions. Clones differ widely in seed-production capacity.[247] Commercial seed is predominately produced through hand crossing. The period from fertilization to ripening is about 120 to 150 days.

Buckwheat (*Fagopyrum sagittatum*)

A rapidly developing annual, buckwheat usually begins to flower 35 to 40 days after sowing. Flowering is indeterminate, and may continue for 25 to 30 days.[92] Having a self-incompatibility system, the crop is naturally cross-pollinated and benefits from pollination by honeybees and other insects. The total growing season required is about 70 to 90 days, making buckwheat a favorite crop for reseeding where a crop has failed early in the season.[12] Even in long-season areas, late planting is practiced in order that flowering should occur after the hottest part of the summer, as the optimum temperature for fertilization is about 20°C.[248] In areas with cool summer climates, this delay is not advantageous.

REFERENCES

1. **Milthorpe, F. L.,** Some physiological principles determining the yield of root crops, in *Proc. Int. Symp. Tropical Root Crops,* Tai, E. A., Ed., University of West Indies, St. Augustine, Trinidad, 1969.
2. **Zadoks, J. C., Chang, T. T., and Konzak, C. F.,** A decimal code for the growth stages of cereals, *Weed Res.,* 14, 415, 1974.
3. **Fehr, W. R., Caviness, C. E., Burmood, D. T., and Pennington, J. S.,** Stages of development descriptions for soybeans, *Glycine max* L. (Merr.), *Crop Sci.,* 11, 929, 1971.
4. **Vanderlys, R. L. and Reeves, H. E.,** Growth stages of sorghum (*Sorghum bicolor* L. [Moench]), *Agron. J.,* 64, 13, 1972.
5. **Monteith, J. L.,** Climate, in *Ecophysiology of Tropical Crops,* Alvim, P. de T. and Kozlowski, T. T., Eds., Academic Press, New York, 1977, chap. 1.
6. **Aitken, Y.,** *Flowering Time, Climate and Genotype,* Melbourne University Press, Melbourne, 1974.
7. **Alvim, P. de T. and Kozlowski, T. T.,** *Ecophysiology of Tropical Crops,* Academic Press, New York, 1977.
8. **Nuttonson, M. Y.,** *Wheat-climate Relationships and the Use of Phenology in Ascertaining the Thermal and Photo-thermal Requirements of Wheat,* American Institute of Crop Ecology, Washington, D.C., 1955.
9. **Nuttonson, M. Y.,** *Rye-climate Relationships and the Use of Phenology in Ascertaining the Thermal and Photo-thermal Requirements of Rye,* American Institute of Crop Ecology, Washington, D.C., 1957.
10. **Nuttonson, M. Y.,** *Barley-Climate Relationships and the Use of Phenology in Ascertaining the Thermal and Photo-thermal Requirements of Barley,* American Institute of Crop Ecology, Washington, D.C., 1957.
11. **Splittstoesser, W. E.,** *Vegetable Growing Handbook,* AVI, Westport, Conn., 1979.
12. **Martin, J. H., Leonard, W. H., and Stamp, D. L.,** *Principles of Field Crop Production,* 3rd ed., Macmillan, New York, 1976.
13. **Borlaug, N. E., Ortega, J. E., Rarvaez, I., Garcia, A., and Rodriguez, R.,** Hybrid wheat in perspective, in Hybrid Wheat Sem. Rep. Crop Qual. Coun., Minneapolis, January 30, 1964, 1.
14. **Busch, R. H. and Chamberlain, D. D.,** Effect of daylength response and semidwarfism on agronomic performance of spring wheat, *Crop Sci.,* 21, 57, 1981.
15. **Russell, W. A. and Hallauer, A. R.,** Corn, in *Hybridization of Crop Plants,* Fehr, W. R. and Hadley, H. H., Eds., American Society of Agronomy and Crop Science Society of America, Madison, 1980, chap. 19.
16. **Allison, J. C. S. and Daynard, T. B.,** Effect of change in time of flowering induced by altering photoperiod or temperature on attributes related to yield in maize, *Crop Sci.,* 19, 1, 1979.
17. **Nanda, K. K., Grover, R., and Chinoy, J. J.,** Factors affecting growth and development of some millets, *Phyton,* 9, 15, 1957.
18. **Oggema, M. W.,** Ann. Rep., National Plant Breeding Station, Njoro, Kenya, 1975.
19. **Hanway, J. J.,** Growth stages of corn (*Zea mays* L.), *Agron. J.,* 55, 487, 1963.
20. **Shaw, R. H.,** Climatic requirement, in *Corn and Corn Improvement,* Sprague, G. F., Ed., American Society of Agronomy, Madison, 1977.
21. **Starzycki, S.,** Diseases, pests and physiology of rye, in *Rye: Production, Chemistry and Technology,* Bushuk, W., Ed., American Association of Cereal Chemists, St. Paul, 1976.
22. **Fowler, D. B. and Gusta, L. V.,** Influence of fall growth and development on cold tolerance of rye and wheat, *Can. J. Plant Sci.,* 57, 751, 1977.

23. **Luh, B. S.**, *Rice: Production and Utilization*, AVI, Westport, Conn., 1980.
24. **Yoshida, S.**, Rice, in *Ecophysiology of Tropical Crops*, Alvim, P. de T. and Kozlowski, T. T., Eds., Academic Press, New York, 1977.
25. **Nuttonson, M. Y.**, *Ecological Crop Geography and Field Practices of Japan*, American Institute of Crop Ecology, Washington, D.C., 1951.
26. **Chinnici, M. F. and Peterson, D. M.**, Temperature and drought effects on blast and other characteristics in developing oats, *Crop Sci.*, 19, 893, 1979.
27. **Friesen, G. and Olson, P. J.**, The effect of 2,4-D on the developmental process in barley and oats, *Can. J. Agric. Sci.*, 33, 315, 1953.
28. **Olson, P. J., Zalik, S., Breakey, W. J., and Brown, D. A.**, Sensitivity of wheat and barley at different stages of growth to treatment with 2,4-D, *Agron. J.*, 43, 77, 1951.
29. **Schlehuber, A. M. and Tucker, B. B.**, Culture of wheat, in *Wheat and Wheat Improvement*, Quisenberry, K. S. and Reitz, L. P., Eds., Agronomy 13, American Society of Agronomy, Madison, 1967, chap. 4.
30. **Hawkins, R. C. and Cooper, P. J. M.**, Effects of seed size on growth and yield of maize in the Kenya highlands, *Exp. Agric.*, 15, 73, 1979.
31. **Lee, K., Lommasson, R. C., and Eastin, J. D.**, Developmental stages on the panicle initiation in sorghum, *Crop Sci.*, 14, 80, 1974.
32. **Ferraris, R.**, *Pearl Millet*, Commonwealth Agricultural Bureau Ser. No. 1, C.A.B. Pastures and Field Crops, Hurley, U.K., 1973.
33. **Ong, C. K. and Everard, A.**, Short-day induction of flowering in pearl millet (*Pennisetum typhoides*) and its effect on plant morphology, *Exp. Agric.*, 15, 401, 1979.
34. **Malm, N. R. and Rachie, K. O.**, The Setaria Millets — a Review of the World Literature, *Nebr. Agr. Exp. Stn. Bull*, SB513, pp. 1-133, 1971.
35. **Curtis, D. L.**, The races of sorghum in Nigeria, *Exp. Agric.*, 3, 275, 1967.
36. **Burton, G. W.**, Photoperiodism in pearl millet, *Pennisetum typhoides*, *Crop Sci.*, 5, 333, 1965.
37. **Williams, C.**, *The Agronomy of the Major Tropical Crops*, Oxford University Press, Oxford, 1967.
38. **Kuhr, S. L., Peterson, C. J., Johnson, V. A., Mattern, P. J., and Schmidt, J. W.**, Results of the Twelfth International Winter Wheat Performance Nursery, 1980, Research Bulletin 303, Agricultural Research Service, United States Department of Agriculture, 1983.
39. **Larter, E. N.**, Personal communication; triticale trial data.
40. **Simmonds, D. H.**, The structure of the developing and mature triticale kernel, in *Triticale: First Man-made Cereal*, Tsen, C. C., Ed., American Association of Cereal Chemists, St. Paul, 1974.
41. **Peterson, R. F.**, *Wheat, Botany, Cultivation and Utilization*, Interscience, New York, 1965.
42. **Schery, R. W.**, *Plants for Man*, Prentice-Hall, Englewood Cliffs, N.J., 1972.
43. **Blackhurst, H. T. and Miller, J. C., Jr.**, Cowpea, in *Hybridization of Crop Plants*, Fehr, W. R. and Hadley, H. H., Eds., American Society Agronomy and Crop Science Society of America, Madison, 1980, chap. 21.
44. **Sharma, D. and Green, J. M.**, Pigeonpea, in *Hybridization of Crop Plants*, Fehr, W. R. and Hadley, H. H., Eds., American Society of Agronomy and Crop Science Society of America, Madison, 1980, chap. 33.
45. **Norden, A. J.**, Peanut, in *Hybridization of Crop Plants*, Fehr, W. R. and Hadley, H. H., Eds., American Society of Agronomy and Crop Science Society of America, Madison, 1980, chap. 31.
46. **Woodroof, J. G.**, *Peanuts: Production, Processing, Product*, AVI, Westport, Conn., 1973.
47. **Rachie, K. O. and Silvestre, P.**, Grain lengumes, in *Food Crops of the Lowland Tropics*, Leakey, C. L. A. and Wills, J. B., Eds., Oxford University Press, Oxford, 1977.
48. **Duke, J. A.**, *Handbook of Legumes of World Economic Importance*, Plenum Press, New York, 1981.
49. **Fehr, W. R. and Hadley, H. H.**, *Hybridization of Crop Plants*, American Society of Agronomy and Crop Science Society of America, Madison, 1980.
50. **Tsukamoto, J. T.**, New Crops Investigations, Manitoba Department of Agriculture Ann. Rep., Winnipeg, 1977—1983.
51. **Crothers, S. E. and Wextermann, D. T.**, Plant population effects on seed yield in *Phaseolus vulgaris*, *Agron. J.*, 68, 958, 1976.
52. **Sandhu, S. S. and Hodges, H. F.**, Effect of photoperiod, light intensity and temperature on vegetative growth, flowering and seed production in *Cicer arietinum* L., *Agron. J.*, 63, 913, 1971.
53. **Auckland, A. K. and van der Maeson, L. J. G.**, Chickpea, in *Hybridization of Crop Plants*, Fehr, W. R. and Hadley, H. H., Eds., American Society of Agronomy and Crop Science Society of America, Madison, 1980, chap. 15.
54. **Purseglove, J. W.**, *Tropical Crops. Dicotyledons, Vol. 1*, Longmans, London, 1968.
55. **Dekhuijzen, H. M., Verkerke, D. R., and Howers, A.**, Physiological aspects of growth and development of *Vicia faba*, in *Vicia faba: Physiology and Breeding*, Thomson, R., Ed., Martinus Nijhoff, The Hague, 1980.
56. **McVetty, P. B. E.**, Personal communication; fababean trial data.

57. **Stanfield, B., Ormrod, D. P., and Fletcher, H. F.,** Response of peas to environment. II. Effects of temperature in controlled-environment cabinets, *Can. J. Plant Sci.,* 46, 195, 1966.
58. **Wallace, D. H. and Enriquez, G. A.,** Daylength and temperature effects on days to flowering in early and late maturing beans (*Phaseolus vulgaris* L.), *J. Am. Soc. Hortic. Sci.,* 105, 583, 1980.
59. **Gutton, E. T.,** Field peas, in *Hybridization of Crop Plants,* Fehr, W. R. and Hadley, H. H., Eds., American Society of Agronomy and Crop Science Society of America, Madison, 1980, chap. 23.
60. **Brigham, R. D.,** Castor, in *Hybridization of Crop Plants,* Fehr, W. R. and Hadley, H. H., Eds., American Society of Agronomy and Crop Science Society of America, Madison, 1980, chap. 14.
61. **Davidson, J. M. and Yermanos, D. M.,** Flowering pattern of flax (*Linum usitatissimum* L.), *Crop Sci.,* 5, 23, 1965.
62. **Appelquist, L.-A. and Ohlson, R.,** *Rapeseed: Cultivation, Composition, Processing and Utilization,* Elsevier, Amsterdam, 1972.
63. **Holmes, M. R. J.,** *Nutrition of the Oilseed Rape Crop,* Applied Science, London, 1980.
64. **Downey, R. K., Klassen, A. J., and Stringham, G. R.,** Rapeseed and mustard, in *Hybridization of Crop Plants,* Fehr, W. R. and Hadley, H. H., Eds., American Society of Agronomy and Crop Science Society of America, Madison, 1980, chap. 35.
65. **Abel, G. H.,** Effects of irrigation regimes, planting dates, nitrogen levels and row spacings on safflower cultivars, *Agron. J.,* 68, 448, 1976.
66. **Knowles, P. F.,** Safflower, in *Hybridization of Crop Plants,* Fehr, W. R. and Hadley, H. H., Eds., American Society of Agronomy and Crop Science of America, Madison, 1980, chap. 38.
67. **Saha, S. N. and Bhargava, S. C.,** Physiological analysis of growth, development and yield of oilseed sesame, *J. Agric. Sci.,* 95, 733, 1980.
68. **Saham, S. N. and Bhargava, S. C.,** Flowering pattern and reproductive efficiency of oil-seed sesame, *Exp. Agric.,* 18, 293, 1982.
69. **Kinman, M. L.,** *Sesame Production, U.S. Dep. Agric. Farmers' Bull.,* No. 2119, 1958, 12 pp.
70. **Hicks, D. R.,** Growth and development, in *Soybean Physiology, Agronomy and Utilization,* Norman, A. G., Ed., Academic Press, New York, 1978, chap. 2.
71. **Fehr, W. R.,** Soybean, in *Hybridization of Crop Plants,* Fehr, W. R. and Hadley, H. H., Eds., American Society of Agronomy and Crop Science Society of America, Madison, 1980, chap. 42.
72. **Scott, W. O. and Aldrich, S. R.,** *Modern Soybean Production,* The Farm Quarterly, Cincinnati, 1970.
73. **Shanmugasiendaram, S.,** Variation in the photoperiod response on several characters in Soybean, *Glycine max* (L.) Merrill, *Euphytica,* 28, 495, 1979.
74. **Major, D. J., Johnson, D. R., Tanner, J. W., and Anderson, I. C.,** Effects of daylength and temperature on soybean development, *Crop Sci.,* 15, 174, 1975.
75. **Zeihn, C., Egli, D. B., Leggett, J. E., and Reikosky, D. A.,** Cultivar differences in N distribution in soybeans, *Agron. J.,* 74, 375, 1982.
76. **Major, D. J., Johnson, D. R., and Leudders, V. D.,** Evaluation of eleven thermal unit methods of predicting soybean development, *Crop Sci.,* 15, 172, 1975.
77. **Parker, M. B., Marchant, W. H., and Mullinize, P. J., Jr.,** Date of planting and row spacing effects on four soybean cultivars, *Agron. J.,* 73, 759, 1981.
78. **Hanway, V. V. and Weber, C. R.,** Dry matter accumulation in eight soybean [*Glycine max* (L) Merrill] varieties, *Agron. J.,* 63, 227, 1971.
79. **Shibles, R., Anderson, I. C., and Gibson, A. H.,** Soybeans, in *Crop Physiology,* Evans, L. F., Ed., Cambridge University Press, Cambridge, 1976.
80. **Robinson, R. G.,** Sunflower phenology — year, variety and date of planting effects on day and growing degree-day summations, *Crop Sci.,* 11, 635, 1971.
81. **Robinson, R. G., Boenat, L. A., Geise, H. A., Johnson, F. K., Kinman, M. L., Mader, E. L., Oswalt, R. M., Putt, E. D., Swallers, C. M., and Williams, J. H.,** Sunflower development at latitudes ranging from 31 to 49 degrees, *Crop Sci.,* 7, 134, 1967.
82. **Robinson, R. G.,** Production and culture, in *Sunflower Science and Technology,* Carter, J. F., Ed., No. 19, Agronomy, American Society of Agronomy, Crop Science Society of America, and Soil Science Society of America, Madison, 1978, chap. 4.
83. **Knowles, P. F.,** Morphology and anatomy, in *Sunflower Science and Technology,* Carter, J. F., Ed., No. 19, Agronomy, American Society of Agronomy, Crop Science Society of America, and Soil Science Society of America, Madison, 1978, chap. 3.
84. **Dedio, W. and Putt, E. D.,** Sunflower, in *Hybridization of Crop Plants,* Fehr, W. R. and Hadley, H. H., Eds., American Society of Agronomy and Crop Science Society of America, Madison, 1980, chap. 45.
85. **Kenaschuk, E. O.,** Flax breeding and genetics, in *Oilseed and Pulse Crops in Western Canada,* Harapiak, J. F., Ed., Western Co-Op Fertilizers Ltd., Calgary, 1975, chap. 7.
86. **Anon.,** Unpublished flax variety trial reports, U.S. and Canada, 1952—1975.

87. **Musnicki, C.,** Investigation on native and foreign winter rape varieties in Poland, *Proc. Int. Rapskongress,* Giessen, West Germany, 1974, 201.
88. **Gross, A. I. H.,** Effect of date of planting on yield, plant height, flowering and maturity of rape and turnip rape, *Agronomy,* 56, 76, 1964.
89. **Anon.,** Unpublished rapeseed variety trial reports, Canada, 1970—1975.
90. **Murray, D. B.,** Coconut palm, in *Ecophysiology of Tropical Crops,* Alvim, P. de T. and Kozlowski, T. T., Eds., Academic Press, New York, 1977, chap. 14.
91. **Chandler, W. H.,** *Evergreen Orchards,* Lea & Febiger, Philadelphia, 1950.
92. **McGregor, S. E.,** *Insect Pollination of Cultivated Crop Plants,* Agricultural Handbook, Agricultural Research Service, United States Department of Agriculture, Washington, D. C., 1976, 496.
93. **Ferwerda, J. D.,** Oil palm, in *Ecophysiology of Tropical Crops,* Alvim, P. de T. and Kozlowski, T. T., Eds., Academic Press, New York, 1977, chap. 13.
94. **Corley, R. H. V., Hardon, J. J., and Wood, B. J., Eds.,** *Oil Palm Research,* Elsevier, Amsterdam, 1976.
95. **Purseglove, J. W.,** *Tropical Crops. Monocotyledons,* Vol. 2, John Wiley & Sons, New York, 1972.
96. **Opitz, K. W.,** Olive Production in California, University of California Circular No. 540, 1966.
97. **Ochse, J. J., Soule, M. J., Jr., Dijkmon, M. J., and Wehlberg, C.,** *Tropical and Subtropical Agriculture,* Vol. 1, Macmillan, New York, 1961.
98. **Potter, G. F.,** The domestic tung industry. I. Production and improvement of the tung tree, *Econ. Bot.,* 13, 328, 1959.
99. **Wilson, L. A.,** Root crops, in *Ecophysiology of Tropical Crops,* Alvim, P. de T. and Kozlowski, T. T., Eds., Academic Press, New York, 1977, chap. 7.
100. **Plaisted, R. L.,** Potato, in *Hybridization of Crop Plants,* Fehr, W. R. and Hadley, H. H., Eds., American Society of Agronomy and Crop Science Society of America, Madison, 1980, chap. 34.
101. **Jones, A.,** Sweet potato, in *Hybridization of Crop Plants,* Fehr, W. R. and Hadley, H. H., Eds., American Society of Agronomy and Crop Science Society of America, Madison, 1980, chap. 46.
102. **Kawano, K.,** Cassava, in *Hybridization of Crop Plants,* Fehr, W. R. and Hadley, H. H., Eds., American Society of Agronomy and Crop Science Society of America, Madison, 1980, chap. 13.
103. **Coursey, D. G. and Booth, R. H.,** Root and tuber crops, in *Food Crops of the Lowland Tropics,* Leakey, C. L. A. and Wills, J. B., Eds., Oxford University Press, Oxford, 1977, chap. 5.
104. **Moorby, J. and Milthorpe, F. L.,** Potato, in *Crop Physiology,* Evans, L. T., Ed., Cambridge University Press, Cambridge, 1975, chap. 8.
105. **Plucknett, D. L.,** Taro production in Hawaii, *World Crops,* 23, 244, 1971.
106. **Gooding, E. G. B. and Hoad, R. M.,** Problems of yam cultivation in Barbardos, in *Proc. Int. Symp. Tropical Root Crops,* Tai, E. A., Ed., University of West Indies, St. Augustine, 1969, 111.
107. **Haynes, P. H., Spence, J. A., and Walter, C. J.,** The use of physiologic studies in the agronomy of root crops, in *Proc. Int. Symp. Tropical Root Crops,* Tai, E. A., Ed., University of West Indies, St. Augustine, 1969, III-1.
108. **James, N. I.,** Sugarcane, in *Hybridization of Crop Plants,* Fehr, W. R. and Hadley, H. H., Eds., American Society of Agronomy and Crop Science Society of America, Madison, 1980, chap. 44.
109. **Nickell, L. G.,** Sugarcane, in *Ecophysiology of Tropical Crops,* Alvim, P. de T. and Kozlowski, T. T., Eds., Academic Press, New York, 1977, chap. 4.
110. **Artschweiger, E., Brandes, E. N., and Starrett, R. C.,** Development of flowers and seed of some varieties of sugarcane, *J. Agric. Res.,* 39, 1, 1929.
111. **Johnson, R. T., Alexander, J. T., Rush, G. E., and Hawkes, G. R.,** *Advances in Sugarbeet Production: Principles and Practices,* Iowa State University Press, Ames, 1971.
112. **Smith, G. A.,** Sugarbeet, in *Hybridization of Crop Plants,* Fehr, W. R. and Hadley, H. H., Eds., American Society of Agronomy and Crop Science Society of America, Madison, 1980, chap. 43.
113. **Coons, C. F.,** Sugar Bush Management for Maple Syrup Producers, Information Branch, Ontario Ministry of Natural Resources, Toronto, 1976.
114. **MacArthur, J. D.,** Curator, Morgan Arboretum, Quebec, personal communication, 1983.
115. **Anon.,** Woody-Plant Seed Manual, Misc. Publ., No. 654, United States Forest Service, Washington, D.C., 1948.
116. **Coleman, O. H.,** Syrup and sugar from sweet sorghum, in *Sorghum Production and Utilization,* Wall, J. S. and Ross, W. M., Eds., AVI, Westport, Conn., 1970, chap. 11.
117. **Broadhead, D. M.,** Sugar production from sweet sorghum as affected by planting date, after-ripe harvesting and storage, *Agron. J.,* 61, 811, 1969.
118. **Alvim, P. deT.,** Cacao, in *Ecophysiology of Tropical Crops,* Alvim, P. de T. and Kozlowski, T. T., Eds., Academic Press, New York, 1977, chap. 9.
119. **Maestri, R. and Barros, S.,** Coffee, in *Ecophysiology of Tropical Crops,* Alvim, P. de T. and Kozlowski, T. T., Eds., Academic Press, New York, 1977, chap. 10.
120. **Browning, G.,** Flower bud dormancy in *Coffea arabica* L., *J. Hortic. Sci.,* 48, 297, 1973.

121. **Haunold, A.,** Hop, in *Hybridization of Crop Plants,* Fehr, W. R. and Hadley, H. H., Eds., American Society of Agronomy and Crop Science Society of America, Madison, 1980, chap. 27.
122. **Fordham, R.,** Tea, in *Ecophysiology of Tropical Crops,* Alvim, P. de T. and Kozlowski, T. T., Eds., Academic Press, New York, 1977, chap. 12.
123. **Eden, T.,** *Tea,* Longmans, London, 1958.
124. **Wernsman, E. A. and Matzinger, D. F.,** Tobacco, in *Hybridization of Crop Plants,* Fehr, W. R. and Hadley, H. H., Eds., American Society of Agronomy and Crop Science Society of America, Madison, 1980, chap. 47.
125. **Garner, W. W.,** *The Production of Tobacco,* McGraw-Hill, New York, 1951.
126. **Wilsie, C. P.,** *Crop Adaptation and Distribution,* W. H. Freeman, San Francisco, 1962.
127. **Lee, J. A.,** Cotton, in *Hybridization of Crop Plants,* Fehr, W. R. and Hadley, H. H., Eds., American Society of Agronomy and Crop Science Society of America, Madison, 1980, chap. 20.
128. **Carns, H. R. and Msuney, J. R.,** Physiology of the cotton plant, in *Advances in Production and Utilization of Quality Cotton; Principles and Practices,* Elliott, F. C., Hoover, M., and Porter, W. K., Jr., Eds., Iowa State University Press, Ames, 1966.
129. **Halevy, J.,** Growth rate and nutrient uptake of two cotton cultivars grown under irrigation, *Agron. J.,* 68, 701, 1976.
130. **Wanjura, D. F. and Newton, O. K.,** Predicting cotton crop boll development, *Agron. J.,* 73, 476, 1981.
131. **Purseglove, J. W.,** *Tropical Crops. Dicotyledons,* Vol. 2, Longman, London, 1968.
132. **Singh, D. P.,** Jute, in *Hybridization of Crop Plants,* Fehr, W. R. and Hadley, H. H., Eds., American Society of Agronomy and Crop Science Society of America, Madison, 1980, chap. 28.
133. **Osborne, J. F. and Singh, D. P.,** Sisal and other long-fibre Agaves, in *Hybridization of Crop Plants,* Fehr, W. R. and Hadley, H. H., Eds., American Society of Agronomy and Crop Science Society of America, Madison, 1980, chap. 40.
134. **Hartmann, H. T., Flocker, W. J., and Kofranek, A. M.,** *Growth, Development and Utilization of Cultivated Plants,* Prentice-Hall, Englewood Cliffs, N.J., 1981.
135. **Gourley, J. H. and Howlett, F. S.,** *Modern Fruit Production,* Macmillan, New York, 1941.
136. **Westwood, M. N.,** *Temperate Zone Pomology,* W. H. Freeman, San Francisco, 1978.
137. **Shoemaker, J. S.,** *Small Fruit Culture,* 5th ed., AVI, Westport, Conn., 1977.
138. **Reuther, W.,** Citrus, in *Ecophysiology of Tropical Crops,* Alvim, P. de T. and Kozlowski, T. T., Eds., Academic Press, New York, 1977, chap. 15.
139. **Teskey, B. J. and Shoemaker, J. S.,** *Tree Fruit Production,* 3rd ed., AVI, Westport, Conn., 1978.
140. **Gardner, V. R., Bradford, F. C., and Hooker, H. D.,** *The Fundamentals of Fruit Production,* 3rd ed., McGraw-Hill, New York, 1952.
141. **Valdeyron, G. and Lloyd, D. G.,** Flowering phenology in the common fig, *Evolution,* 33, 673, 1979.
142. **Childers, N. A.,** *Modern Fruit Science,* 8th ed., Rutgers University Press, New Brunswick, N.J., 1978.
143. **Tai, E. A.,** Banana, in *Ecophysiology of Tropical Crops,* Alvim, P. de T. and Kozlowski, T. T., Eds., Academic Press, New York, 1977, chap. 16.
144. **Bartholomew, D. P. and Kadzimin, S. B.,** Pineapple, in *Ecophysiology of Tropical Crops,* Alvim, P. de T. and Kozlowski, T. T., Eds., Academic Press, New York, 1977, chap. 5.
145. **Samson, J. A.,** *Tropical Fruits,* Longman, London, 1980.
146. **Brown, D. S.,** Use of temperature records to predict the time of harvest of apricots, *Proc. Am. Soc. Hortic. Sci.,* 60, 197, 1952.
147. **Fisher, D. V.,** Time of blossom induction in apricots, *Proc. Am. Soc. Hortic. Sci.,* 58, 19, 1951.
148. **Anon.,** Growing Cherries East of the Rocky Mountains, *U.S. Dep. Agric. Farmers' Bull.,* No. 2185, 1962.
149. **Albert, D. W. and Hilgeman, R. H.,** *Date Growing in Arizona, Univ. of Ariz. Bull.,* No. 149, University of Arizona, Tucson, 1935.
150. **Nixon, R. W.,** Growing dates in the United States, *U.S. Dep. Agric. Inform. Bull.,* No. 207, 1959.
151. **Fletcher, W. A.,** Growing Chinese gooseberries, Bull. No. 349, New Zealand Department of Agriculture, 1971.
152. **Stevens, I.,** Flower power; kiwifruit pollination, Seminar, Tauranga, New Zealand Ministry of Agriculture and Fisheries, September 28, 1982.
153. **Nagy, S., Shaw, P. E., and Veldhuis, N. K.,** *Citrus Science and Technology,* AVI, Westport, Conn., 1977.
154. **Singh, L. B.,** Mango, in *Ecophysiology of Tropical Crops,* Alvim, P. de T. and Kozlowski, T. T., Eds., Academic Press, New York, 1977, chap. 18.
155. **Ruehle, G. D. and Ledin, R. B.,** Mango growing in Florida, *Fla. Agric. Exp. Stn. Bull.,* 574, 1955.
156. **Bailey, L. H.,** Hortus III, *Manual of Cultivated Plants Most Commonly Grown in the United States and Canada,* Macmillan, New York, 1949.
157. **Harkness, R. W.,** Papaya Growing in Florida, Circ. 133, Florida Agricultural Extension Service, 1955.

158. **Young, E. and Houser, J.,** Influence of Siberian C. rootstock on peach bloom delay, water potential and pollen meiosis, *J. Am. Soc. Hortic. Sci.,* 105, 242, 1980.
159. **Roberts, R. H. and Struckmeyer, B. E.,** Growth and fruiting of the cranberry, *Am. Soc. Hortic. Sci., Proc.,* 40, 373, 1942.
160. **Craig, D. L.,** Growing red raspberries in eastern Canada, Publication 1196, Agriculture Canada, 1974.
161. **Waldo, G. F.,** Fruit bud development in strawberry varieties and species, *J. Agric. Res.,* 40, 393, 1930.
162. **Woodroof, J. G.,** *Tree Nuts: Production, Processing, Products,* Vol. 1, AVI, Westport, Conn., 1967.
163. **Nambiar, N. C.,** Cashew, in *Ecophysiology of Tropical Crops,* Alvim, P. de T. and Kozlowski, T. T., Eds., Academic Press, New York, 1977, chap. 17.
164. **Nuttonson, M. Y.,** *The Physical Environment and Agriculture in Australia,* American Institute of Crop Ecology, Washington, D.C., 1958.
165. **Gould, W. A.,** *Tomato Production, Processing and Quality Evaluation,* AVI, Westport, Conn., 1974.
166. **Hawthorn, L. R.,** Growing vegetable seeds for sale, in *Seeds, Yearbook of Agriculture,* Steffeurd, A., Ed., U.S. Department of Agriculture, U.S. Government Printing Office, Washington, D.C., 1961.
167. **Hebblethwaite, P. D.,** *Seed Production,* Butterworth, London, 1980.
168. **Burrage, A. C.,** *Burrage on Vegetables,* Houghton Mifflin, Boston, 1975.
169. **Carleton, R. M.,** *Vegetables for Today's Gardens,* D Van Nostrand, Princeton, 1967.
170. **Luther, G.,** personal communication, 1983.
171. **Purseglove, J. W., Brown, E. G., Green, C. L., and Robbins, S. R. J.,** *Spices,* Vol. 1, Longman, New York, 1981.
172. **Purseglove, J. W., Brown, E. G., Green, C. L., and Robbins, S. R. J.,** *Spices,* Vol. 2, Longman, New York, 1981.
173. **Rosengarten, F., Jr.,** *The Book of Spices,* Livingston, Wynnewood, Oklahoma, 1969.
174. **Calder, D. M.,** Inflorescence induction and initiation in the Gramineae, in *The Growth of Cereals and Grasses,* Milthorpe, F. L. and Ivins, J. D., Eds., Butterworths, London, 1966, sec. II, 59.
175. **Walton, P. D.,** *Production and Management of Cultivated Forages,* Prentice-Hall, Reston, Va., 1983.
176. **Lamp, H. F.,** Reproductive activity in *Bromus inermis* in relation to phases of tiller development, *Bot. Gaz.,* 113, 413, 1952.
177. **Evans, M. W. and Wilsie, J. M.,** Flowering of bromegrass, *Bromus inermis,* in the greenhouse, as influenced by length of day, temperature, and level of fertility, *J. Am. Soc. Agron.,* 38, 923, 1946.
178. **Newell, L. C.,** Controlled life cycles of bromegrass, *Bromis inermis* Leyss., used in improvement, *Agron. J.,* 43, 417, 1951.
179. **Gardner, E. P. and Loomis, W. E.,** Floral induction and development in orchard grass, *Plant Physiol.,* 28, 201, 1983.
180. **Evans, M. W.,** Relation of latitude to certain phases of growth in timothy, *Am. J. Bot.,* 26, 212, 1939.
181. **Cooper, J. P. and Calder, D. M.,** The inductive requirements for flowering of some temperate grasses, *J. Br. Grassl. Soc.,* 19, 6, 1964.
182. **Cooper, J. P.,** Studies on growth and development in *Lolium.* III. Influence of season and latitude on ear emergence, *J. Ecol.,* 40, 352, 1952.
183. **Evans, L. T.,** Reproduction, in *Grasses and Grasslands,* Barnard, C., Ed., Macmillan, London, 1964.
184. **Allard, H. A. and Evans, M. W.,** Growth and flowering of some tame and wild grasses in response to different photoperiods, *J. Agric. Res.,* 62, 193, 1941.
185. **Hanson, A. A. and Sprague, V. G.,** Heading of perennial grasses under greenhouse conditions, *Agron. J.,* 45, 248, 1953.
186. **Knowles, R. P.,** Seed production of perennial grasses in the greenhouse, *Can. J. Plant Sci.,* 41, 1, 1961.
187. **Sachs, R. M.,** Inflorescence induction and initiation, in *The Biology and Utilization of Grasses* Youngner, V. B. and McKell, C. M., Eds., Academic Press, New York, 1972, chap. 25.
188. **Latting, J.,** Differentiation in the grass inflorescence, in *The Biology and Utilization of Grasses,* Youngner, V. B. and McKell, C. M., Eds., Academic Press, New York, 1972, chap. 26.
189. **Davies, I.,** The Influence of Management on Tiller Development and Herbage Growth, Tech. Bull. No. 3, Welsh Plant Breeding Station, Plas Gogerddan, Aberystwyth, 1969.
190. **Evans, L. T., Wardlaw, I. F., and Williams, C. N.,** Environmental control of plant growth, in *Grasses and Grasslands,* Barnard, C., Ed., Macmillan, London, 1964, chap. 7.
191. **Bogdan, A. V.,** *Tropical Pasture and Fodder Plants,* Longman, London, 1977.
192. **Knight, W. E. and Bennett, H. W.,** Preliminary report of the effect of photoperiod and temperature on the flowering and growth of several southern grasses, *Agron. J.,* 45, 268, 1953.
193. **Youngner, V. B.,** Physiology of growth and development, in *Turfgrass Science,* Hanson, A. A. and Juska, F. V., Eds., American Society of Agronomy, Madison, 1969.
194. **Burton, G. W.,** The adaptability and breeding of suitable grasses for the southeastern states, *Adv. Agron.,* 3, 197, 1951.
195. **Boonman, J. G.,** Seed production of tropical grasses in Kenya, in *Seed Production,* Hebblethwaite, P. D., Ed., Butterworths, London, 1980, chap. 15.

196. **Wheeler, W. A. and Hill, D. D.,** *Grassland Seeds,* Part 2, D. van Nostrand, Princeton, 1957.
197. **Håbjørg, A.,** Effects of photoperiod and temperature on floral differentiation, development and seed yield of different latitudinal ecotypes of *Poa pratensis,* in *Seed Production,* Hebblethwaite, P. D., Ed., Butterworths, London, 1980, chap. 5.
198. **Hovin, A. W., Berg, C. C., Bashaw,. E. C., Buckner, R. C., Dewey, D. R., Dunn, G. M., Hoveland, C. S., Rincker, C. M., and Wood, G. M.,** Effects of geographic origin and seed production environments on apomixis in Kentucky bluegrass, *Crop Sci.,* 16, 635, 1976.
199. **Olmstead, C. E.,** Growth and development in range grasses. V. Photoperiodic response of clonal divisions of three latitudinal strains of side-oats grama, *Bot. Gaz.,* 106, 382, 1945.
200. **Abe, J. and Kawabata, S.,** Heading behaviour in Japan of Turkish populations of cocksfoot, *Dactylis glomerata* L., *Euphytica,* 28, 643, 1979.
201. **Cooper, H. W., Smith, J. E., Jr., and Atkins, M. D.,** Producing and Harvesting Grass Seed in the Great Plains, *U.S. Dep. Agric. Farmers' Bull.,* No. 2112, 1957.
202. **Griffiths, D. J., Roberto, H. M., Lewis, J., Stoddart, J. L., and Bean, E. W.,** Principles of Herbage Seed Production, Tech. Bull. No. 1, Welsh Plant Breeding Station, Plas Gogerddan, Aberystwyth, 1967.
203. **Anon.,** Crested Wheatgrass, Publication 1295, Information Division, Canada Department of Agriculture, Ottawa, 1971.
204. **Knowles, R. P. and Horner, W. H.,** Methods of selfing and crossing crested wheatgrass, *Agropyron cristatum* (L.) Beauv., *Sci. Agric.,* 23, 10, 1943.
205. **Jung, G. A., Kocher, R. E., Gross, C. F., Berg, C. C., and Bennett, O. L.,** Nonstructural carbohydrates in the spring herbage of temperate grasses, *Crop Sci.,* 16, 353, 1976.
206. **Canode, C. I. and Van Keuren, R. W.,** Seed Production Characteristics of Selected Grass Species and Varieties, Bull. 647, Washington Agricultural Experiment Station, Institute of Agricultural Science, Washington State University, Pullman, 1963.
207. **Smith, D. C.,** *Forage Management in the North,* 4th ed., Kendall/Hunt, Dubuque, Iowa, 1975.
208. **Templeton, W. C., Jr., Mott, G. O., and Bula, R. J.,** Some effects of temperature and light on growth and flowering in tall fescue, *Festuca arundinacea* Schreb. II. Floral development, *Crop Sci.,* 1, 283, 1961.
209. **Bean, E. W.,** Short-day and low-temperature control of floral induction in *Festuca, Ann. Bot.,* 34, 57, 1970.
210. **Elliott, C. R. and Baenziger, H.,** *Creeping Red Fescue,* Publication 1122, Canada Department of Agriculture, Ottawa, 1967.
211. **Murray, J., Wilton, A. A., and Powell, J. B.,** Floral induction and development in *Festuca rubra* L. — differential clonal response to environmental conditions, *Crop Sci.,* 13, 645, 1973.
212. **Elling, L. J. and McGraw, R. L.,** Progress Report of Seed Production Research, Grass-Legume Seed Institute, Department of Agriculture and Plant Genetics, University of Minnesota, Minneapolis, 1975—1983.
213. **Najda, H. G.,** Forage Cultivar Trials, Publication No. 83-16B, Canada Agriculture Research Station, Beaverlodge, Alberta, in cooperation with Alberta Agriculture, Edmonton, 1983.
214. **Silsbery, J. H.,** Interrelationships in the growth and development of *Lolium* I. Some effects of vernalization of growth and development, *Aust. J. Agric. Res.,* 16, 903, 1965.
215. **Hebblethwaite, P. D., Wright, D., and Noble, A.,** Some physiological aspects of seed yield in *Lolium perenne* L. (perennial ryegrass), in *Seed Production,* Hebblethwaite, P. D., Ed., Butterworths, London, 1980, chap. 6.
216. **Andersen, S. and Andersen, K.,** The relationship between seed maturation and seed yield in grasses, in *Seed Production,* Hebblethwaite, P. D., Ed., Butterworths, London, 1980, chap. 11.
217. **Knight, W. E.,** The influence of photoperiod and temperature on growth, flowering and seed production of dallisgrass, *Paspalum dilatatum,* Poir., *Agron. J.,* 47, 555, 1955.
218. **Taylor, R. W. and Allison, D. W.,** Response of three warm-season grasses to varying fertility levels on fine soils, *Can. J. Plant Sci.,* 62, 657, 1982.
219. **Mason, W. and Lachance, L.,** Effects of initial harvest date on dry matter yield, in-vitro dry matter digestibility and protein in timothy, tall fescue, reed canarygrass and Kentucky bluegrass, *Can. J. Plant Sci.,* 63, 675, 1983.
220. **Wolf, D. D., Brown, R. H., and Blaser, R. E.,** Physiology of growth and development, in *Tall Fescue,* Buckner, R. C. and Bush, L. P., Eds., No. 20, Agronomy Series, American Society of Agronomy and Crop Science Society of America, Madison, 1979.
221. **Bonin, S. G. and Goplen, B. P.,** A histological study of seed shattering in reed canary grass, *Can. J. Plant Sci.,* 43, 200, 1963.
222. **Evans, M. W.,** The Life History of Timothy, *U.S. Dep. Agric. Bull. No.,* 1450, 1927, 56 pp.
223. **Evans, M. W.,** Relations of length of day to growth of timothy, *J. Agric. Res.,* 48, 571, 1934.
224. **Langer, R. H. M. and Ryle, G. J. A.,** The effect of time of sowing on flowering and fertile tiller production in S-48 timothy, *J. Agric. Sci.,* 53, 145, 1959.

225. **Hanson, C. H.,** *Alfalfa Science and Technology,* No. 15, Agronomy Ser., American Society of Agronomy, Madison, 1972.
226. **Bula, R. J. and Massengale, M. A.,** Environmental physiology, in *Alfalfa Science and Technology,* Hanson, C. A., Ed., American Society of Agronomy, Madison, 1972, chap. 8.
227. **Christian, K. R.,** Effects of the environment on the growth of alfalfa, *Adv. Agron.,* 29, 183, 1977.
228. **Barnes, D. K.,** Alfalfa, in *Hybridization of Crop Plants,* Fehr, W. R. and Hadley, H. H., Eds., American Society of Agronomy and Crop Science Society of America, Madison, 1980, chap. 9.
229. **Turkington, R. and Franks, G. D.,** The biology of Canadian weeds. 41. *Lotus corniculatus* L., *Can. J. Plant Sci.,* 60, 965, 1980.
230. **Turkington, R., Covers, P. B., and Rempel, E.,** The biology of Canadian weeds. 29. *Melilotus alba* Desr. and *Melilotus officinalis* (L.) Lam., *Can. J. Plant Sci.,* 58, 523, 1978.
231. **Hawkins, R. P.,** Investigations on local strains of herbage plants. II. Types of red clover and their identification, *J. Br. Grassl. Soc.,* 8, 213, 1953.
232. **Bird, J. N.,** Early and late types of red clover, *Sci. Agric.,* 28, 444, 1948.
233. **Fergus, E. N. and Hollowell, E. A.,** Red clover, *Adv. Agron.,* 12, 365, 1960.
234. **Stoddart, J. L.,** Floral initiation and its relationship to growth stage in red clover (*Trifolium pratense* L.), *Nature (London),* 184, 559, 1959.
235. **Knight, W. E. and Hollowell, E. A.,** Crimson clover, *Adv. Agron.,* 25, 47, 1973.
236. **Turkington, R. and Burdon, J. J.,** The biology of Canadian weeds, 57. *Trifolium repens* L., *Can. J. Plant Sci.,* 63, 243, 1983.
237. **Bula, R. J.,** Vegetative and floral development in red clover as affected by duration and intensity of illumination, *Agron. J.,* 52, 74, 1960.
238. **Conje, A. M. and Carlson, I. T.,** Performance of crosses within and between two diverse sources of birdsfoot trefoil, *Lotus corniculatus* L., *Crop Sci.,* 13, 357, 1973.
239. **Winch, J. E. and Macdonald, H. A.,** Flower, pod and seed development relative to the timing of the seed harvest of Viking birdsfoot trefoil (*Lotus corniculatus*), *Can. J. Plant Sci.,* 41, 523, 1961.
240. **Willard, C. J.,** An experimental study of sweetclover, *Ohio Agric. Exp. Stn.,* Bull. No. 405, 1927.
241. **Sinskaya, E. N.,** *Flora of Cultivated Plants of the U.S.S.R. XIII. Perennial Leguminous Plants,* Israel Prog. Scient. Translation, Jerusalem, 1961.
242. **Townsend, C. E.,** Forage legumes, in *Hybridization of Crop Plants,* Fehr, W. R. and Hadley, H. H., Eds., American Society of Agronomy and Crop Science Society of America, Madison, 1980, chap. 25.
243. **Hanna, M. R., Cooke, D. A., Smoliak, S., and Goplen, B. P.,** Sainfoin for Western Canada, *Agric. Canada Publ.,* No. 1470, Information Canada, Ottawa, 1974.
244. **Dubbs, A. L.,** Sainfoin as a honey crop, in *Sainfoin Symposium,* Cooper, C. S. and Carleton, A. E., Eds., Artcraft Printers, Bozeman, 108.
245. **Morley, F. H. W.,** Subterranean clover, *Adv. Agron.,* 13, 57, 1961.
246. **Bates, R. P.,** Effects of photoperiods on plant growth, flowering, seed production and tannin content of *Lespedeza cuneata* Don, *Agron. J.,* 47, 564, 1955.
247. **Chin, H. F. and Lassim, M. B. M.,** Plantation crop seed production in Malaysia, in *Seed Production,* Hebblethwaite, P. D., Ed., Butterworths, London, 1980, chap. 29.
248. **DeJong, H.,** Buckwheat, *Field Crop Abstr.,* 25, 389, 1972.

BIOLOGICAL NITROGEN FIXATION IN FIELD CROPS*

Gudni Hardarson, Seth K. A. Danso, and Felipe Zapata

INTRODUCTION

Nitrogen (N), one of the main building blocks of protein, is ultimately derived from soil and atmosphere through plant or microbial assimilation of this element. Enormous reserves of N occur in rocks, with some 1.9×10^{17} metric tons occurring in primary rocks alone (Table 1). However, most N in rock is bound in forms which are not immediately available for plant uptake. The small proportion of this vast resource that is released annually is often insufficient to meet the demands for plant growth, and this in part explains why crop growth is often limited by the availability of N. In addition to N in rocks, N_2 is the most abundant gas (78%) in the atmosphere. N_2 is, however, so inert that it can not be utilized by plants alone and it can therefore only enter biological systems when it has been fixed or bonded to other atoms, such as hydrogen and oxygen.

The strong triple bond in N_2 can be artificially broken by the application of high temperature and pressure. This process, known as the Haber-Bosch process shown in Equation 1, is the basis for the manufacture of nitrogenous fertilizers.

$$N_2 + 3H_2 \xrightarrow[450°, 200 \text{ atm}]{\text{catalyst}} 2\,NH_3 \qquad (1)$$

Although many natural ecosystems do not receive nitrogenous fertilizers, N balance studies have shown that soil nitrogen losses due to crop uptake, leaching, erosion, denitrification etc., are apparently replenished to varying extents by a process of biological nitrogen fixation. The additions from this process are significant in many soils and other habitats and therefore constitute an important source of N for crop growth and protein production. Biologically fixed N_2 is not only inexpensive, but also does not involve many of the disadvantages, such as pollution hazards, associated with excessive use of inorganic N fertilizers. It appears to be the most promising alternative or supplement to inorganic fertilizer N. An intelligent manipulation and full exploitation of biological nitrogen fixation is therefore necessary. Although there are different associations and organisms involved in biological nitrogen fixation, only organisms which possess the enzyme nitrogenase are able to catalyze the biological reduction of N_2 to NH_3 according to the following equation:

$$N_2 + 6H^+ + xATP \xrightarrow[\text{nitrogenase}]{6\,e^-} 2\,NH_3 + xADP + xPi \qquad (2)$$

Thus both electrons and energy in the form of ATP are needed for biological reductions of N_2 to NH_3.

In nature the possession of the nitrogenase enzyme and thus the ability to carry out biological nitrogen fixation are confined exclusively to the prokaryotes. These are predominantly bacteria, including the blue-green algae which are classified as cyanobacteria.[1] No higher organisms have developed the capability to fix N_2, although several participate in-

---

* The mention of specific companies or their products or brand names does not imply any endorsement or recommendation on the part of the International Atomic Energy Agency or the Food and Agricultural Organization of the United Nations.

Table 1
GEOCHEMICAL DISTRIBUTION OF NITROGEN

Source	Total mass × 10^{20} g	% of N
Fundamental rocks	1,930.0	97.82
Atmosphere	38.646	1.96
Ancient sedimentary rocks	4.0	0.2
Terrestrial humus	0.0082	Negligible
Sea-bottom organic compounds	0.0054	Negligible
Living organisms	0.00028	Negligible

Reproduced from *Soil Nitrogen*, Agronomy Monograph No. 10, 1965, pages 1—42 by permission of the American Society of Agronomy, Crop Science Society of America, and Soil Science Society of America.

directly by forming symbiotic associations with N_2-fixing bacteria, or supplying substrates for their growth.

The first N_2-fixing bacteria were isolated by Winogradsky in 1893, followed by the isolation of *Azotobacter* by Beijerinck in 1901. Since then, various N_2-fixing organisms with various nutritional requirements and belonging to different ecological and trophic levels have been identified (Table 2). Examples of N_2 fixers identified include the obligate anaerobe *Clostridium*, obligate aerobe *Azotobacter*, facultative anaerobe *Klebsiella*, heterotroph *Rhizobium*, photoautotroph *Rhodomicrobium*, chemoautotroph *Thiobacillus*, and algal N_2 fixers such as *Anabaena* and *Nostoc*. These microbes may fix N_2 when alone, or when in association with higher organisms, mainly plants.

NONSYMBIOTIC NITROGEN FIXATION

Many microorganisms that fix N_2 asymbiotically have been identified. They range from those that fix N_2 completely independent of plants to those that fix N_2 when associated with the roots of certain plants, without the formation of any recognizable anatomical structures, such as the nodules of legumes.

In the initial stages, the potential of free-living bacteria to fix N_2 was demonstrated by their ability to multiply in what was considered to be nitrogen-free media or by determinations of increases in total N in incubated soils. However, using these methods, many unconvincing claims were reported.[1] The acetylene reduction technique and $^{15}N_2$ have recently been used to prove the N_2-fixing ability of many microorganisms. Examples of free-living N_2-fixing bacteria and algae that have been established to be capable of fixing N_2 are listed in Table 2.

High N_2 fixation by free-living microorganisms is strongly dependent upon the presence of substantial amounts of readily oxidizable organic matter in soil. Jensen[2] has questioned the importance of free-living N_2 fixation in soils, since, according to him, the amounts of organic matter needed to carry out substantial N_2 fixation may not be available in many soils. However, many unfertilized soils of low N status have continued to support sustained and moderate grain yields in the absence of symbiotic N_2 fixation, with no apparent decline in soil N. This, together with the high estimates of free-living N_2 fixation reported in the literature from N balance studies, has provided strong evidence that the contribution from this source of N could be important in many soils. According to Greenland,[3] it is not unusual to observe 20 to 50 kg N/ha/year of N fixed by free-living organisms. Data provided by Bartholomew et al.,[4] as well as Greenland and Nye[5] show that the contribution of nonsymbiotic N_2 fixers can be high in many soils. Odu[6] has suggested that in tropical soils the high organic-matter turnover and the high flushes of mineralization associated with the wetting

Table 2
ORGANISMS KNOWN TO FIX ATMOSPHERIC NITROGEN

Order	Family	Genus	Species
Eubacteriales	Azotobacteraceae	*Azotobacter*	*beijerinckii*
			chroococcum
			paspali
			vinelandii
		Azomonas	*insignis*
			macrocytogenes
			agilis
		Beijerinckia	*indica*
			fluminensis
			derxii
		Derxia	*gummosa*
	Rhizobiaceae[a]	*Rhizobium*	species (cowpea)
			japonicum
			leguminosarum
	Bacillaceae	*Bacillus*	*macerans*
			polymyxa
		Clostridium	*butyricum*
			pasteurianum
			saccharobutyricum
			acetobutyricum
			beijerinckia
			tyrobutyricum
			acetobutylicum
			felsineum
			kluyverii
			lactoacetophilum
			madisonii
			pectinovorum
			tetanomorphum
			butylicum
	Enterobacteriaceae	*Klebsiella*	*pneumoniae*
			aerogenes
		Enterobacter	*aerogenes*
			cloacae
			agglomerans
		Erwinia	*herbicola*
		Citrobacter	*freundii*
			intermedius
		Escherichia	*coli*
			intermedia
Actinomycetales	Mycobacteriaceae	*Mycobacterium*	*flavum*
			roseo-album
			azotabsorptum
	Corynebacteriaceae	*Xanthobacter*	*autotrophicus*
Hyphomicrobiales	Hyphomicrobiaceae	*Rhodomicrobium*	*vannielii*
Pseudomonadales	Athiorhodaceae	*Rhodopseudomonas*	*palustris*
			capsulata
			gelatinosa
			spheroides
		Rhodospirillum	*rubrum*
	Thiorhodaceae	*Chromatium*	*vinosum*
			minutissimum
	Chlorobacteriaceae	*Chlorobium*	*thiosulfatophilum*
	Ectothiorhodaceae	*Ectothiorhodospira*	*shaposhnikovii*
	Methanomonadaceae	*Methylosinus*	*trichosporium*
			sporium

Table 2 (continued)
ORGANISMS KNOWN TO FIX ATMOSPHERIC NITROGEN

Order	Family	Genus	Species
Pseudomonadales (cont.)		*Methylococcus*	*capsulatus*
	Thiobacteriaceae	*Thiobacillus*	*ferro-oxidans*
	Spirillaceae	*Azospirillum*	*lipoferum*
			brasilense
		Aquaspirillum	*peregrinum*
			fasciculus
		Desulfovibrio	*desulfuricans*
			vulgaris
			gigas
		Desulfotomaculum	*orientis*
			ruminis
Nostocales	Nostocaceae	*Anabaena*	
		Anabaenopsis	
		Aulosira	
		Cylindrospermum	
		Nostoc	
	Rivulariaceae	*Calothrix*	
	Scytonemataceae	*Scytonema*	
		Tolypothrix	
	Oscillatoriaceae	*Trichodesmium*	
		Lyngbya	
		Phormidium	
		Plectonema	
Stigonematales	Stigonemataceae	*Fischerella*	
		Hapalosiphon	
		Mastigocladus	
		Stigonema	
		Westiellopsis	
Chroococcales		*Gloeocapsa*	

[a] The species of *Rhizobium* listed have been reported to fix N_2 in culture. Other species (*R. trifolii*, *R. meliloti*, *R. lupini*, and *R. phaseoli*) fix N_2 in symbiosis with appropriate host legumes.

From Dalton, H., in *Methods for Biological Nitrogen Fixation*, Bergersen, F. J., Ed., copyright 1980, John Wiley & Sons. Reprinted by permission of John Wiley & Sons, Ltd.

and drying cycles in soils could make substantial quantities of energy available for nonsymbiotic N_2 fixation to proceed at a high rate in soil.

More recent findings have shown that many plants are capable of stimulating N_2 fixation by free-living bacteria which live in their rhizospheres, without entering into symbiotic relationship with the plant, or forming any recognizable anatomical structures as a result of the N_2 fixation.[7] This has aroused further interest in the potential of free-living microorganisms as sources of N in soils. This process has been called associative N_2 fixation.[7] According to Balandreau,[8] this enhancement in N_2 fixation is due to the photosynthate exuded from these plants, which then support the growth and N_2-fixing activity of the bacteria. Varieties of plants with enhanced capacity to exude photosynthate through their roots may thus be selectively suited for high associative N_2 fixation.[8] Rennie[9] has provided evidence showing that different varieties of the same plant species can indeed support associative N_2 fixation to different extents. Furthermore, it has been reported that tropical plants with the C_4 photosynthetic pathway stimulate rhizospheric N_2 fixation more than those with the C_3 pathway.[7] This could be because the plants with the C_4 pathway can use light of intensities twice that which saturates other plants and use water much more efficiently.[10] These plants therefore present an attractive possibility for trapping considerable light energy for increased

N_2 fixation. The possibility exists that many of the high values reported for free-living N_2 fixation could be attributable largely to associative N_2 fixation in the rhizosphere of plants.

There is still insufficient data to prove conclusively that the benefit derived by many plants in the presence of free-living N_2 fixers is the result of biological N_2 fixation and not any synergistic effects, due to the presence of these bacteria, such as increased availability of growth hormones. Unequivocal evidence for associative N_2 fixation can only be obtained from the use of ^{15}N, but these types of studies have unfortunately been few. Data reported by Rennie[9] using ^{15}N in experiments with potted plants have shown that one variety of wheat is capable of satisfying 60% of its N requirement from associative N_2 fixation, while Boddey and Chalk[11] have similarly used ^{15}N to estimate nonsymbiotic N_2 fixation in *Paspalum notatum* varieties and found that 8 to 23% of the total plant N was derived from fixation. More field studies, using ^{15}N, are needed to establish the contribution of free-living N_2 fixers to plant growth.

In addition, many species of blue-green algae have long been established to fix N_2. Blue-green algae, being photoautotrophic, do not need any external supply of available organic matter, since sunlight provides the needed energy for nitrogen fixation. Thus, subject to the availability of light, free-living N_2-fixing blue-green algae represent a potential source of inexpensive N for plants. Blue-green algae have been suggested to be important in flooded soils, where they may contribute between 10 to 30 kg N/ha/year or more to the soil.[12] However, there have been very few valid estimates of N_2 fixed by blue-green algae growing in the field. Much of the evidence has been of an indirect nature, such as through increased rice yields due to inoculation of these organisms in paddy soils. For example, Venkataraman[13] has reported 10 to 20% increase in grain yield as a result of algal application in the absence of any added chemical N fertilizer. Algal N_2 fixation is however strongly inhibited in the presence of inorganic N fertilizers, and this therefore limits the potential of blue-green algae as supplements to inorganic fertilizers, unless strains are eventually selected which can fix N_2 in the presence of moderate to high N fertilizers.

SYMBIOTIC NITROGEN FIXATION

By Legumes

Symbiotic nitrogen fixation by the legumes is the oldest known and most studied of all biological N_2-fixing systems, and one which offers great promise for alleviating world protein and food shortage. Legumes are indeed second only to the cereals in terms of global consumption, and are of extremely high protein content, in comparison to the cereals and most other food crops (Table 3). Although in terms of dry matter production, legumes account for only 9% of the combined yield of cereals and legumes, they constitute as much as 24% of the total protein yield of the two crops, because of the high protein content of the latter. The importance of legumes as a source of proteins is appreciated more if it is recognized that most of these legumes are capable of achieving optimum yields even on soils of low fertility, with very little or no nitrogenous fertilizer input, once the right conditions exist for nodulation and N_2 fixation.

Legumes belong to the family of plants classified as Leguminosae. These are divided into three subfamilies, Caesalpinioideae, Papilionoideae, and Mimosoideae. The Leguminosae are of worldwide distribution and have been reported in all geographical regions of the world. A detailed review of leguminous species has recently been published.[14]

The microsymbiont partner of the legume is a rod-shaped bacterium of genus *Rhizobium*, belonging to the family *Rhizobiaceae*. *Rhizobium* is aerobic, non-spore-forming, and usually a gum-producing heterotroph, which can grow and multiply on many carbon substrates. Many factors can affect the growth and survival of the root nodule bacterium, and this has been reviewed by Danso.[15] Unlike many other bacteria which have been classified into

Table 3
CONTRIBUTION OF CEREALS AND GRAIN LEGUMES TO PROTEIN PRODUCTION
(1976—1978 YEARLY AVERAGE)

	Production (million tonnes)	Average protein content (%)	Protein production (million tonnes)
Cereals			
Wheat	415.6	12.3	51.1
Rice	365.8	8.3	30.4
Maize	348.7	8.9	31.0
Barley	187.1	8.6	16.1
Sorghum	67.5	11.0	7.4
Oats	50.1	11.0	5.5
Others	80.0	10.7	8.6
Total (cereals)	1514.8	—	150.1
Grain Legumes			
Soybean	74.2	38.0	28.2
Groundnut	18.2	25.6	4.7
Dried beans	16.4	22.3	3.7
Dried peas	16.0	24.1	3.9
Broad beans	11.3	25.1	2.8
Chickpeas	7.0	24.0	1.7
Lentils	1.2	28.0	0.3
Others	9.0	24.2	2.2
Total (legumes)	153.3	—	47.5
Total	1668.1		197.6

From Crabbe, D. and Lawson, S., *The World Food Book*, Kogan Page, London, 1981. With permission.

species by classical microbiological or biochemical tests, classification of this organism into species has been based on the host plants it is able to nodulate. It has been established that different types of *Rhizobium* form nodules on distinct groups of legume species, and not on others. Six species of *Rhizobium* have been delineated. These are *R. meliloti, R. trifolii, R. leguminosarum, R. lupini, R. japonicum,* and *R. phaseoli*. There is yet a seventh group of species in this genus whose host range is so wide that further classification into definite species has not been possible. It has been designated simply as *Rhizobium* sp., or cowpea *Rhizobium*. It is most likely that future research efforts will further subdivide this group into more distinct species. Burton[16] has summarized the different *Rhizobium* species and their cross-inoculation groups in Table 4.

Estimates of amounts of N_2 fixed by legumes have been varied, and some figures have been compiled for various grain and forage legumes in Tables 5 and 6, respectively. These measurements are, on the other hand, obtained under very different environmental and experimental conditions and these factors can greatly affect amounts of N fixed (Table 7). Any comparisons between species in Tables 5 and 6 should therefore be done with caution. It is also likely that the potential of each species to fix nitrogen can be higher than the maximum value recorded.

By Nonlegumes

Several plants in addition to the legumes are known to be associated with prokaryotes in a symbiotic relationship which results in the incorporation of N_2 into plant tissues. The

Table 4
LEGUMINOUS PLANTS WHICH TEND TO RESPOND SIMILARLY WHEN INOCULATED WITH THE SAME STRAIN OF *RHIZOBIUM*

Rhizobium species	Group	Leguminous species
Rhizobium meliloti	1	*Medicago sativa, M. falcata, M. minima, M. tribuloides, Melilotus denticulata, M. alba, M. officinalis, M. indica*
	2	*Medicago arabica, M. hispida, M. lupulina, M. orbicularis, M. praecox, M. truncatula, M. scutellata, M. polymorpha, M. rotata, M. rigidula, Trigonella foenum-graecum*
	3	*Medicago laciniata*
	4	*Medicago rugosa*
R. trifolii	5	*Trifolium incarnatum, T. subterraneum, T. alexandrinum, T. hirtum, T. glomeratum, T. arvense, T. angustifolium*
	6	*Trifolium pratense, T. repens, T. hybridum, T. fragiferum, T. procumbens, T. nigrescens, T. glomeratum*
	7	*T. vesiculosum, T. berytheum, T. bocconi, T. boissiere, T. compactum, T. leucanthum, T. mutabile, T. vernum, T. physodes, T. dasydrum*
	8	*Trifolium rueppellianum, T. tembense, T. usambarense, T. steudneri, T. burchellianum* var. *burchellianum, T. burchellianum* var. *johnstonii, T. africanum, T. pseudostriatum*
	9	*T. semipilosum* var. *kilimanjaricum, T. masaiense, T. cheranganiense, T. rueppellianum* var. *lanceolatum*
	10	*T. medium, T. sarosience, T. alpestre*
	11	*T. ambiguum*
	12	*T. heldreichianum*
	13	*T. mosaiense*
	14	*T. reflexum*
	15	*T. rubens*
	16	*T. semipilosum*
R. leguminosarum	17	*Pisum sativum, Vicia villosa, V. hirsuta, V. tenuifolia, V. tetrasperma, Lens esculenta, Lathyrus aphaca, L. cicera, L. hirsutus, L. odoratus, Lathyrus sylvestris*
	18	*Lathyrus ochrus, L. tuberosus, L. szenitzii*
	19	*Lathyrus sativus, L. clymenum, L. tingitanus*
	20	*Vicia faba, V. narbonensis*
	21	*Vicia sativa, V. amphicarpa*
R. phaseoli	22	*Phaseolus vulgaris, P. coccineus, P. angustifolia*
R. lupini	23	*Lupinus albicaulis, L. albifrons, L. albus, L. angustifolius, L. arboreus, L. argenteus, L. benthami, L. formosus, L. luteus, L. micranthus, L. perennis, L. sericus, Lotus ulliginosus, L. americanus, L. pedunculatus, L. strictus, L. strigosus*
	24	*L. densiflorus, L. vallicola*
	25	*L. nanus*
	26	*L. polyphyllus*
	27	*L. subcarnosus*
	28	*L. succulentus*
R. japonicum	29	*Glycine max*
Rhizobium spp. (cowpea type)	30	*Vigna unguiculata, V. sesquipedalis, V. luteola, V. cylindrica, V. angularis, V. radiata, V. mungo, Desmodium* sp., *Alysicarpus vaginalis, Crotalaria* sp., *Macroptilium lathyroides, M. atropurpureus, Psophocarpus* sp., *Lespedeza striata, L. stipulaceae, Indigofera* sp., *Cajanus cajan*

Table 4 (continued)
LEGUMINOUS PLANTS WHICH TEND TO RESPOND SIMILARLY WHEN INOCULATED WITH THE SAME STRAIN OF *RHIZOBIUM*

Rhizobium species	Group	Leguminous species
Rhizobium spp. (cowpea type)	30	*Vigna unguiculata, V. sesquipedalis, V. luteola, V. cylindrica V. angularis, V. radiata, V. mungo, Desmodium* sp., *Alysicarpus vaginalis, Crotalaria* sp., *Macroptilium lathyroides, M. atropurpureus, Psophocarpus* sp., *Lespedeza striata, L. stipulaceae, Indigofera* sp., *Cajanus cajan*
	31	*Phaseolus limensis, P. lunatus, P. aconitifolius, Canavalia ensiformis, Canavalia lineata*
	32	*Arachis hypogaea, A. glabrata, Cyamopsis tetragonoloba, Lespedeza sericea, L. japonica, L. bicolor*
	33	*Centrosema pubescens, Galactia* sp.
	34	*Lotononis bainesii*
	35	*Lotononis angolensis*
Rhizobium spp. (*Lotus*)	36	*Lotus corniculatus, L. tenuis, L. angustissimus, L. tetragonolobus, Dorycnium hirsutum, D. rectum, D. suffruticosum, L. caucasieus, L. crassifolius, L. creticus, L. edulus, L. frondosus, L. subpinnatus, L. weilleri, Anthyllis vulneraria, A. latoides*
	37	*Lotus uliginosus, L. americanus, L. scoparius, L. angustissimus, L. pedunculatus, L. strictus, L. strigosus, Ornithopus sativum, Lupinus angustifolius, L. albus, L. luteus*
Rhizobium spp. (*Coronilla, Petalostemum-Onobrychis*)	38	*Coronilla varia, Onobrychis viciaefolia, Petalostemum purpureum, P. candidum, P. microphyllum, P. multiflorum, P. villosum, Leucaena leucocephala, L. retusa*
Rhizobium spp. (various)	39	*Dalea alopecuroides*
	40	*Strophostyles helvola*
	41	*Robinia pseudoacacia, R. hispida*
	42	*Amorpha canescens*
	43	*Caragana arborescens, C. frutesceus*
	44	*Oxytropis sericea*
Rhizobium spp. *Astragalus* sp.	45	*A. cicer, A. falcata, A. canadensis, A. mexicanus, A. orbiculatus*

From Burton, J. C., in *Recent Advances in Biological Nitrogen Fixation*, Subba Rao, N. S., Ed., Edward Arnold, London, 1980. With permission.

association is intimate in nature, and specific between particular micro- and macrosymbiont partners. All rhizosphere and phylosphere prokaryotes which lack this intimacy and specificity are therefore not considered as symbiotic.

The most studied nodulated nonlegume plant is *Alnus*. Field-grown *Alnus* typically possess root nodules that approach the size of tennis balls.[1] Ten- to fifteen-year-old *Hippophae*, another nonlegume species which forms nodules on its roots, have been reported to bear about 30 kg/ha of nodules.[17] Some other nonlegume plants bearing root nodule are *Casuarina, Cercocarpus, Coriaria, Discaria, Dryas, Eleagnus, Myrica, Purshia, Shepherdia*, and *Podocarpus*. Microscopic observations of nodule exudates suggest that the microsymbiont is an actinomycete, and has been designated as *Frankia*.[1]

Many of these nodulated nonleguminous plants have been shown to incorporate $^{15}N_2$ as evidence of N_2 fixation, and have been found to fix significant amounts of N_2. Many of the

Table 5
ESTIMATES OF NITROGEN FIXATION BY GRAIN LEGUMES

Species	Maximum fixation (kg/ha)	Ref.
Arachis hypogaea (groundnut)	124	13, 48—50
Cajanus cajan (pigeon pea)	280	51
Calapogonium mucunoides	450	52
Canavalia ensiformis	49	53
Cyanopsis tetragonolobus (guar)	220	50, 54
Glycine max (soybean)	180	48, 55—68
Lupinus spp.	208	64
Phaseolus aureus (green gram)	342	52
P. aureus (mung bean)	61	69
P. vulgaris (common bean)	64	70
Pisum sativum (peas)	85	64, 71
Vicia faba	552	48, 64, 72, 73
Vigna sinensis (cowpea)	354	50, 52, 53

Table 6
ESTIMATES OF NITROGEN FIXATION BY FORAGE LEGUMES

Species	Maximum fixation (kg/ha)	Ref.
Centrosema pubescens	395	74—77
Lespedeza spp.	193	78
Lotus corniculatus	116	79, 80
Medicago lupulina	209	81
M. sativa (alfalfa)	463	41, 64, 78, 80, 82—88
Melilotus alba (sweet clover)	183	64, 89
Stylosanthes spp.	220	50, 76, 90
Trifolium spp. (clovers)	673	64, 65, 78, 80, 84, 87, 88, 91—109
Trigonella foenum-graecum	573	65, 73, 110
Vicia villosa (vetch)	184	64, 78

available figures of N_2 fixation were, however, obtained from indirect estimates. Silver[18] quotes N_2-fixation values of 218 kg N/ha/year for *Casuarina equisetifolia* and 192 kg N/ha/year for *Coriaria*. In addition, it has been reported that these nodulated nonlegumes may contribute significant amounts of N to associated crops through direct release into the soil medium.[18]

It is now well established that several plant species possess nodules on their leaves that contain bacterial symbionts. The present evidence that significant N_2 fixation occurs in these nodules is, however, weak. Foliar nodulation is limited to three angiosperm families — the Rubiaceae, Myrsinaceae, and Dioscoreaceae.[19] Nodulated species of these plants are confined mainly to tropical and subtropical regions, and the three most-studied species are *Pavetta zimmermanniana*, *Psychotria punctata*, and *Ardisia crispa*.[19] A variety of ill-identified bacteria have been isolated from these leaf nodules, but the best identified is probably *Klebsiella*.

Table 7
EFFECT OF SOME EXPERIMENTAL CONDITIONS ON THE AMOUNT OF NITROGEN FIXED BY SEVERAL LEGUME CROPS AS MEASURED BY ^{15}N SUBSTRATE LABELING METHODOLOGY[48,111]

Crop and location	Labeled starter N applied (kg/ha)	Other treatments	N fixed (kg/ha)
Soybean; Hungary	40	0 kg P/ha at planting	7
	40	35 kg P/ha at planting	42
	40	70 kg P/ha at planting	62
	40	105 kg P/ha at planting	71
Soybean; Sri Lanka	20	—	3
	20	5000 kg straw at planting	41
Soybean; India	20	Inoculated with *Rhizobium*	102
	20	Not inoculated	79
Groundnut; Ghana	15	No inoculum	82
	15	Inoculum on seed	94
	15	Inoculum in soil	100
	15	Inoculum in soil + B + Mo	106
Soybean varieties; Austria	20	Ada	16
	20	Dunadja	18
	20	Chippewa	45
	20	Evans	50

A cyanobacterium, *Anabaena azollae*, has been found to form a symbiotic association with a fern plant, *Azolla*. The *Anabaena* occupies cavities in the dorsal lobe of fern leaves, where it fixes N_2 and makes it possible for the fern to grow in waters deficient in inorganic N. *Azolla* is widely distributed on the surface of paddy fields and water bodies in both temperate and tropical regions. The *Azolla-Anabaena* association is of high agronomic importance, and can contribute significant amounts of N for plant growth. Research at the International Rice Research Institute in the Philippines has shown that *Azolla* has the ability to fix as much as 1.5 kg N/ha/day or 500 kg N/ha/year.[20]

METHODS FOR ESTIMATING N_2 FIXATION

For proper management and a full realization of the benefits of this plant-microbial association, it is necessary to estimate how much nitrogen is fixed under different conditions in the field. It is only after this is known that various factors can be manipulated so as to increase the amount and proportion of N a plant derives from biological fixation. A suitable method for accurately measuring the amount of N crops derive from fixation is therefore an important requirement in any program aimed at maximizing biological nitrogen fixation.

Increment in N Yield and Plant Growth

The quantity of N_2 fixed by crops can be estimated by comparing the total nitrogen in a nonfixing plant with the nitrogen yield in the N_2-fixing plant of interest.[21,22] Any increase in N in the N_2-fixing plant over the control plant is considered to be due to N_2 fixation. This method has been used mostly in studies of nitrogen fixation by leguminous crops. The selection of an appropriate control crop is crucial for this methodology, since the method assumes that both fixing and nonfixing plants take up equal amounts of N from soil. The following nonfixing crops have been used to measure the amount of N derived from soil:

1. A nonnodulating legume isoline.
2. An uninoculated legume, where indigenous rhizobia are absent.
3. A legume, heavily inoculated with ineffective rhizobia capable of precluding infection by effective rhizobia.
4. A nonlegume with similar rooting pattern and N uptake to the legume.

The main advantages of this method are that it is relatively simple and inexpensive, compared to other methods; it is very useful in N-deficient soil or medium and also, where only single treatment differences are being investigated. It is often used in greenhouse or light-room studies to evaluate effectiveness of different strains of *Rhizobium*.[21]

The assumption that both control plant and N_2 fixer absorb the same quantity of soil N is not always valid.[23] The accuracy of the method is also greatly affected in soils of high N content, or where additional fertilizer N has been applied.

Total plant dry weight and leaf color have also been used to assess N_2 fixed when nitrogen is the main limiting factor. However, this method is only relative, and cannot be used to quantify N_2 fixed. The higher the N_2 fixed in this case, the higher is the total plant biomass, and the greener the leaf color in the absence of limiting factors other than N.

Nitrogen Balance Method

This method involves measurements of the gains and losses of N in a soil-plant system. The contribution made by microorganisms to the nitrogen content of soil-plant systems may be derived from the following equation:[3]

$$\Delta N = (F + S + M + D + R) - (C + L + V + E) \qquad (3)$$

where ΔN = the change in N content of a given mass of soil between times t_1 and t_2.

Inputs
F = nitrogen derived from N_2 fixation
S = nitrogen returned to soil surface from subsurface by plant roots or upward movement of N
M = nitrogen additions from manure, seeds, fertilizers, etc.
D = nitrogen added from dust and precipitation
R = nitrogen added from rainfall
Losses
C = nitrogen removed from soil by the standing vegetation
L = nitrogen lost by leaching
V = gaseous nitrogen losses
E = nitrogen lost by erosion

From the above equation, the total N_2 fixed (F) can be quantified if all the other terms are measured. Under unfertilized systems where the contribution from V, E, R, S, and D are small (such as with unfertilized pastures) and may therefore be ignored, the equation may be rewritten as

$$F = \Delta N + L + C \qquad (4)$$

Nitrogen balance studies provide only an indirect evidence for N_2 fixed. Since this method for N_2 fixation estimation involves several independent and unrelated measurements, the precision and accuracy of the value obtained are strongly influenced by the errors associated with each of these independent methods of assay.[24] For this reason N balance studies tend to be of long duration so as to be able to detect significant changes which have to be large. In addition, some of these estimates, such as gaseous N or leaching losses, cannot be made easily in a field area, and may only be assessed on small plots or lysimeters, or only in pot studies, thus limiting the usefulness of this method for field estimations of N_2 fixed.

The Acetylene Reduction Technique

The acetylene (C_2H_2) reduction technique was developed as a result of the finding by Schollhorn and Burris[25] that N_2 fixation is inhibited by acetylene, and it was also found to be converted by nitrogenase to ethylene (C_2H_4).[26] Ethylene has been identified as the sole product of nitrogenase-catalyzed acetylene reduction and no further reduction products have been found.[27] The conversion of acetylene to ethylene is represented by the following equation:

$$C_2H_2 + 2H^+ \xrightarrow[\text{nitrogenase}]{2e^-} C_2H_4 \qquad (5)$$

The conversion is nitrogen to ammonia, the first identified product of N_2 fixation, is represented in Equation 2.

The ethylene produced in the acetylene reduction technique is retrieved readily as a gas sample which is then analyzed quantitatively by a gas chromatograph. Ethylene gas is also stable during storage, thereby permitting the analysis to be done at a later period or date after sampling.

There is no universally established technique used for acetylene reduction assay. Instead, various improvised methods have been adopted for in vitro as well as in vivo assays. However, the procedures involved could be summarized into:

1. Sample preparation
2. Incubation of sample in acetylene-enriched air
3. Termination of reaction or sample collection
4. Gas chromatographic analysis
5. Assessment of N_2 fixed based on ethylene produced

These steps have been discussed in detail by Hardy et al.[27] and Masterson and Murphy.[28]

The technique is relatively simple, inexpensive, and very sensitive. The main disadvantage, however, is that this is a short-term enzyme activity measurement, often conducted in environments different from field situations in which crops are grown. Extrapolating results of such short-term measurements involving diurnal, daily, and seasonal variation to a growing season is therefore of doubtful validity. In addition, based on the electron requirements for nitrogenase-catalyzed C_2H_2 (Equation 5) and N_2 reduction (Equation 2) a conversion factor of 3 is normally used to obtain N_2 fixed from acetylene reduction figures. This conversion factor has, however, been found to differ under many conditions and may therefore often give erroneous N_2 fixation values.[27]

Use of Isotopes of N to Measure Nitrogen Fixation

Nitrogen has six isotopes, varying in atomic mass from 12 to 17 (Table 8). Of these, ^{14}N and ^{15}N are stable isotopes, while the rest are radioactive. The longest-lived radioactive isotope of N is ^{13}N, with a half-life of only 10.05 min, which therefore severely restricts the use of this radioisotope to studies of short duration. ^{13}N is thus of little use in N_2 fixation studies lasting beyond 2 hr. In contrast, the stable isotopes of N can be used in experiments

Table 8
ISOTOPES OF NITROGEN

Mass number	Natural abundance (%)	Half-life
12	—	0.0125 sec
13	—	10.05 min
14	99.634	(stable)
15	0.366	(stable)
16	—	7.36 sec
17	—	4.14 sec

of any duration, and are also safe to handle. $^{15}N_2$ as well as ^{15}N enriched or depleted inorganic or organic fertilizers have therefore been widely used for quantifying nitrogen fixed.

The use of ^{15}N as a tracer is based on the occurrence in nature of ^{14}N and ^{15}N in an almost constant ratio of 272 ± 0.3 to 1 yielding 0.3663 ± 0.0004 atom % ^{15}N. The amount of ^{15}N in a sample is conveniently expressed as % ^{15}N atom excess over the natural abundance in atmospheric nitrogen of 0.3663. In N isotopic tracer studies, the system under investigation is supplied with materials containing $^{15}N/^{14}N$ ratios measurably different from the ^{15}N natural abundance. It is also essential that the nitrogen isotope ratio should again be measurably different from ^{15}N natural abundance at the time the system under investigation is sampled. In the case of plants, for example, the uptake of ^{15}N-enriched fertilizer added to soil will result in a $^{15}N/^{14}N$ ratio greater than 0.3663% within the plant. The extent of increase in the $^{15}N/^{14}N$ ratio in the plant over that of natural abundance is a reflection of the extent of uptake of the labeled ^{15}N fertilizer, while a decrease in the % ^{15}N atom excess of the fertilizer nitrogen within the plant is an indication of the extent to which the plant took up N from available sources of unlabeled N.

The method for the determination of $^{15}N/^{14}N$ ratios in soil and plant samples have been reviewed by Fiedler and Proksch.[29]

Isotopes of N may be used in the gaseous N_2 form, or as labeled substrates in solid or liquid forms.

Use of $^{15}N_2$

The earliest application of $^{15}N_2$ in N-fixation studies was by Burris and Miller in 1941.[30] This method has been used to provide direct evidence for N_2 fixation since the ^{15}N concentration in plants exposed to $^{15}N_2$ is greater than the 0.3663% natural abundance after analysis. The extent to which ^{15}N is detected in the plant also provides an estimate of the proportion of the N of the plant that was derived from fixation, and is thus a direct method for quantifying N_2 fixed. The use of $^{15}N_2$ involves the enclosure of plants in chambers supplied with the enriched gas.[31] The environment within the chamber is, however, different from that in a field situation. Also, the plants cannot be confined in these chambers for long periods. Results obtained from such studies therefore tend to be instantaneous and subject to errors associated with extrapolating data from short-term studies to a growing season which involves diurnal, daily, and seasonal variations.[32]

Use of ^{15}N-Enriched Fertilizers or Substrates

The method involves the growth of N_2-fixing and nonfixing (reference) plants in soil fertilized with ^{15}N-enriched inorganic or organic fertilizers. The basic assumption in this technique is that when plants which take up nutrients from a similar volume of soil are confronted with two or more sources of N, they will take up N from each of these sources in direct proportion to the amounts available.[33]

The accuracy and precision of this method depends to a great extent on selecting a suitable non-N_2-fixing reference crop. The selection of the appropriate reference plant is therefore crucial, and it is essential to observe the following:

1. That the reference crop does not itself fix nitrogen. This can, if necessary, be checked very quickly, using the acetylene reduction assay.
2. The rooting depths of both reference and fixing crops should be similar, or both crops should derive all of their N from the same zone.
3. Both N_2-fixing and standard crops should go through similar growth or physiological stages, and mature about the same time.
4. Both N_2-fixing and standard crops should be planted and harvested at the same time.
5. Both crops should be affected in similar fashion by changes in environmental conditions, such as temperature and water during growth period.

For estimating N_2 fixed in grain legumes, the following standards have been used:

1. A nonlegume, nonfixing plant
2. A nonnodulating legume plant
3. An uninoculated legume plant in soils devoid of the appropriate *Rhizobium* strains

There are three main variations in the technique: (1) where both N_2-fixing and nonfixing plants are sampling the same $^{15}N/^{14}N$ pool; (2) where they are sampling different $^{15}N/^{14}N$ pools; and (3) the "single-treatment" substrate labeling methodology.

Sampling from the Same $^{15}N/^{14}N$ Pool

This is the case when both fixing and reference plants are growing on soil to which the same quantity and ^{15}N-enriched substrates have been applied. Thus in the absence of any supply of N other than soil and ^{15}N fertilizer, a fixing plant and a nonfixing reference plant will contain the same ratio of $^{15}N/^{14}N$ since they are taking N of similar $^{15}N/^{14}N$ composition, but not necessarily the same total quantity of N. In both plants, the $^{15}N/^{14}N$ ratio within the plant is lowered by the N absorbed from the unlabeled soil. However, in the presence of N_2, the fixing plant further lowers the ratio of $^{15}N/^{14}N$ due to incorporation of N from unlabeled air, while this does not occur in the nonfixing plant. The extent to which the $^{15}N/^{14}N$ ratio in the fixing crop is decreased relative to the nonfixing plant is therefore an indication of N_2-fixing ability, and can be used to estimate N fixed in the field.

The determination of N_2 fixation using this approach is depicted in Figure 1 by a fictitious example. By using ^{15}N-labeled fertilizer, 50% of the N in the reference crop (plant I) can be seen to have been derived from the applied fertilizer. Since there are only two sources of N available to this crop,

$$\% \text{ Ndffert} + \% \text{ Ndfsoil} = 100 \qquad (6)$$

where % Ndffert and % Ndfsoil stand for the percentage of nitrogen in the plant derived from fertilizer and soil, respectively.

Referring again to Figure 1, it follows from Equation 6 that the other half or 50% of the N in plant came from soil. This then establishes that the ratio of soil to fertilizer N available to the fixing and nonfixing plants was 1:1 in this example.

For the N_2-fixing crop (plant II in Figure 1), there is a third source of N available to the plants; i.e., N_2 from the atmosphere. The total N in plant can therefore be represented by the following equation:

$$\% \text{ Ndffert} + \% \text{ Ndfsoil} + \% \text{ Ndffix} = 100 \qquad (7)$$

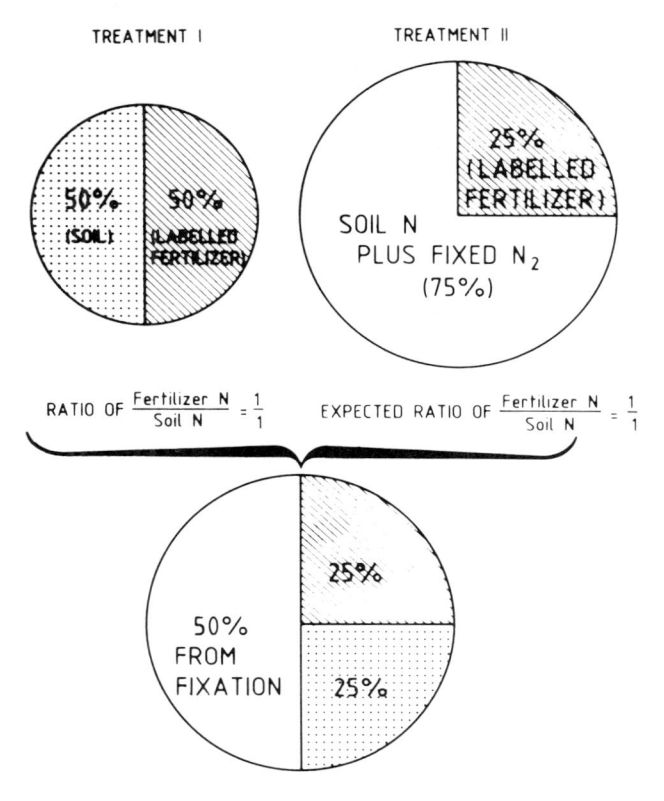

FIGURE 1. A fictitious example showing how percentage of N derived from fixation can be determined by measuring percentage of N derived from fertilizer in nonfixing (I) and fixing (II) plants based on the assumption that both plants take up N from each source in direct proportion to the amounts available.

where % Ndffix is the percentage of nitrogen in the plant derived from nitrogen fixation. The nonfixing reference crop took up N from soil and fertilizer in the ratio 1:1, and this is assumed also to be the case for the fixing crop since:

$$\frac{\text{Ndffert}_{(nonfixing)}}{\text{Ndfsoil}_{(nonfixing)}} = \frac{\text{Ndffert}_{(fixing)}}{\text{Ndfsoil}_{(fixing)}} \qquad (8)$$

In the example, % Ndffert in the fixing crop was 25%. Therefore, according to Equation 8 the % Ndfsoil in the fixing crop must also be 25%. The rest of the N taken up (50%) must be derived from fixation, since according to Equation 7, % Ndffix = 100 − (% Ndffert + % Ndfsoil).

Based on the above concepts, assumptions, and equations, Fried and Middleboe[33] derived the following equation to estimate N_2 fixation:

$$\% \text{ Ndffix} = \left(1 - \frac{\% \text{ Ndffert}_{(fixing)}}{\% \text{ Ndffert}_{(nonfixing)}}\right) \times 100 \qquad (9)$$

But since

$$\% \text{ Ndffert} = \frac{\% \ ^{15}\text{N atom excess of sample}}{\% \ ^{15}\text{N atom excess of fertilizer}} \times 100 \qquad (10)$$

Table 9
DATA RECORDED FOR NODULATING (FIXING) AND NONNODULATING (NONFIXING) SOYBEAN, BOTH OF VAR 'CHIPPEWA' AT THE SEIBERSDORF LABORATORY, AUSTRIA, IN 1981; 20 kg N/ha OF N-15-LABELED FERTILIZER WAS APPLIED TO THE FIXING AND NONFIXING CROPS

Example Showing Calculation

Fixing crop	Dry-matter yield(1)[a] (kg/ha)	N(2) (%)	N yield(4) (kg/ha)	N-15 atom excess(3) (%)	Ndffert(5) (%)	N fert yield(6) (kg/ha)	Ndffix(8) (%)	Fixed N(9) (kg/ha)
Stems	4478	0.63	28.2	0.152	3.16	0.89		
Leaves	2743	1.90	52.1	0.158	3.28	1.71		
Pods	1867	2.58	48.2	0.132	2.74	1.32		
Total			128.5		3.05(7)	3.92	26	33

Note: Two other values are needed for the calculation: (1) % ^{15}N atom excess of fertilizer = 4.81; and (2) % Ndffert$_{(nonfixing)}$ = 4.14 (calculated by the same method as % Ndffert for fixing crop). Equations (1) through (9) are given in Table 10.

And since %^{15}N atom excess of fertilizer is the same for both fixing and reference crop, Equation 9 can be rewritten as

$$\% \text{ Ndffix} = \left(1 - \frac{\%\ ^{15}\text{N atom excess in fixing crop}}{\%\ ^{15}\text{N atom excess in nonfixing crop}}\right) \times 100 \qquad (11)$$

And the amount of N_2 fixed can be derived from:

$$N_2 \text{ fixed} = \frac{\% \text{ Ndffix} \times \text{total N in fixing crop}}{100} \qquad (12)$$

The use of these formulas is demonstrated in Tables 9 and 10.

Sampling from Different $^{15}N/^{14}N$ Pools

Often it is necessary to apply different doses of N to fixing and nonfixing plants because high levels of inorganic N can depress N_2 fixation. This therefore necessitates the application of low amounts of labeled N fertilizer to the fixing crop in order to estimate N_2 fixed. However, such amounts may be too low to support the proper growth of the reference plants, especially in soils of low fertility. For these reasons, it is practical to give a reasonable dose of ^{15}N-labeled fertilizer to the reference crop, while the fixing crop receives a low quantity. Using the A-value concept of Fried and Broeshart,[34] it is still possible to determine N_2 fixed under these conditions.

According to the A-value concept,

$$\frac{\% \text{ Ndffert}}{A_{\text{fert}}} = \frac{\% \text{ Ndfsoil}}{A_{\text{soil}}} = \frac{\% \text{ Ndsoil} + \% \text{ Ndffix}}{A_{\text{soil+fix}}} = \frac{\% \text{ Ndffix}}{A_{\text{fix}}} \qquad (13)$$

where A_{fert} is amount of fertilizer applied, and A_{soil} and A_{fix} are nitrogen available from soil and fixation, respectively, as expressed in fertilizer units.

For the reference crop which derives its N from only soil and fertilizer, Equation 13 may be rewritten as:

Table 10
MEASUREMENTS MADE AND THE FORMULAS USED TO CALCULATE % NDFFIX AND AMOUNT OF FIXED N IN TABLE 9

Measured values

(1) Dry matter yield (D.M.) of pods, straws, and leaves (kg/ha)
(2) % N of each plant part in (1)
(3) % ^{15}N atom excess of each plant part in (1) and of fertilizer applied

Calculated values

(4) N yield (kg/ha) of each plant part $= \dfrac{\text{D.M. of each plant part} \times \% \text{ N}}{100}$

(5) % Nddfert $= \dfrac{\% \ ^{15}\text{N atom excess of sample}}{\% \ ^{15}\text{N atom excess of fertilizer}} \times 100$

(6) N-fert yield (kg/ha) $= \dfrac{\text{N yield (kg/ha)} \times \% \text{ Ndffert}}{100}$

(7) % Ndffert (weighed average) $= \dfrac{\text{Total N fert. yield}}{\text{total N yield}} \times 100$

(8) % Ndffix $= \left(1 - \dfrac{\% \text{ Ndffert (fixing)}}{\% \text{ Ndffert (nonfixing)}}\right) \times 100$

or

% Ndffix $= \left(1 - \dfrac{\% \ ^{15}\text{N atom excess (fixing)}}{\% \ ^{15}\text{N atom excess (nonfixing)}}\right) \times 100$

(9) N fixed (kg/ha) $= \dfrac{\% \text{ Ndffix} \times \text{total N in fixing crop}}{100}$

$$\frac{\% \text{ Ndffert}}{A_{\text{fert}}} = \frac{\% \text{ Ndfsoil}}{A_{\text{soil}}} \tag{14}$$

From Equation 6, % Ndfsoil = 100 − % Ndffert.
Equation 14 can therefore be rewritten as:

$$A_{\text{soil}} = \frac{100 - \% \text{ Ndffert}}{\% \text{ Ndffert}} \times A_{\text{fert}} \tag{15}$$

Since % Ndffert is determined experimentally by the use of ^{15}N, and the amount of fertilizer added (A_{fert}) is known, the available amount of soil N (A_{soil}) in added fertilizer units can then be determined from Equation 15.

However, for the fixing plant, a third source of N, i.e., N_2, is available for plant growth, and therefore

$$\frac{\% \text{ Ndffert}}{A_{\text{fert}}} = \frac{\% \text{ Ndfsoil} + \% \text{ Ndffix}}{A_{\text{soil}+\text{fix}}} \tag{16}$$

From Equation 7, % Ndfsoil + % Ndffix = 100 − % Ndffert.
Equation 16 can therefore be rewritten as

Table 11
DATA RECORDED FOR *VICIA FABA* (FIXING CROP) AND BARLEY (NONFIXING CROP) AT THE SEIBERSDORF LABORATORY, AUSTRIA, IN 1979; 20 kg N/ha OF ^{15}N LABELED FERTILIZER WAS APPLIED TO THE *V. FABA* AND 100 kg N/ha TO THE BARLEY

Example Showing Calculation

	Total N yield(4)[a] (kg/ha)	Ndffert(7) (%)	A_{soil}(8)	$A_{soil+fix}$(9)	A_{fix}(10)	Ndffix(11) (%)	Fixed N(12) (kg/ha)
Fixing crop[b]	151.7	0.877		2260	1810	79	120
Nonfixing crop[c]		18.17	450				

[a] Equations for (4) through (12) are given in Table 12.
[b] N-fert applied: 20 kg N/ha, 5.64 % ^{15}N atom excess; A_{fert} = 20.
[c] N-fert applied: 100 kg N/ha, 1.00 % ^{15}N atom excess; A_{fert} = 100.

$$A_{soil+fix} = \frac{100 - \% \text{ Ndffert}}{\% \text{ Ndffert}} \times A_{fert} \quad (17)$$

Again, % Ndffert is determined experimentally, and A_{fert} is known. Equations 15 and 17 can then be used to establish A_{fix} as follows:

$$A_{fix} = (A_{soil+fix}) - A_{soil} \quad (18)$$

And from Equation 13,

$$\% \text{ Ndffix} = A_{fix} \times \frac{\% \text{ Ndffert}}{A_{fert}}$$

The amount of fixed N can be calculated according to Equation 12. The use of these formulas in different situations is shown in Tables 11 and 12.

In studies of nitrogen fixation by various grain legumes in the field, 20 kg of N per hectare of 5% ^{15}N atom excess fertilizer and 100 kg N per ha of 1% ^{15}N atom excess fertilizer have generally been applied to fixing and nonfixing crops, respectively. Where there is good reason to believe that 20 kg N/ha may inhibit N fixation, it is recommended that 10 kg of N per hectare of 10% atom excess ^{15}N should be applied to the fixing crop. In order to ensure uniform application of N, especially when low quantities are used, the fertilizer is desolved in water and distributed uniformly over each plot, taking precautions to avoid waterlogging of the soil as well as cross-contamination.

"Single-Treatment" Substrate Labeling Methodology

This involves a combination of two isotope methods[35]; i.e., the substrate labeling technique for nitrogen-fixation estimation described previously and the "single-treatment" experiment without interaction.[36]

In the "single-treatment" experiment all treatments are the same as far as the plant and soil are concerned, but only one treatment in each treatment combination has been labeled with isotopes as explained in Table 13. Therefore nitrogen fixation is identical in the matching treatments within the limit of experimental error. By this approach, time of application of

Table 12
MEASUREMENTS WHICH WERE MADE AS WELL AS THE FORMULAS USED TO CALCULATE % NDFFIX AND AMOUNT OF FIXED N IN TABLE 11

Measured values

(1)–(3) Same as in Table 10

Calculated values

(4)–(7) Same as in Table 10

(8) $A_{soil} = A_{fert} \times \dfrac{100 - \%\,\text{Ndffert}}{\%\,\text{Ndffert}}$

(9) $A_{soil + fix} = A_{fert} \times \dfrac{100 - \%\,\text{Ndffert}}{\%\,\text{Ndffert}}$

(10) $A_{fix} = (A_{soil + fix}) - A_{soil}$

(11) $\%\,\text{Ndffix} = A_{fix} \times \dfrac{\%\,\text{Ndffert}}{A_{fert}}$

(12) $\text{N fixed (kg/ha)} = \dfrac{\%\,\text{Ndffix} \times \text{total N in fixing crop}}{100}$

fertilizer, and fertilizer source and placements, etc. can be studied at the same time as N_2 fixation is measured.

The calculations for this type of experiment are similar to the calculations using the A-value approach as shown above, where different amounts of labeled fertilizer are applied to the fixing and the nonfixing crops. However, correction has to be made for the proportion of nitrogen taken up from the other unlabeled fertilizer applied.

The same assumption holds as for the previous approaches in the use of ^{15}N-labeled fertilizer to estimate nitrogen fixation.

The following equations show the sources of nitrogen for the fixing and the nonfixing crops:

Nonfixing crop: % Ndfsoil + % Ndffert* + % Ndf other fert = 100   (20)

Fixing crop: % Ndfsoil + % Ndffix + % Ndffert* + % Ndf other fert = 100   (21)

where Ndffert* is nitrogen derived from labeled fertilizer, whereas other fertilizers are unlabeled.

The following formula can be used to calculate A_{soil} and $A_{soil + fix}$ as derived from Equations 13, 20, and 21.

For nonfixing crops:

$$A_{soil} = A_{fert*} \times \left(\dfrac{100 - \%\,\text{Ndffert*} - \%\,\text{Ndf other fert}}{\%\,\text{Ndffert*}} \right) \quad (22)$$

For fixing crops:

$$A_{soil + fix} = A_{fert*} \times \left(\dfrac{100 - \%\,\text{Ndffert*} - \%\,\text{Ndf other fert}}{\%\,\text{Ndffert*}} \right) \quad (23)$$

where A_{fert*} is the amount of labeled fertilizer applied.

A_{fix} is calculated according Equation 18 and % Ndffix according to the following equation:

$$\% \text{ Ndffix} = A_{fix} \times \frac{\% \text{ Ndffert*}}{A_{fert*}} \quad (24)$$

Fixed N is calculated according to Equation 12. The use of the above formula in different situations is shown in Tables 13 and 14.

Advantages and Disadvantages of the Use of ^{15}N Substrate Labeling Methodology

The ^{15}N substrate labeling method has the advantage of giving the best estimates of N_2 fixation under field condition, because it is possible to measure N derived from fixation, distinct from all other sources of N available to the plant. This methodology requires a relatively high cost expenditure for ^{15}N and instruments needed. However, the cost of ^{15}N fertilizers is at the time of writing of this article far lower than it was a decade or so ago, and with the introduction of relatively cheap emission spectrometers, the need for expensive mass spectrometers can often be avoided. It is now possible to conduct a proper field experiment with $500 to $3000 worth of ^{15}N fertilizers. Some sources of ^{15}N-labeled fertilizers and emission and mass spectrometers are listed in Tables 15, 16, and 17, respectively. Table 18 lists some institutes which analyze ^{15}N samples.

Use of Differences in N Natural Abundance of ^{15}N to Measure Nitrogen Fixation

As a result of isotope discrimination effects occurring during soil formation, most soils have been found to have a slightly higher ^{15}N abundance than the atmosphere.[37-39] As an outcome of this difference in ^{15}N abundance between soil and atmospheric N_2, N_2-fixing plants have been found to have a lower ^{15}N enrichment than non-N_2-fixing plants, and this has therefore been used as evidence and measure of N_2 fixation.[39,40] According to LaRue and Patterson,[22] the following equation is used to calculate N fixed, using variations in ^{15}N natural abundance:

$$\% \text{ N fixed} = 100 \times \left(\frac{\% ~^{15}N \text{ control} - \% ~^{15}N \text{ fixing system}}{\% ~^{15}N \text{ control} - (\% ~^{15}N \text{ air/b})} \right)$$

where b is an isotope discrimination factor of fixation.

The main disadvantages of this method are the rather small differences in ^{15}N abundance being traced,[41] and the high variability of ^{15}N abundance in soils,[42,43] which therefore raises doubts as to the suitability of the method to estimate N_2 fixation.[44]

ACKNOWLEDGMENT

Thanks are due to Drs. H. Broeshart, M. Fried, C. G. Lamm, and K. Reichardt at the FAO/IAEA Division, Vienna for reviewing the article, and to Mrs. M. Kneissl and Ms. G. Dreger for typing the manuscript.

Table 13
DATA RECORDED AT THE SEIBERSDORF LABORATORY, AUSTRIA IN 1980 FOR *VICIA FABA* (FIXING) AND WHEAT (NONFIXING) USING "SINGLE-TREATMENT" EXPERIMENTAL APPROACH

Example Showing Calculations

		Amount of N applied (kg/ha) Broadcast	Banded	Total N yield (kg/ha)	Ndffert (%)	$A_{soil}(8)$[a]	$A_{soil\ +\ fix}(9)$	$A_{fix}(10)$	Ndffix(11) (%)	Fixed N(12) (kg/ha)
V. faba	1a	10[b]	10	243.3	1.3		745	538	70	170
V. faba	1b	10	10[b]	225.4	1.9		509	367	70	158
Wheat	2a	20[b]	20		7.8	207				
Wheat	2b	20	20[b]		11.4	142				

[a] Equations for (8) through (12) are given in Table 14.
[b] Labeled with ^{15}N.

Table 14
MEASUREMENTS WHICH WERE MADE AS WELL AS THE FORMULAS USED TO CALCULATE VALUES IN TABLE 13

Measured values

(1)–(3) Same as in Table 10

Calculated values

(4)–(7) Same as in Table 10

(8) $A_{soil} = A_{fert^a} \times \left(\dfrac{100 - \% \text{ Ndffert}^a - \% \text{ Ndf other fert}}{\% \text{ Ndffert}^a} \right)$

% Ndf other fertilizer is measured by the isotope application; e.g., for treatment wheat 2a in Table 13:

$$A_{soil} = 20 \times \left(\dfrac{100 - 7.8 - 11.4}{7.8} \right) = 207$$

(9) $A_{soil + fix} = A_{fert^a} \times \left(\dfrac{100 - \% \text{ Ndffert}^a - \% \text{ Ndf other fert}}{\% \text{ Ndffert}^a} \right)$

(10) $A_{fix} = A_{soil + fix} - A_{soil}$

(11) $\% \text{ Ndffix} = A_{fix} \times \dfrac{\% \text{ Ndffert}^a}{A_{fert^a}}$

(12) $\text{N fixed (kg/ha)} = \dfrac{\% \text{ Ndffix} \times \text{total N in fixing crop}}{100}$

[a] Labeled fertilizer.

Table 15
SOURCES OF ^{15}N-ENRICHED COMPOUNDS

Company	Address
Amersham Buchler GmbH	Braunschweig, Gieselweg 1 Postfach 1120 3300 Braunschweig F.R.G.
Azote Products Chimique	40 Avenue Hoche Paris 8eme France
Bio-Rad Laboratories	32nd and Griffin Avenue Richmond, CA 94804
Cambridge Isotope Laboratories	20 Commerce Way Woburn, MA 01801
Commissariat a l'Energie Atomique	91191 Gif-Sur-Ivette France
Isocommerz GmbH	DDR 1115 Berlin Lindenberger Weg 70 D.R.G.
Icon Services	19 Ox Bow Lane Summit, N.J. 07901
Isomet Corp.	Palisades Park, N.Y. 07650
Isotec Inc.	7542 McEwen Road Dayton, OH 45459
Merck, Sharp and Dohme, Canada Ltd.	Isotope Division P.O. Box 899 Pointe Claire/Dorval 700 Quebec H9R 4P7 Canada
Ministerio de Industria	Junta de Energia Nuclear Apartado 3055, C. Universitaria Madrid 3 Spain
Monsanto Research Corp.	P.O. Box 32 Miamisburg, OH 45342
Rohstoff-Einfur GmbH	4000 Düsseldorf Faunastrasse 61 F.R.G.
Stohler Isotope Chemicals	49 Jones Road Waltham, MA 02154

Table 16
SOURCES OF EMISSION SPECTROMETERS FOR ^{15}N/^{14}N RATIO DETERMINATION

Company	Address
JASCO (Japan Spectroscopic)	2967-5, Ishikawa-Lho Hachioji City Tokyo 192 Japan
ISOCOMMERZ GmbH	705 Leipzig Permoserstr. 15 D.R.G.

Table 17
SOME SOURCES OF MASS SPECTROMETERS

Company	Address
Hewlett-Packard Intercontinental	3495 Deer Creek Rd. Palo Alto, CA 94304
Hitachi Scientific Instruments, Nissei Sangyo Co., Ltd.	Mori 17th Bldg. 2 Sakuragawa-Cho, Shiba Nishikubo, Minato-Ku Tokyo 105 Japan
VG Isotopes Ltd., Micromass	Ion Path, Road Three Winsford, Cheshire CW7 3BX U.K.
Finnigan MAT	355 River Oaks Parkway San Jose, Calif. 95134

Table 18
SOME INSTITUTES WHICH OFFER ANALYSIS OF ^{15}N SAMPLES

Institute	Address
Isotope Services, Inc.	329 Potrillo Drive Los Alamos, NM 87544
Bureau of Stable Isotope Analysis	BSIA Limited Le Mercure 13, Avenue Albert 1 21000 Dijon France
FAO/IAEA Agricultural Biotechnology Laboratory[a]	IAEA Wagramerstrasse 5 P.O. Box 100 A-1400 Vienna Austria

[a] For contractors only.

REFERENCES

1. **Alexander, M.,** *Introduction to Soil Microbiology,* John Wiley & Sons, New York, 1961.
2. **Jensen, H. L.,** A survey of biological nitrogen fixation in relation to the world supply of nitrogen, *Trans. 4th Int. Congr. Soil Sci.,* 1, 1950, 165.
3. **Greenland, D. J.,** Contribution of microorganisms to the nitrogen status of tropical soils, in *Biological Nitrogen Fixation in Farming Systems of the Tropics,* Ayanaba, A. and Dart, P. J., Eds., John Wiley & Sons, New York, 1977, 13.
4. **Bartholomew, W. V., Mayer, J., and Landelout, H.,** Mineral nutrient immobilization under forest and grass fallow in the Yangambi region, INAEC, Serie Scientifique No. 57, International Atomic Energy Agency, Vienna, 1953.
5. **Greenland, D. J. and Nye, P. H.,** Increases in the carbon and nitrogen contents of tropical soils under natural fallows, *J. Soil Sci.,* 9, 284, 1959.
6. **Odu, C. T. I.,** Contribution of free-living bacteria to the nitrogen status of humid tropical soils, in *Biological Nitrogen Fixation in Farming Systems of the Tropics,* Ayanaba, A. and Dart, P. J., Eds., John Wiley & Sons, New York, 1977, 257.

7. **Dobereiner, J.,** Present and future opportunities to improve the nitrogen nutrition of crops through biological fixation, in *Biological Nitrogen Fixation in Farming Systems of the Tropics,* Ayanaba, A. and Dart, P. J., Eds., John Wiley & Sons, New York, 1977, 3.
8. **Balandreau, J.,** Is it possible to breed new rice varieties for N_2 fixation?, Research Coordination Meet. Use of Isotopes in Studies of Dinitrogen Fixation, International Atomic Energy Agency, Vienna, 1981.
9. **Rennie, R. J.,** Potential use of induced mutations to improve symbioses of crop plants with N_2-fixing bacteria, in Induced Mutations — A Tool in Plant Research, Proc. Symp. International Atomic Energy Agency, Vienna, March 9 to 13, 1981.
10. **Hatch, M. D. and Slack, C. R.,** Photosynthetic CO_2 fixation pathways, *Annu. Rev. Plant Physiol.,* 21, 141, 1970.
11. **Boddey, R. M., Chalk, P. M., Victoria, R., and Matsui, E.,** The ^{15}N isotope technique applied to the estimation of biological nitrogen fixation associated with *Paspalum notatum* cv. *Batatais* in the field, *Soil Biol. Biochem.,* 15, 25, 1983.
12. **Roger, P. A. and Kulasooriya, S. A.,** *Blue-Green Algae and Rice,* The International Rice Research Institute, Manila, Philippines, 1980.
13. **Venkataraman, G. S.,** Non-symbiotic nitrogen fixation, 12th Int. Congr. Soil Science, Vol. 2, New Delhi, February 8 to 16, 1982.
14. **Allen, O. N. and Allen, E. K.,** *The Leguminosae, A Source Book of Characteristics, Uses and Nodulation,* Macmillan, New York, 1981.
15. **Danso, S. K. A.,** The ecology of *Rhizobium* and recent advances in the study of the ecology of *Rhizobium,* in *Biological Nitrogen Fixation in Farming Systems of the Tropics,* Ayanaba, A. and Dart, P. J., Eds., John Wiley & Sons, New York, 1977, 115.
16. **Burton, J. C.,** New development in inoculating legumes, in *Recent Advances in Biological Nitrogen Fixation,* Subba Rao, N. S., Ed., Edward Arnold, London, 1980.
17. **Akkermans, A. D. L.,** Nitrogen Fixation and Nodulation of *Alnus* and *Hippophae* under Natural Conditions, Ph.D. thesis, University of Leiden, Leiden, 1971.
18. **Silver, W. S.,** Dinitrogen fixation by plant associations excluding legumes, in *A Treatise on Dinitrogen Fixation,* Section IV, Hardy, R. W. F. and Gibson, A. H., Eds., John Wiley & Sons, New York, 1977, 141.
19. **Silver, W. S.,** Foliar associations in higher plants, in *Treatise on Dinitrogen Fixation,* Section III, Hardy, R. W. F. and Silver, W. S., Eds., John Wiley & Sons, New York, 1977, 153.
20. **Watanabe, I., Ke-Zhi, B., Berja, N. S., Espinas, C. R., Ito, O., and Subudhi, B. P. R.,** The *Azolla-Anabaena* complex and its use in rice culture, *IRRI Res. Pap.,* Ser. 69, 1981, 1.
21. **Vincent, J. M.,** *A Manual for the Practical Study of Root-Nodule Bacteria,* Blackwell Scientific, Oxford, 1970.
22. **LaRue, T. A. and Patterson, T. G.,** How much nitrogen do legumes fix?, *Adv. Agron.,* 34, 15, 1981.
23. **Ruschel, A. P., Vose, P. B., Victoria, R. L., and Salati, E.,** Comparison of isotope techniques and non-nodulating isolines to study the effect of ammonium fertilization on dinitrogen fixation in soybean, *Glycine Max., Plant Soil,* 53, 513, 1979.
24. **Greenland, D. J. and Watanabe, I.,** The continuing nitrogen enigma, 12th Int. Congr. Soil. Science, Vol. 5, New Delhi, February 8 to 16, 1982.
25. **Schollhorn, R. and Burris, R. H.,** Acetylene as a competitive inhibitor of N_2 fixation, *Proc. Natl. Acad. Sci. U.S.A.,* 58, 213, 1967.
26. **Dilworth, M. J.,** Acetylene reduction by nitrogen-fixing preparations from *Clostridium pasteurianum, Biochem. Biophys. Acta,* 127, 285, 1966.
27. **Hardy, R. W. F., Burns, R. C., and Holsten, R. D.,** Applications of the acetylene-ethylene assay for measurement of nitrogen fixation, *Soil Biol. Biochem.,* 5, 47, 1973.
28. **Masterson, C. L. and Murphy, P. M.,** The acetylene reduction technique, in *Recent Advances in Biological Nitrogen Fixation,* Subba Rao, N. S., Ed., Edward Arnold, London, 1981, 8.
29. **Fiedler, R. and Proksch, G.,** The determination of nitrogen-15 by emission and mass spectrometry in biochemical analysis: a review, *Anal. Chim. Acta,* 78, 1, 1975.
30. **Burris, R. H. and Miller, C. E.,** Application of ^{15}N to the study of biological nitrogen fixation, *Science,* 93, 114, 1941.
31. **Witty, J. F. and Day, J. M.,** Use of $^{15}N_2$ in evaluating asymbiotic N_2 fixation, in *Isotopes in Biological Dinitrogen Fixation,* International Atomic Energy Agency, Vienna, 1978.
32. **Knowles, R.,** The measurement of nitrogen fixation, *Curr. Perspectives in Nitrogen Fixation,* Gibson, E. H. and Newton, W. E., Eds., Australian Academy of Science, 1981, 327.
33. **Fried, M. and Middelboe, V.,** Measurement of amount of nitrogen fixed by a legume crop, *Plant Soil,* 47, 713, 1977.
34. **Fried, M. and Broeshart, H.,** An independent measurement of the amount of nitrogen fixed by legume crops, *Plant Soil,* 43, 707, 1975.

35. **Fried, M. and Broeshart, H.,** A further extension of the method for independently measuring the amount of nitrogen fixed by a legume crop, *Plant Soil,* 61, 331, 1981.
36. **Fried, M., Soper, R. J., and Broeshart, H.,** [15]N labelled single-treatment fertility experiments, *Agron. J.,* 67, 393, 1975.
37. **Chang, H. H., Bremner, J. M., and Edwards, A. P.,** Variations of nitrogen-15 abundance in soils, *Science,* 146, 1574, 1964.
38. **Bremner, J. M. and Tababai, M. A.,** Nitrogen-15 enrichment of soils and soil derived nitrate, *J. Environ. Qual.,* 2, 363, 1973.
39. **Kohl, D. H., Shearer, G., and Harper, J. E.,** Estimates of N_2 fixation based on differences in natural abundance of ^{15}N in nodulating and non-nodulating isolines of soybeans, *Plant Physiol.,* 66, 61, 1980.
40. **Amarger, N., Mariotti, A., Mariotti, F., Durr, J. C., Bourguignon, C., and Lagacherie, B.,** Estimate of symbiotic fixed nitrogen in field grown soybean using variations in ^{15}N natural abundance, *Plant Soil,* 52, 269, 1979.
41. **Lee, C. and Smith, D.,** Influence of nitrogen fertilizer on stands, yield of herbage and protein and nitrogenous fractions of field grown alfalfa, *Agron. J.,* 64, 527, 1972.
42. **Shearer, G., Kohl, D. H., and Harper, J. E.,** Distribution of ^{15}N among plant parts of nodulating and non-nodulating isolines of soybean, *Plant Physiol.,* 66, 57, 1980
43. **Broadbent, F. E., Rauschkolb, R. S., Lewis, K. A., and Chang, G. Y.,** Spatial variability of nitrogen-15 and total nitrogen in some virgin and cultivated soils, *Soil Sci. Soc. Am. Proc.,* 44, 524, 1980.
44. **Owens, L.,** Use of ^{15}N-enriched soil to study N_2 fixation in grasses, in *Genetic Engineering for Nitrogen Fixation,* Hollander, A., Ed., Plenum Press, New York, 1977.
45. **Stevenson, F. J.,** Origin and distribution of nitrogen in soil, in *Soil Nitrogen,* Vol. 10, Bartholomew, W. V. and Clcoh, F. E., Eds., American Society of Agronomy, Madison, 1965.
46. **Dalton, H.,** The cultivation of diazotrophic microorganisms, in *Methods for Biological Nitrogen Fixation,* Bergersen, F. J., Ed., John Wiley & Sons, Chichester, 1980, 13.
47. **Crabbe, D. and Lawson, S.,** *The World Food Book,* Kogan Page, London, 1981.
48. **Fried, M.,** Direct quantitative assessment in the field of fertilizer management practices, 11th Int. Congr. Soil Science, Alberta, Canada, June 19 to 27, 1978, 103.
49. **Seeger, J. R.,** Effect of nitrogen fertilizing on the nodulation and yield of groundnut, *Bull. Inst. Agron. Stn. Rech. Gambloux,* 29, 197, 1961.
50. **Wetselaar, R.,** Estimation of nitrogen fixation by four legumes in a dry monsoonal area of north-western Australia, *Aust. J. Exp. Agric. Anim. Husb.,* 7, 518, 1968.
51. **Sen, A. N.,** Nitrogen economy of soil under rahar *(Cajanus cajan), Indian Soc. Soil Sci.,* 6, 171, 1958.
52. **Agboola, A. A. and Fayemi, A. A. A.,** Fixation and excretion of nitrogen by tropical legumes, *Agron. J.,* 64, 409, 1972.
53. **Gargantini, H. and Wutke, A. C. P.,** Fixacao do nitrogenio do ar pelas bacterias que vivem associadas as raizes do feijao de porco e do feijao biano, *Bragantia,* 19, 639, 1960; *Field Crops Abstr.,* 14, 191.
54. **Sanderson, K. W.,** Guar (*Cyanopsis tetragonaloba*) in the Rhodesian low veld, *Rhod. Agric. J.,* 71, 17, 1974.
55. **Bezdicek, D. F., Evans, D. W., Abede, B., and Witters, R. E.,** Evaluation of peat and granular inoculum for soybean yield and N fixation under irrigation, *Agron. J.,* 70, 865, 1978.
56. **Bhangoo, M. S. and Albritton, D. J.,** Nodulating and non-nodulating Lee soybean isolines response to applied nitrogen, *Agron. J.,* 68, 642, 1976.
57. **Deibert, E. J., Bijeriego, M., and Olson, R. A.,** Utilization of ^{15}N fertilizer by nodulating and non-nodulating soybean isolines, *Agron. J.,* 71, 717, 1979.
58. **Ham, G. E.,** Use of ^{15}N in evaluating symbiotic N_2-fixation in field grown soybean, in *Isotopes in Biological Dinitrogen Fixation,* International Atomic Energy Agency, Vienna, 1978, 151.
59. **Ham, G. E. and Caldwell, A. C.,** Fertilizer placement effects on soybean seed yield, N_2 fixation and ^{33}P uptake, *Agron. J.,* 70, 779, 1978.
60. **Ham, G. E., Lawn, R. J., and Brun, W. A.,** Influence of inoculation, nitrogen fertilizers and photosynthetic source-sink manipulations on field grown soybeans, in *Symbiotic Nitrogen Fixation in Plants,* Nutman, P. S., Ed., Cambridge University Press, Cambridge, 1976, 239.
61. **Johnson, H. S. and Hume, D. J.,** Effects of nitrogen sources and organic matter on nitrogen fixation and yield of soybeans, *Can. J. Plant Sci.,* 52, 991, 1972.
62. **Lagacherie, B. and Obaton, M.,** L'inoculation du soja — resultats d'essais et orientation du travail, Academie d'Agriculture de France, seance 67-79, 1973.
63. **Legg, J. O. and Sloger, C.,** A tracer method for determining symbiotic nitrogen fixation in field studies, in Proc. 2nd Int. Conf. Stable Isotopes, Klein, E. R. and Klein, P. D., Eds., Oak Brook, Ill., October 20 to 23, 1975, 661.
64. **Lyon, T. L. and Bizzel, J. A.,** A comparison of several legumes with respect to nitrogen accretion, *J. Am. Soc. Agron.,* 26, 657, 1934.

65. **Rizk, S. G.,** Atmospheric nitrogen fixed by legumes under Egyptian conditions. I. Egyptian clover *(Trifolium alexandrinum), J. Soil Science U. A. R.,* 2, 253, 1962.
66. **Vest, G.,** Nitrogen increases in a non-nodulating soybean genotype grown with nodulating genotypes, *Agron. J.,* 63, 356, 1971.
67. **Weber, C. R.,** Nodulating and non-nodulating soybean isolines. I. Agronomic and chemical attributes, *Agron. J.,* 58, 43, 1966.
68. **Weber, C. R.,** Nodulating and non-nodulating soybean isolines. II. Response to applied nitrogen and modified soil conditions, *Agron. J.,* 58, 46, 1966.
69. **Singh, P. and Chonbey, S. D.,** Inoculation — a cheap source of nitrogen to legumes, *Indian Farming,* 20, 33, 1971.
70. **Ruschel, A. P., Vose, P. B., Matsui, E., Victoria, R. L., and Tsai Saito, S. M.,** Field evaluation of N_2 fixation and N-utilization by *Phaseolus* bean varieties determined by ^{15}N isotope dilution, *Plant Soil,* 65, 397, 1982.
71. **Mahler, R. L., Bezdicek, D. F., and Witters, R. E.,** Influence of slope position on nitrogen fixation and yield of dry peas, *Agron. J.,* 71, 348, 1979.
72. **McEwen, J.,** Fertilizer nitrogen and growth regulators for field beans (*Vicia faba* L.). I. Effect of seed bed applications of large dressings of fertilizer nitrogen and residual effects on following winter wheat, *J. Agric. Sci.,* 74, 61, 1970.
73. **Yankovitch, L.,** Essai de la determination de la fixation de l'azote atmospherique par les legumineuses, *Ann. Serv. Bot. Agron. Tunis.,* 16, 303, 1940.
74. **Akhurst, C. G.,** Report of the Rubber Research Institute (Soil Division) for 1939, Kuala Lumpur, Malaysia, p. 61.
75. **Moore, A. W.,** Symbiotic nitrogen fixation in a grazed tropical grass-legume pasture, *Nature (London),* 185, 638, 1960.
76. **Odu, C. T. I., Fayemi, A. A., and Ogunwale, J. A.,** Effect of pH on the growth, nodulation and nitrogen fixation of *Centrosema pubescens* and *Stylosanthes gracilis, J. Sci. Food Agric.,* 22, 57, 1971.
77. **Watson, G. A.,** Nitrogen fixation by *Centrosema pubescens, J. Rubber Res. Inst. Malays.,* 15, 2, 1957.
78. **Dawson, R. C.,** Potential for increasing protein production by legume inoculation, *Plant Soil,* 32, 655, 1970.
79. **Allos, H. F. and Bartholomew, W. V.,** Replacement of symbiotic nitrogen by available nitrogen, *Soil Sci.,* 87, 61, 1959.
80. **Washko, J. B. and Marriott, L. F.,** Yield and nutritive value of grass herbage as influenced by nitrogen fertilization in the northeastern United States, in *Proc. 8th Int. Grassland Congr., Reading, U.K.,* Alden Press, Oxford, 1960, 137.
81. **Radulovic, V. and Nutman, P. S.,** Field experiments on legume inoculation, *Rep. Rothamsted Exp. Stn. 1962* (Part I), 79.
82. **Bear, F. E.,** Making the most of our nitrogen resources, *Proc. Soil Sci. Soc. Am.,* 7, 294, 1942.
83. **Heichel, G. H., Barnes, D. K., and Vance, C. P.,** Nitrogen fixation of alfalfa in the seeding year, *Crop Sci.,* 21, 330, 1981.
84. **Lyon, T. L. and Bizzel, J. A.,** Nitrogen accumulation in soil as influenced by the cropping system, *J. Am. Soc. Agron.,* 25, 266, 1933.
85. **Nutman, P. S.,** IBP field experiments on nitrogen fixation by nodulated legumes, in *Symbiotic Nitrogen Fixation in Plants,* Nutman, P. S., Ed., Cambridge University Press, Cambridge, 1976, 211.
86. **Souza, D. I. A. de,** Legume nodulation and nitrogen fixation studies in Kenya, *East Afr. Agric. For. J.,* 34, 299, 1969.
87. **Wedin, W. F.,** Legume and inorganic nitrogen for pasture swards in subhumid, microthermal climates of the United States, *Proc. 9th Int. Grassland Congr., Sao Paulo,* Vol. 2, Edicoes Limitada, 1965, 1163.
88. **Williams, R. J. B. and Cooke, G. N.,** Experiments on herbage crops at Saxmundham, 1967—1971, *Rep. Rothamsted Exp. Stn. 1971* (Part 2), 95.
89. **Ashford, R. and Bolton, J. L.,** Effects of sulphur and nitrogen fertilization and inoculation with *Rhizobium meliloti* on the growth of sweet clover (*Melilotus alba* Dear.), *Can. J. Plant Sci.* 41, 81, 1961.
90. **Crack, B. J.,** Changes in soil nitrogen following different establishment procedures for Townsville stylo on a solodic soil in north eastern Queensland, *Aust. J. Exp. Agric. Anim. Husb.,* 12, 274, 1972.
91. **Bland, B. F.,** The effect of cutting frequency and root segregation on the yield from perennial ryegrass-white clover associations, *J. Agric. Sci.,* 69, 391, 1967.
92. **Butler, G. W. and Bathurst, N. O.,** The underground transfer of nitrogen from clover to associated grass, Proc. Int. Grassland Cong., Palmerston North, New Zealand, 1956, 168.
93. **Castle, M. E. and Reid, D.,** Nitrogen and herbage production, *J. Br. Grassl. Soc.,* 18, 1, 1963.
94. **Clarke, A. L.,** Nitrogen accretion by an impoverished red-brown earth soil under short-term leys, *Proc. 11th Int. Grassland Cong.,* Queensland, Norman, M. J. T., Ed., University of Queensland Press, Queensland, 1970, 461.

95. **Cowling, D. W. and Lockyer, D. R.**, A comparison of the reaction of different grass species to fertilizer nitrogen and to growth in association with white clover. II. Yield of nitrogen, *J. Br. Grassl. Soc.*, 22, 53, 1967.
96. **Donald, C. M.**, The impact of cheap nitrogen, *J. Inst. Agric. Sci.*, 26, 319, 1960.
97. **Edmeades, D. C. and Goh, K. M.**, Symbiotic nitrogen fixation in a sequence of pastures of increasing age measured by a ^{15}N dilution technique, *N. Z. J. Agric. Res.*, 21, 623, 1978.
98. **Green, J. O. and Cowling, D. W.**, The nitrogen nutrition of grassland, in *Proc. 8th Int. Grassland Cong.*, Reading, U.K., Alden Press, Oxford, 1960, 126.
99. **Haystead, A. and Lowe, A. G.**, Nitrogen fixation by white clover in hill pasture, *J. Br. Grassl. Soc.*, 32, 57, 1977.
100. **Haystead, A. and Marriott, C.**, Fixation and transfer of nitrogen in a white-grass sward under hill conditions, *Ann. Appl. Biol.*, 88, 453, 1978.
101. **Jones, M. B., Delwiche, C. C., and Williams, W. A.**, Uptake and losses of ^{15}N applied to annual grass and clover in lysimeters, *Agron. J.*, 69, 1019, 1977.
102. **McAuliffe, C., Chamblee, D. S., Uribe-Arango, H., and Woodhouse, W. W.**, Influence of inorganic nitrogen on nitrogen fixation by legumes as revealed by ^{15}N, *Agron. J.*, 50, 334, 1958.
103. **Melville, J. and Sears, P. O.**, Pasture growth and soil fertility. II. The influence of red and white clovers, superphosphate, lime and dung and urine on the chemical composition of pasture, *N. Z. J. Sci. Technol.*, Suppl. I, 30, 1953.
104. **Munro, J. M. M.**, The role of white clover in hill areas, *J. Brit. Grassl. Soc., White Clover Research, Occasional Symp.*, 6, 259, 1970.
105. **Phillips, D. A. and Bennet, J. P.**, Measuring symbiotic nitrogen fixation in rangeland plots of *Trifolium subterraneum* L. and *Bromus mollis* L., *Agron. J.*, 70, 671, 1978.
106. **Royset, S.**, Trials with nodule bacteria for white clover, *Medd. det Norske Myrselsk*, 56, 114, 1958.
107. **Sau, A.**, Legumes and fertilizers and sources of grassland nitrogen in temperate regions of the USSR, *Proc. Int. 11th Grassland Congr., Queensland*, Norman, M. J. T., Ed., University of Queensland Press, Queensland, 1970, 416.
108. **Sears, P. O.**, Pasture growth and soil fertility. I. The influence of red and white clovers, superphosphate, lime and sheep grazing on pasture yields and botanical composition, *N. Z. J. Sci. Technol.*, A-35(Suppl. 1), 1, 1953.
109. **Watson, E. R.**, The influence of subterranean clover pastures on soil fertility. III. The effect of applied phosphorus and sulphur, *Aust. J. Agric. Res.*, 20, 447, 1969.
110. **Essafi, A.**, Remarques sur l'importance et sur l'estimation de la fixation d'azote atmospherique par les legumineuses sous le climat de la Tunisie, *C. R. Acad. Agric. Fr.*, 46, 611, 1960.
111. **Hardarson, G., Zapata, F., and Danso, S. K. A.**, Symbiotic nitrogen fixation by soybean varieties as affected by nitrogen fertilizer application, in *Advances in Nitrogen Fixation Research*, Veeger, C. and Newton, W. E., Eds., Martinus Nijhoff/Dr. W. Junk, The Hague, 1984, 34.

Environmental Factors and Plant Growth

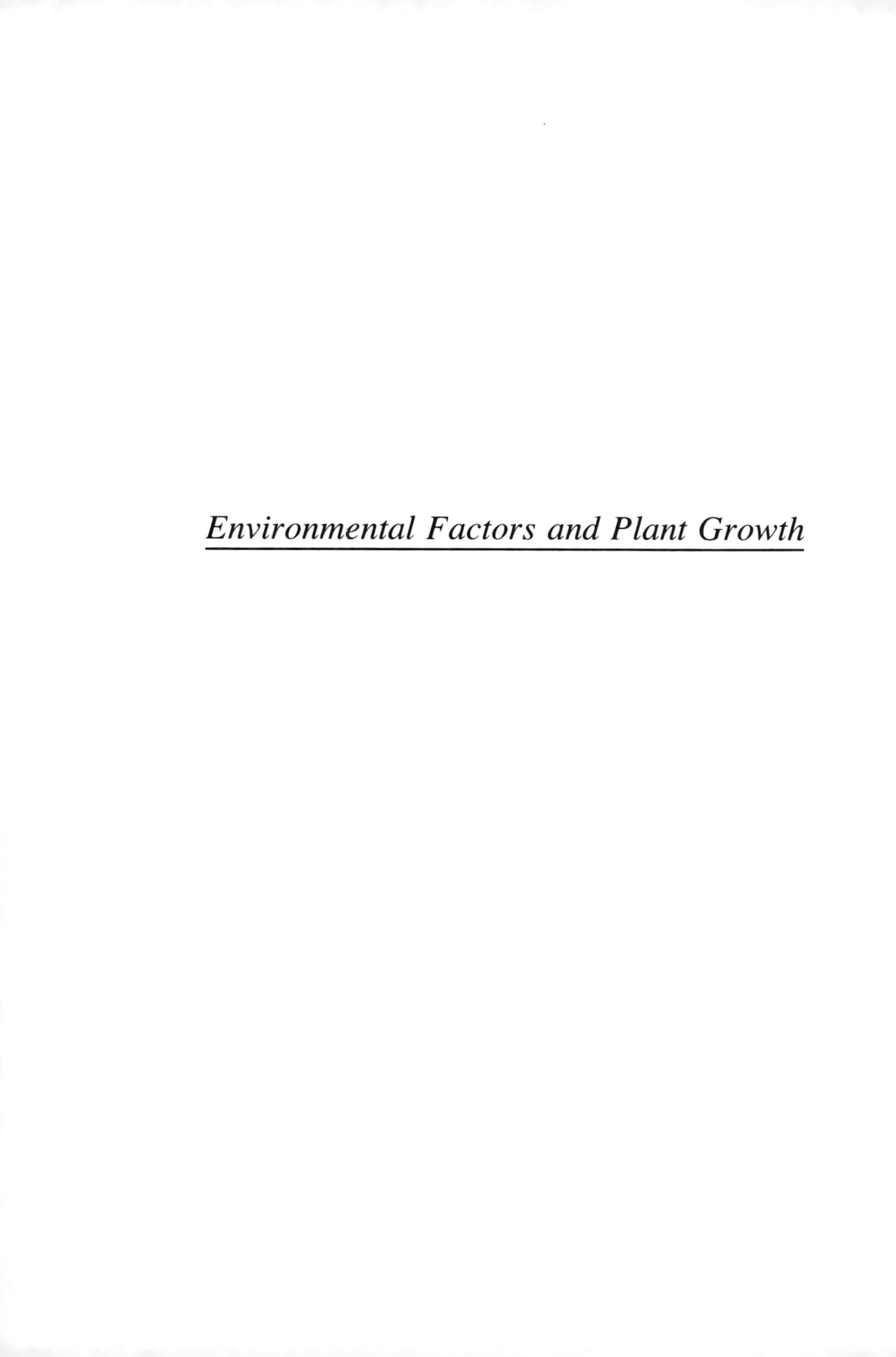

LIGHT AND PHOTOPERIOD*

David J. Major

The intensity, duration, and quality of radiant energy from the sun combine directly and indirectly to control the growth and development of plants. Large amounts of photosynthetically active radiation (PAR), between 400 and 700 nm, are used directly in photosynthesis by chlorophyll for the assimilation of carbon dioxide which, in turn, is used for growth and dry-weight increase. Small amounts of energy with wavelengths between 670 and 760 nm are used by the photoreversible pigment phytochrome to measure length of day. In addition to the direct effect of sunlight on photosynthesis, the radiant energy from the sun affects air temperature, which in turn has profound effects on rates of growth and development. These effects are discussed in the Chapter on "Plant Temperature Stress".

The use of controlled-environment facilities has resulted in a better understanding of artificial light requirements for plant growth and has increased the awareness by researchers of its effects. The measurement of light energy should be considered on the basis that light is a physical entity,[1] and therefore qualitative measurements such as illumination should not be used. Radiant energy can be defined as the amount of energy required to do work. The rate of energy flow with time is an expression of power and the expression of power applied over area is a measure of intensity. For example, an erg, the expression of energy, is the force of 1 dyne applied through a distance of 1 centimeter. When this is integrated over time it becomes an expression of power (erg/sec), and power per unit area is intensity (ergs sec^{-1} cm^{-2}). Radiant-energy measurements in plant science reports are usually expressed as energy, power, or intensity. The acceptable International System (SI) units are the joule for total radiant energy, the watt for power, and watts per square meter (W/m^2) for intensity. Conversions among irradiance terms are presented in Table 1. Illuminance is not generally considered an acceptable measurement of light because it is a measurement of visible radiation; that is, it is measured using a sensor with spectral responses similar to those of the human eye. It is usually used to describe lighting where the human eye is the sensor. Therefore, measurement of photosynthetically active radiation is preferred. Some conversions among illumination terms are presented in Table 2 and an estimate of the conversion factor for kilolux to W/m^2 for various light sources is presented in Table 3. It should be kept in mind that each source of light or combination of sources of light will require a different conversion factor when transforming illuminance to irradiance. Many plant scientists use equipment that measures the photon flux density of PAR with units of μE m^{-2} sec^{-1}. This is the number of moles of photons incident upon a unit area per unit time. Thus, Avogadro's number of photons (6.022 × 10^{23}) is the number of photons in 1 mol. The mathematical conversion from μE m^{-2} sec^{-1} to W/m^2 is possible, but complicated. Some approximate conversions are provided in Table 4.

The irradiance of artificial light sources changes with the age of the lamps, so it is also important to know the average age of lamps in a controlled-environment cabinet and the irradiance should be routinely measured. An example of the deterioration of various types of light bulbs is shown in Figure 1.

The quality of light for plant growth is assessed by determining how the source satisfies the requirements of the plant over the visible spectrum. The action spectra of four measurable plant processes — photosynthesis, phytochrome responses, high energy reaction, and phototropism — are shown in Figure 2. Photosynthesis is the process whereby light energy is converted to chemical energy and it is responsible for increases in dry weight of the plant.

---

* All tables follow text.

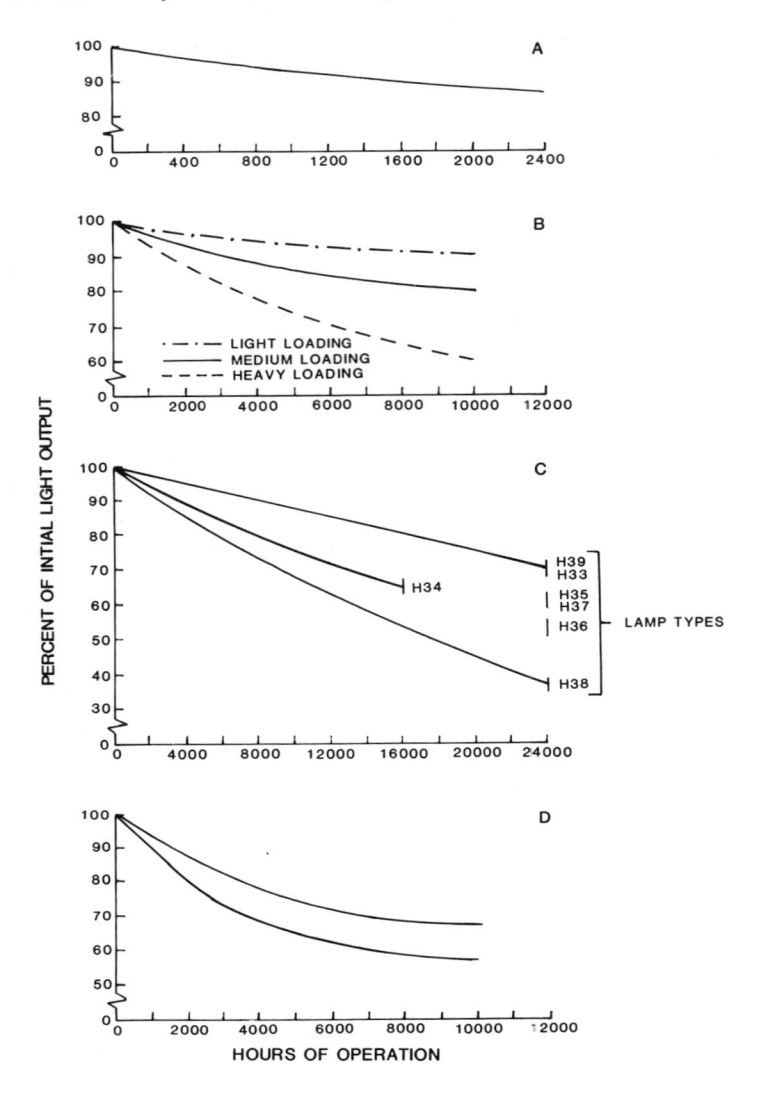

FIGURE 1. Lamp outputs as they change throughout lamp life. (A) Incandescent, (B) fluorescent; (C) mercury-fluorescent (H33 = 400 W, H34 = 1000 W low current, H35 = 700 W, H36 = 1000 W high current, H37 = 250 W, H38 = 100 W, H39 = 175 W); and (D) 400-W metal halide (upper line) and 400-W phosphor-coated metal halide (lower line). (Adapted from Bickford, E. D. and Dunn, S., *Lighting for Plant Growth*, Kent State University Press, Kent, Oh., 1972. With permission.)

Phytochrome is photoreversible pigment which runs the biological clock and controls the rate of development. The high-energy reaction is active in the blue and far-red regions but the receptor pigment is as yet unknown. Phototropism is influenced by blue light but again the pigment is not known. It controls the movement toward light, but protoplasmic streaming, protein synthesis, and stomatal opening are also involved. In order to obtain satisfactory growth the light source must supply sufficient light in these wavelengths. The spectral distribution of energy of commonly used light sources is shown in Figure 3. A comparison of the spectral distribution of energy of incandescent lamps with the action spectra of plants shows that maximum emission occurs in the region where phytochrome is most active. Only

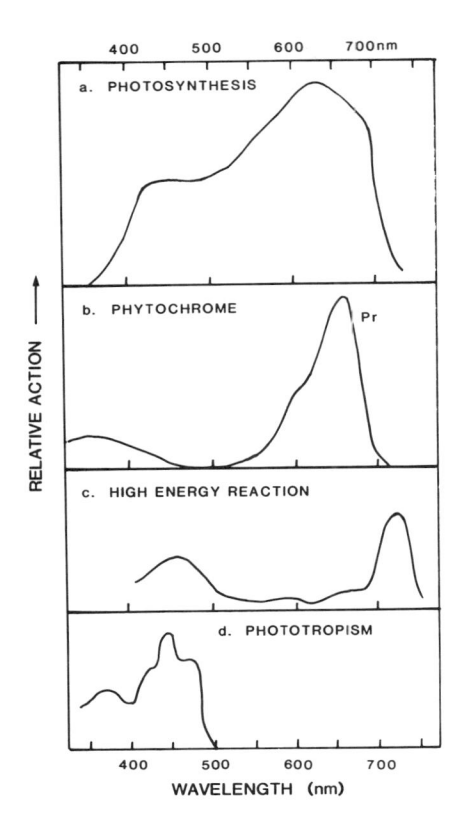

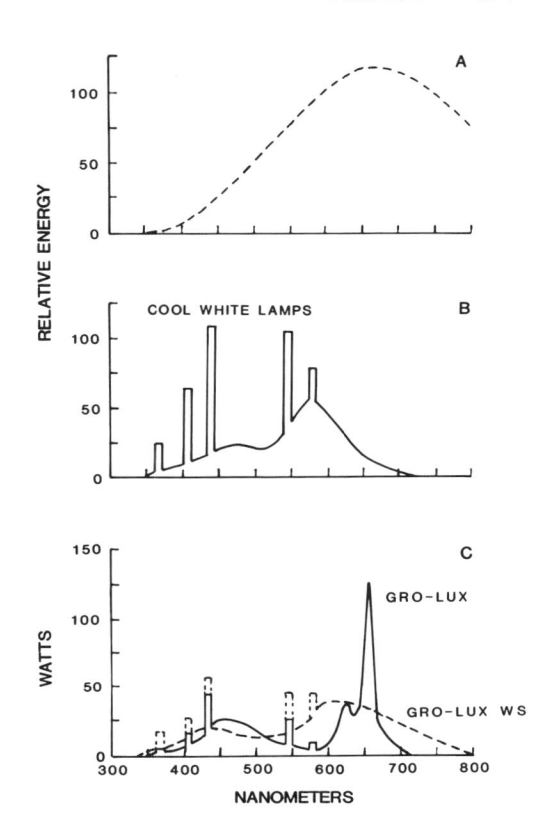

FIGURE 2. Action spectra. (a) Of photosynthesis under equal incident energy; (b) of phytochrome transformations; (c) of the high-energy reaction; and (d) of phototropism. (From Evans, L. T., in *Plant Response to Climatic Factors*, Slatyer, R. O., Ed., UNESCO, Paris, 1973. With permission.)

FIGURE 3. Approximate spectral energy distribution. (A) Of a 150-W parabolic aluminized reflector incandescent lamp; (B) of cool-white fluorescent lamps, and (C) of Standard Gro-Lux® and Gro-Lux® wide spectrum lamps. (Adapted from Bickford, E. D. and Dunn, S., *Lighting for Plant Growth*, Kent State University Press, Kent, Oh., 1972. With permission.)

10 to 20% of the total energy emitted by incandescent lamps is physiologically active. Since only low light intensities in the region of 600 to 750 nm are required for photoperiod extensions, incandescent lamps are commonly used for this purpose. The fluorescent lamp is used in most controlled-environment chambers to provide the energy needed for photosynthesis. Incandescent lamps are usually used in conjunction with cool or warm white fluorescent lamps. Newer types of plant-growth lamps, such as the Gro-Lux® and the Gro-Lux® Wide Spectrum lamps, have been designed specifically to meet the needs of plants. The Gro-Lux® Wide Spectrum has more energy in the far-red and blue regions than any other type of lamp used alone, so its use probably makes incandescent bulbs unnecessary.

Natural lighting and its measurement is frequently of interest to plant scientists, particularly those in the fields of water and energy balances, photosynthesis, and remote sensing. As with artificial light, the intensity and duration are usually of paramount importance. For remote sensing, the solar angle is also of considerable importance.

There are no currently accepted plant science standards for defining sunrise and sunset nor the duration of the length of day (photoperiod). If sunrise and sunset are chosen as the time when the upper limit of the sun is just visible on the horizon then the center of the sun is 50 min below the horizon (List[2]). But even when the center of the sun is 50 min below the horizon, the intensity is still high in terms of plant responses. To adjust to a lower intensity some plant researchers have extended the photoperiod determined when the sun is 50' below the horizon by 1 hr[3] or by 21% of astronomical twilight,[4] assuming that this will more accurately reflect the photoperiod experienced by field-grown plants. Civil twilight is the time it takes the center of the sun to pass from $-50'$ to $-6°00'$. At $6°00'$ below the horizon normal outdoor activities are just possible (List[2]). Nautical twilight is when the center of the sun is $-12°00'$, and astronomical twilight is $-18°00'$. The duration of daylight for sunrise and sunset at solar altitude of $-50'$ for latitudes from 0 to 65 at 5-day intervals is included in Table 5. The procedure for calculating solar angle, sunrise, sunset, photoperiod, and solar radiation at the surface of the atmosphere is included in Tables 6 through 9. A computer program describing the method of calculating sunrise and sunset prepared by Robertson and Russelo[5] is presented in Table 10.

The annual and diurnal cycles of photoperiod are so precise that plants have evolved a diurnal and seasonal timing mechanism using the duration of day or night. The development of all crop plants, short- or long-day types, is influenced primarily by photoperiod and temperature, but the extent to which these variables interact is not known. A short-day species flowers most rapidly when photoperiod is short, and a long-day species flowers most rapidly when photoperiod is long. In natural daylight conditions, multicollinearity makes simultaneous study of the effects of the temperature and photoperiod difficult. At high latitudes photoperiod and temperature vary too much to allow the use of sequential planting dates in field studies of genotypic responses to photoperiod. Nevertheless, some investigators have made use of the effects of naturally changing photoperiods in serial plantings in the field (see Major[6]). Controlled-environment chambers or rooms are capable of precise control of many variables but they are too expensive for plant breeders or crop physiologists and modelers who need to screen large numbers of genotypes. The greenhouse is relatively inexpensive and offers some temperature control but its limitation is that there is no control of photoperiod. The natural photoperiod changes continually, so if a model explaining photoperiod response is realistic, it should be capable of operating in any of these conditions. The photoperiod response of any genotype can be determined if the basic vegetative phase (BVP), the maximum or minimum optimal photoperiod (MOP), and photoperiod sensitivity are known (Figure 4). Optimal photoperiods of short- or long-day plants are defined as those that have no delaying effect on floral development. Under such conditions the number of days to flowering, which occurs in a constant number of days, is an estimate of length of BVP. The BVP is a period of juvenility through which the plant must pass before it can respond to the photoperiod stimulus. Since, by definition, the plant must reach a minimum size before it can flower, the BVP is a constant length regardless of the photoperiod, but will vary among genotypes. The BVP of rice cultivars, for example, varies from 10 to 63 days from germination, and its length is controlled by two genes.

The length of the BVP can be measured, in optimal photoperiod conditions, as the number of days to a readily measured event associated with flowering, such as heading in cereals. By its definition, the length of the BVP is measured by determination of the date of floral initiation. In rice, the reproductive phase (floral initiation to flowering) and the ripening phase (flowering to physiological maturity) are relatively constant at 35 days and 30 to 35 days.

The maximum or minimum optimal photoperiod separates the optimal photoperiods from the nonoptimal photoperiods in short- and long-day species, respectively. Nonoptimal photoperiods are defined as those that have a delaying effect on floral development. Since flowering is delayed beyond the BVP, this phase is the photoperiod-induced phase (PIP). The PIP is the duration from the end of the BVP to floral initiation and the duration of this phase is controlled by photoperiod. Rice cultivars of low-latitude origin have been found to have shorter MOP than those from higher latitudes.

For some genotypes, a critical photoperiod also exists. The mathematical definition of a critical point is the point on the graph of a function where the derivative is zero or infinity.

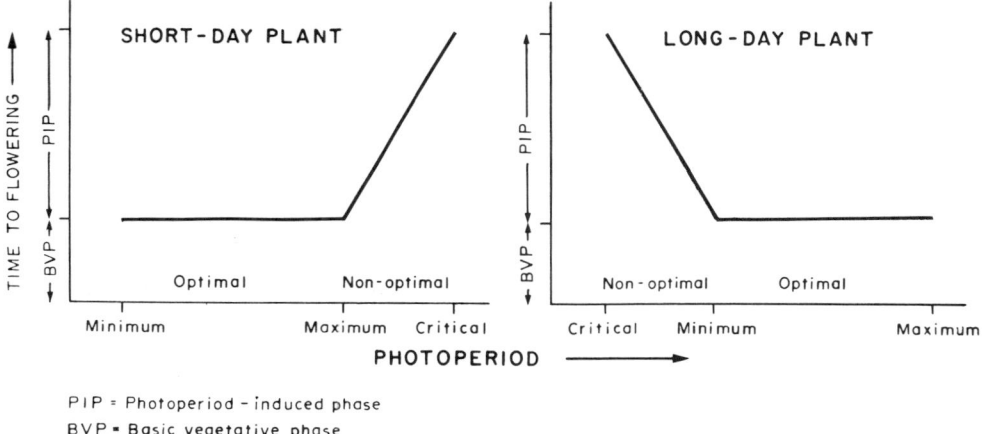

FIGURE 4. Responses of short- and long-day plants to photoperiod.

For a short-day species, a critical photoperiod would indicate that, as photoperiod increased, either flowering would never occur at or above the critical photoperiod (i.e., derivative is infinite) or flowering would occur in a constant number of days (i.e., derivative is zero). Photoperiod sensitivity is defined as the derivative of the function relating time to flowering in nonoptimal photoperiod conditions and will have units of days delay of flowering per hour increase of daylength. There is evidence that genetic variability exists for photoperiod sensitivity in rice, oats, and sorghum, but these estimates of photoperiod sensitivity were obtained by subtracting the minimum growth duration from the maximum growth duration.

The response of perennial crops to photoperiod has not received much attention. Alfalfa, for example, is clearly a long-day plant but it also requires vernalization. It is not known if species such as alfalfa require floral induction once in their lifetime or each year. A great many perennial species are probably short-day species. A list of field crop species obtained from Fehr and Hadley[9] is presented in Table 11 to illustrate the type of response to photoperiod. Within a crop, genotypes will be found which range from no measurable response to photoperiod to very sensitive.

The two principal plant processes controlled by light are photoperiodism and photosynthesis. The first determines the timing of flowering and maturity and the second determines the rate of dry-matter production. Photoperiodism is a timing mechanism that allows the plant to determine the length of the day and requires only low intensities of red and far-red wavelengths. Daily dry-matter production is the result of rate of photosynthesis, which is dependent upon intensity of PAR, as well as duration.

When the rates of photosynthesis of crop species are compared in terms of net carbon dioxide fixation it is apparent that a number of species, including corn, have a rate of CO_2 fixation much higher than that of other species. The differences in photosynthetic rate cannot be attributed to differences in chlorophyll content or to the proportionality between the quanta of light absorbed and the CO_2 assimilated. These efficient species have the C-4 photosynthetic pathway. The CO_2-fixation pathway involving C-4 species has been found to be operative in tropical grasses as well as in certain broadleaves. Of significance to the C-4 pathway is that it is linked to a number of unique characteristics, such as leaf anatomy, chloroplast ultrastructure, low CO_2 compensation points, and a lack of photorespiration.

In the Calvin cycle, (C-3 pathway), found in species such as wheat, CO_2 is added to the C-2 of an intermediate compound and then two molecules of 3-phosphoglyceric acid (3-PGA) are formed. In the C-4 pathway the acceptor is a four-carbon compound. The important enzymes in the two systems are RuDP carboxylase in the C-3 plant and phosphoenolpyruvate

(PEP)-carboxylase in the C-4 plant. The PEP-carboxylase has a much greater affinity for CO_2 than does RuDP-carboxylase, allowing the C-4 plants to photosynthesize at much lower CO_2 concentrations than C-3 plants. Another important difference among these types is that C-3 plants exhibit photorespiration, a process whereby recently fixed CO_2 is released in the light. Photorespiration increases as light intensity and temperature increase.

Plants can be divided into two groups based on either high or low compensation points. This point represents the equilibrium between the rates of photosynthesis and respiration. It is about 50 ppm for C-3 plants and about 5 ppm for C-4 plants.

C-4 plants also differ from C-3 plants anatomically. The vascular bundles of C-4 plants are surrounded by a bundle sheath of relatively large parenchyma cells containing large chloroplasts that lack grana but accumulate starch. Surrounding this bundle sheath are one or more layers of smaller mesophyll cells. The chloroplasts in these cells differ from those in the bundle sheath in that they are smaller, randomly arranged, possess well-formed grana and do not accumulate starch. Presumably, the C-4 anatomy facilitates unloading of photosynthates into the vascular bundles.

Within the grass family, the distribution of the C-4 pathway is found in those lines that evolved during the Cretaceous period in the tropics. The ability of these types of plants to maintain maximum photosynthesis under high light intensities and temperatures may be related to their leaf anatomy and chloroplast structure.

The measurement of the rate of net photosynthesis is subject to a great deal of environmental and genetic disturbance. The group that has the C-4 pathway includes the species such as corn, sorghum, and sugarcane. The group that has the C-3 pathway includes the temperate cereal grains, rice, soybeans, etc. Some estimates of rates of photosynthesis and maximal growth rates of some crop species[11] are presented in Table 10.

Table 1
CONVERSION FACTORS OF RADIANT ENERGY, POWER AND INTENSITY FROM NON-SI TO ACCEPTABLE SI UNITS

	Non-SI units Multiply	By	SI units To obtain
Energy	Calorie	4.19	Joule
	Erg	10^{-7}	Joule
Power	Cal/min	6.98×10^{-2}	Watt
	Erg/sec	10^{-7}	Watt
Intensity	Erg sec^{-1} cm^{-2}	10^{-3}	Watt/m^2
	Cal min^{-1} cm^{-2}	6.98×10^2	Watt/m^2

Table 2
CONVERSION FACTORS FOR ILLUMINATION

Multiply	By	To obtain
Brightness		
Footlambert	1.08×10^{-3}	Lambert
Cal/mm^2	314	Lambert
Cal/mm^2	2.92×10^5	Footlambert
Illuminance		
Lux, meter candle	10^{-4}	Lumen/cm^2
Footcandle	10.75	Lux, meter candle

Table 3
CONVERSION FACTORS FOR CONVERTING KILOLUX (klx) READINGS INTO SI UNITS (W/m^2) FOR COMMON LIGHT SOURCES

Light sources	300—800 nm[a]	400—700 nm[a]	400—700 nm[b]
	Multiply by		
Incandescent	—	4.2	4.0
Incandescent (500 W)	7.0	—	—
Fluorescent lamps			
White	2.8	—	2.7
Warm-white	2.7	2.8	—
De-lux Warm-white	—	3.2	—
Cool-white	3.0	3.1	—
De-lux cool white	—	3.4	—
Daylight	—	3.4	4.0
Blue	7.2	5.8	—
Green	2.4	2.1	—
Red (magnesium-gumanate)	11.4	—	—
Red	—	8.6	—
Gold	3.2	2.3	—
Gro-Lux®	8.9	—	—
Mercury	3.3	—	3.0
Sunlight	—	4.0	—

To convert klx to W/m^2

Table 3 (continued)
CONVERSION FACTORS FOR CONVERTING KILOLUX (klx) READINGS INTO SI UNITS (W/m²) FOR COMMON LIGHT SOURCES

Light sources	300—800 nm[a]	400—700 nm[a]	400—700 nm[b]
	\multicolumn{3}{c}{Multiply by}		
Metal halide	—	—	3.1
Sodium (H.P.)	—	—	2.8

To convert klx to W/m²

[a] From Bickford, E. D. and Dunn, S., *Lighting for Plant Growth*, Kent State University Press, Kent, Ohio, 1972. With permission.

[b] From Biggs, W. W. and Hansen, M. C., in Li-COR Instrumentation for Biological and Environmental Sciences, Li-COR, Inc., Lincoln, 1981, 42. With permission.

Table 4
APPROXIMATE CONVERSION FACTORS TO µE m⁻² sec⁻¹ FOR VARIOUS PAR LIGHT SOURCES

To convert to µE m⁻² sec⁻¹	Daylight	White fluorescent	Incandescent	Metal halide	Sodium (H.P.)	Mercury
			Multiply by			
W/m²	4.6	4.6	5.0	4.6	5.0	4.7
Kilolux	18	12	20	14	14	14

Adapted from Biggs and Hansen.[7]

Table 5
DAY LENGTH IN HOURS

LATITUDE 0

DATE	JAN	FEB	MAR	APR	MAY	JUNE	JULY	AUG	SEPT	OCT	NOV	DEC
5	12.0	12.0	12.0	12.0	12.0	12.0	12.0	12.0	12.0	12.0	12.0	12.0
10	12.0	12.0	12.0	12.0	12.0	12.0	12.0	12.0	12.0	12.0	12.0	12.0
15	12.0	12.0	12.0	12.0	12.0	12.0	12.0	12.0	12.0	12.0	12.0	12.0
20	12.0	12.0	12.0	12.0	12.0	12.0	12.0	12.0	12.0	12.0	12.0	12.0
25	12.0	12.0	12.0	12.0	12.0	12.0	12.0	12.0	12.0	12.0	12.0	12.0
30	12.0		12.0	12.0	12.0	12.0	12.0	12.0	12.0	12.0	12.0	12.0

LATITUDE 5

DATE	JAN	FEB	MAR	APR	MAY	JUNE	JULY	AUG	SEPT	OCT	NOV	DEC
5	11.6	11.7	11.8	12.0	12.1	12.2	12.2	12.1	12.0	11.8	11.7	11.6
10	11.6	11.7	11.8	12.0	12.1	12.2	12.2	12.1	11.9	11.8	11.7	11.6
15	11.6	11.7	11.9	12.0	12.1	12.2	12.1	12.1	11.9	11.8	11.7	11.6
20	11.6	11.8	11.9	12.0	12.1	12.2	12.1	12.0	11.9	11.8	11.6	11.6
25	11.6	11.8	11.9	12.0	12.1	12.2	12.1	12.0	11.9	11.7	11.6	11.6
30	11.7		11.9	12.1	12.1	12.2	12.1	12.0	11.9	11.7	11.6	11.6

Table 5 (continued)
DAY LENGTH IN HOURS

LATITUDE 10

DATE	JAN	FEB	MAR	APR	MAY	JUNE	JULY	AUG	SEPT	OCT	NOV	DEC
5	11.3	11.5	11.7	12.0	12.3	12.4	12.4	12.3	12.1	11.8	11.5	11.3
10	11.3	11.5	11.8	12.1	12.3	12.4	12.4	12.3	12.0	11.7	11.5	11.3
15	11.4	11.6	11.8	12.1	12.3	12.5	12.4	12.2	12.0	11.7	11.4	11.3
20	11.4	11.6	11.9	12.2	12.4	12.5	12.4	12.2	11.9	11.6	11.4	11.3
25	11.4	11.7	11.9	12.2	12.4	12.5	12.4	12.1	11.9	11.6	11.4	11.3
30	11.4		12.0	12.2	12.4	12.5	12.3	12.1	11.8	11.6	11.3	11.3

LATITUDE 15

DATE	JAN	FEB	MAR	APR	MAY	JUNE	JULY	AUG	SEPT	OCT	NOV	DEC
5	11.0	11.3	11.7	12.1	12.5	12.7	12.7	12.5	12.1	11.7	11.3	11.0
10	11.0	11.3	11.7	12.2	12.5	12.7	12.7	12.5	12.1	11.7	11.3	11.0
15	11.1	11.4	11.8	12.2	12.6	12.8	12.7	12.4	12.0	11.6	11.2	11.0
20	11.1	11.5	11.9	12.3	12.6	12.8	12.7	12.3	11.9	11.5	11.1	11.0
25	11.2	11.5	11.9	12.4	12.7	12.8	12.6	12.3	11.9	11.5	11.1	11.0
30	11.2		12.0	12.4	12.7	12.8	12.6	12.2	11.8	11.4	11.1	11.0

LATITUDE 20

DATE	JAN	FEB	MAR	APR	MAY	JUNE	JULY	AUG	SEPT	OCT	NOV	DEC
5	10.7	11.1	11.6	12.2	12.7	13.0	13.0	12.7	12.2	11.7	11.1	10.7
10	10.7	11.2	11.7	12.3	12.8	13.1	13.0	12.7	12.1	11.6	11.0	10.7
15	10.8	11.2	11.8	12.3	12.8	13.1	13.0	12.6	12.0	11.5	11.0	10.7
20	10.8	11.3	11.9	12.4	12.9	13.1	12.9	12.5	11.9	11.4	10.9	10.7
25	10.9	11.4	12.0	12.5	12.9	13.1	12.9	12.4	11.9	11.3	10.8	10.7
30	11.0		12.1	12.6	13.0	13.1	12.8	12.3	11.8	11.2	10.8	10.7

LATITUDE 25

DATE	JAN	FEB	MAR	APR	MAY	JUNE	JULY	AUG	SEPT	OCT	NOV	DEC
5	10.4	10.8	11.5	12.2	12.9	13.3	13.4	13.0	12.3	11.6	10.9	10.4
10	10.4	10.9	11.6	12.4	13.0	13.4	13.3	12.9	12.2	11.5	10.8	10.4
15	10.5	11.1	11.7	12.5	13.1	13.4	13.3	12.8	12.1	11.4	10.7	10.3
20	10.5	11.2	11.8	12.6	13.2	13.4	13.2	12.7	12.0	11.2	10.6	10.3
25	10.6	11.3	12.0	12.7	13.2	13.4	13.2	12.6	11.8	11.1	10.5	10.3
30	10.7		12.1	12.8	13.3	13.4	13.1	12.5	11.7	11.0	10.5	10.3

LATITUDE 30

DATE	JAN	FEB	MAR	APR	MAY	JUNE	JULY	AUG	SEPT	OCT	NOV	DEC
5	10.0	10.6	11.4	12.3	13.1	13.7	13.7	13.2	12.4	11.5	10.6	10.0
10	10.1	10.7	11.5	12.5	13.3	13.7	13.7	13.1	12.3	11.4	10.5	10.0
15	10.1	10.8	11.7	12.6	13.4	13.8	13.6	13.0	12.1	11.2	10.4	9.9
20	10.2	11.0	11.8	12.9	13.5	13.8	13.6	12.9	12.0	11.1	10.3	9.9
25	10.3	11.1	12.0	12.9	13.6	13.8	13.5	12.7	11.8	10.9	10.2	9.9
30	10.4		12.1	13.0	13.6	13.8	13.4	12.6	11.7	10.8	10.1	9.9

LATITUDE 35

DATE	JAN	FEB	MAR	APR	MAY	JUNE	JULY	AUG	SEPT	OCT	NOV	DEC
5	9.6	10.3	11.3	12.4	13.4	14.1	14.1	13.5	12.5	11.4	10.4	9.6
10	9.6	10.5	11.5	12.6	13.5	14.1	14.1	13.4	12.4	11.3	10.2	9.6
15	9.7	10.6	11.6	12.8	13.7	14.2	14.0	13.2	12.2	11.1	10.1	9.5
20	9.8	10.8	11.8	12.9	13.8	14.2	13.9	13.1	12.0	10.9	9.9	9.5
25	10.0	11.0	12.0	13.1	13.9	14.2	13.8	12.9	11.8	10.7	9.8	9.5
30	10.1		12.2	13.3	14.0	14.2	13.7	12.7	11.6	10.6	9.7	9.5

LATITUDE 40

DATE	JAN	FEB	MAR	APR	MAY	JUNE	JULY	AUG	SEPT	OCT	NOV	DEC
5	9.1	10.0	11.1	12.5	13.7	14.5	14.6	13.9	12.7	11.4	10.1	9.2
10	9.2	10.2	11.4	12.7	13.9	14.6	14.5	13.7	12.4	11.1	9.9	9.1
15	9.3	10.4	11.6	12.9	14.0	14.7	14.4	13.5	12.2	10.9	9.7	9.0
20	9.4	10.6	11.8	13.1	14.2	14.7	14.3	13.3	12.0	10.7	9.5	9.0
25	9.6	10.8	12.0	13.3	14.3	14.7	14.2	13.1	11.8	10.5	9.4	9.0
30	9.7		12.2	13.5	14.4	14.6	14.0	12.9	11.6	10.3	9.3	9.0

Table 5 (continued)
DAY LENGTH IN HOURS

LATITUDE 42

DATE	JAN	FEB	MAR	APR	MAY	JUNE	JULY	AUG	SEPT	OCT	NOV	DEC
5	8.9	9.8	11.1	12.5	13.8	14.7	14.8	14.0	12.7	11.3	9.9	8.9
10	9.0	10.0	11.3	12.8	14.0	14.8	14.7	13.8	12.5	11.1	9.7	8.9
15	9.1	10.2	11.6	13.0	14.2	14.9	14.6	13.6	12.3	10.8	9.5	8.8
20	9.2	10.5	11.8	13.2	14.4	14.9	14.5	13.4	12.0	10.6	9.3	8.8
25	9.4	10.7	12.0	13.4	14.5	14.9	14.4	13.2	11.8	10.4	9.2	8.8
30	9.6		12.3	13.6	14.6	14.9	14.2	13.0	11.6	10.2	9.1	8.8

LATITUDE 44

DATE	JAN	FEB	MAR	APR	MAY	JUNE	JULY	AUG	SEPT	OCT	NOV	DEC
5	8.6	9.7	11.0	12.6	14.0	15.0	15.0	14.2	12.8	11.3	9.8	8.7
10	8.7	9.9	11.3	12.8	14.2	15.0	14.9	13.9	12.5	11.0	9.6	8.6
15	8.9	10.1	11.5	13.1	14.4	15.1	14.8	13.7	12.3	10.8	9.3	8.5
20	9.0	10.4	11.8	13.3	14.5	15.1	14.7	13.5	12.0	10.5	9.2	8.5
25	9.2	10.6	12.0	13.5	14.7	15.1	14.5	13.3	11.8	10.3	9.0	8.5
30	9.4		12.3	13.8	14.8	15.1	14.4	13.1	11.5	10.0	8.8	8.6

LATITUDE 46

DATE	JAN	FEB	MAR	APR	MAY	JUNE	JULY	AUG	SEPT	OCT	NOV	DEC
5	8.4	9.5	11.0	12.6	14.1	15.2	15.3	14.3	12.8	11.2	9.6	8.5
10	8.5	9.7	11.2	12.9	14.4	15.3	15.2	14.1	12.6	11.0	9.4	8.4
15	8.6	10.0	11.5	13.2	14.6	15.3	15.1	13.9	12.3	10.7	9.1	8.3
20	8.8	10.3	11.8	13.4	14.7	15.4	14.9	13.6	12.0	10.4	8.9	8.3
25	9.0	10.5	12.0	13.7	14.9	15.4	14.7	13.4	11.8	10.2	8.8	8.3
30	9.2		12.3	13.9	15.1	15.3	14.6	13.1	11.5	9.9	8.6	8.3

LATITUDE 48

DATE	JAN	FEB	MAR	APR	MAY	JUNE	JULY	AUG	SEPT	OCT	NOV	DEC
5	8.1	9.3	10.9	12.7	14.3	15.4	15.5	14.5	12.9	11.2	9.4	8.2
10	8.2	9.6	11.2	13.0	14.5	15.5	15.4	14.3	12.6	10.9	9.2	8.1
15	8.4	9.9	11.5	13.3	14.8	15.6	15.3	14.0	12.3	10.6	8.9	8.0
20	8.6	10.1	11.8	13.5	15.0	15.6	15.1	13.8	12.0	10.3	8.7	8.0
25	8.8	10.4	12.1	13.8	15.1	15.6	15.0	13.5	11.8	10.0	8.5	8.0
30	9.0		12.3	14.1	15.3	15.6	14.8	13.2	11.5	9.8	8.3	8.0

LATITUDE 50

DATE	JAN	FEB	MAR	APR	MAY	JUNE	JULY	AUG	SEPT	OCT	NOV	DEC
5	7.8	9.1	10.8	12.7	14.5	15.7	15.8	14.7	13.0	11.1	9.2	7.9
10	7.9	9.4	11.1	13.1	14.7	15.8	15.7	14.4	12.7	10.8	9.0	7.8
15	8.1	9.7	11.4	13.4	15.0	15.9	15.6	14.2	12.4	10.5	8.7	7.7
20	8.3	10.0	11.8	13.6	15.2	15.9	15.4	13.9	12.1	10.2	8.5	7.6
25	8.5	10.3	12.1	13.9	15.4	15.9	15.2	13.6	11.7	9.9	8.2	7.6
30	8.8		12.4	14.2	15.6	15.9	15.0	13.3	11.4	9.6	8.0	7.7

LATITUDE 52

DATE	JAN	FEB	MAR	APR	MAY	JUNE	JULY	AUG	SEPT	OCT	NOV	DEC
5	7.4	8.9	10.7	12.8	14.7	16.0	16.1	14.9	13.0	11.1	9.0	7.6
10	7.6	9.2	11.1	13.1	15.0	16.1	16.0	14.6	12.7	10.7	8.7	7.4
15	7.8	9.5	11.4	13.5	15.2	16.2	15.9	14.3	12.4	10.4	8.4	7.3
20	8.0	9.9	11.7	13.8	15.4	16.3	15.7	14.0	12.1	10.1	8.2	7.3
25	8.3	10.2	12.1	14.1	15.7	16.3	15.5	13.7	11.7	9.7	7.9	7.3
30	8.5		12.4	14.4	15.9	16.2	15.2	13.4	11.4	9.4	7.7	7.3

LATITUDE 54

DATE	JAN	FEB	MAR	APR	MAY	JUNE	JULY	AUG	SEPT	OCT	NOV	DEC
5	7.1	8.6	10.6	12.9	14.9	16.4	16.5	15.1	13.1	11.0	8.8	7.2
10	7.2	9.0	11.0	13.2	15.2	16.5	16.3	14.8	12.8	10.6	8.5	7.0
15	7.4	9.3	11.4	13.6	15.5	16.6	16.2	14.5	12.4	10.3	8.2	6.9
20	7.7	9.7	11.7	13.9	15.7	16.6	16.0	14.2	12.1	9.9	7.9	6.9
25	8.0	10.1	12.1	14.3	16.0	16.6	15.7	13.9	11.7	9.6	7.6	6.9
30	8.3		12.4	14.6	16.2	16.6	15.5	13.5	11.4	9.2	7.4	6.9

Table 5 (continued)
DAY LENGTH IN HOURS

LATITUDE 56

DATE	JAN	FEB	MAR	APR	MAY	JUNE	JULY	AUG	SEPT	OCT	NOV	DEC
5	6.6	8.4	10.5	12.9	15.1	16.8	16.9	15.4	13.2	10.9	8.5	6.7
10	6.8	8.7	10.9	13.3	15.5	16.9	16.7	15.1	12.8	10.5	8.2	6.6
15	7.0	9.1	11.3	13.7	15.8	17.0	16.5	14.7	12.5	10.1	7.8	6.4
20	7.3	9.5	11.7	14.1	16.1	17.1	16.3	14.4	12.1	9.8	7.5	6.4
25	7.6	9.9	12.1	14.4	16.3	17.0	16.1	14.0	11.7	9.4	7.2	6.4
30	7.9		12.5	14.8	16.6	17.0	15.8	13.7	11.3	9.0	7.0	6.5

LATITUDE 58

DATE	JAN	FEB	MAR	APR	MAY	JUNE	JULY	AUG	SEPT	OCT	NOV	DEC
5	6.1	8.1	10.4	13.0	15.4	17.2	17.4	15.7	13.3	10.8	8.3	6.2
10	6.3	8.5	10.8	13.4	15.8	17.4	17.2	15.4	12.9	10.4	7.9	6.0
15	6.6	8.9	11.3	13.8	16.1	17.5	17.0	15.0	12.5	10.0	7.5	5.9
20	6.9	9.3	11.7	14.2	16.4	17.5	16.7	14.6	12.1	9.6	7.1	5.8
25	7.2	9.7	12.1	14.6	16.7	17.5	16.4	14.2	11.7	9.2	6.8	5.8
30	7.6		12.5	15.0	17.0	17.5	16.1	13.8	11.3	8.7	6.5	5.9

LATITUDE 60

DATE	JAN	FEB	MAR	APR	MAY	JUNE	JULY	AUG	SEPT	OCT	NOV	DEC
5	5.4	7.7	10.3	13.1	15.7	17.8	17.9	16.1	13.4	10.7	7.9	5.6
10	5.7	8.2	10.8	13.6	16.1	17.9	17.7	15.7	13.0	10.3	7.5	5.4
15	6.0	8.6	11.2	14.0	16.5	18.1	17.5	15.2	12.5	9.8	7.1	5.2
20	6.4	9.1	11.7	14.4	16.9	18.1	17.2	14.8	12.1	9.4	6.6	5.1
25	6.8	9.5	12.1	14.9	17.2	18.1	16.9	14.4	11.7	8.9	6.3	5.1
30	7.2		12.6	15.3	17.5	18.1	16.5	14.0	11.2	8.5	5.9	5.2

LATITUDE 61

DATE	JAN	FEB	MAR	APR	MAY	JUNE	JULY	AUG	SEPT	OCT	NOV	DEC
5	5.1	7.5	10.2	13.2	15.9	18.1	18.2	16.2	13.5	10.7	7.7	5.3
10	5.4	8.0	10.7	13.6	16.3	18.3	18.0	15.8	13.0	10.2	7.3	5.0
15	5.7	8.5	11.2	14.1	16.7	18.4	17.8	15.4	12.6	9.7	6.8	4.8
20	6.1	9.0	11.7	14.6	17.1	18.5	17.4	14.9	12.1	9.3	6.4	4.7
25	6.5	9.4	12.1	15.0	17.4	18.5	17.1	14.5	11.6	8.8	6.0	4.7
30	7.0		12.6	15.5	17.8	18.4	16.7	14.0	11.2	8.3	5.6	4.8

LATITUDE 62

DATE	JAN	FEB	MAR	APR	MAY	JUNE	JULY	AUG	SEPT	OCT	NOV	DEC
5	4.6	7.3	10.1	13.2	16.1	18.4	18.6	16.4	13.6	10.6	7.5	4.9
10	5.0	7.8	10.6	13.7	16.5	18.6	18.4	16.0	13.1	10.1	7.0	4.6
15	5.3	8.3	11.1	14.2	17.0	18.8	18.1	15.5	12.6	9.6	6.5	4.4
20	5.8	8.8	11.6	14.7	17.4	18.9	17.7	15.1	12.1	9.1	6.1	4.3
25	6.2	9.3	12.1	15.1	17.7	18.8	17.4	14.6	11.6	8.6	5.6	4.3
30	6.7		12.6	15.6	18.1	18.8	17.0	14.1	11.1	8.1	5.2	4.4

LATITUDE 63

DATE	JAN	FEB	MAR	APR	MAY	JUNE	JULY	AUG	SEPT	OCT	NOV	DEC
5	4.1	7.1	10.1	13.3	16.3	18.8	19.0	16.7	13.6	10.6	7.3	4.4
10	4.5	7.6	10.6	13.8	16.7	19.0	18.7	16.2	13.1	10.1	6.8	4.1
15	4.9	8.1	11.1	14.3	17.2	19.2	18.4	15.7	12.6	9.5	6.3	3.8
20	5.4	8.7	11.6	14.8	17.6	19.3	18.1	15.2	12.1	9.0	5.7	3.7
25	5.9	9.2	12.1	15.3	18.1	19.3	17.7	14.7	11.6	8.5	5.3	3.7
30	6.4		12.7	15.8	18.4	19.2	17.2	14.2	11.1	8.0	4.8	3.8

LATITUDE 64

DATE	JAN	FEB	MAR	APR	MAY	JUNE	JULY	AUG	SEPT	OCT	NOV	DEC
5	3.5	6.8	10.0	13.3	16.5	19.2	19.5	16.9	13.7	10.5	7.1	3.8
10	4.0	7.4	10.5	13.9	17.0	19.5	19.2	16.4	13.2	10.0	6.5	3.4
15	4.5	8.0	11.1	14.4	17.5	19.7	18.8	15.9	12.7	9.4	5.9	3.1
20	5.0	8.5	11.6	14.9	18.0	19.8	18.4	15.4	12.1	8.9	5.4	3.0
25	5.5	9.1	12.2	15.5	18.4	19.8	18.0	14.9	11.6	8.3	4.8	3.0
30	6.1		12.7	16.0	18.8	19.7	17.5	14.3	11.1	7.8	4.3	3.2

Table 5 (continued)
DAY LENGTH IN HOURS

LATITUDE 65

DATE	JAN	FEB	MAR	APR	MAY	JUNE	JULY	AUG	SEPT	OCT	NOV	DEC
5	2.8	6.5	9.9	13.4	16.7	19.8	20.0	17.2	13.8	10.5	6.8	3.2
10	3.3	7.1	10.5	14.0	17.3	20.1	19.7	16.6	13.2	9.9	6.2	2.6
15	3.9	7.7	11.0	14.5	17.8	20.3	19.3	16.1	12.7	9.3	5.6	2.2
20	4.5	8.4	11.6	15.1	18.3	20.4	18.8	15.6	12.1	8.7	5.0	2.0
25	5.1	8.9	12.2	15.6	18.8	20.4	18.3	15.0	11.6	8.1	4.3	2.0
30	5.8		12.7	16.2	19.3	20.3	17.8	14.5	11.0	7.5	3.7	2.2

Table 6
SYMBOLS USED IN THE ASTRONOMICAL EQUATIONS FOR SUNRISE AND SUNSET

a	= solar altitude (measured from center of the sun)
$a_o, a_1...b_1, b_2$, etc.	= constants in an harmonic equation
D	= day of year (1 January = 1)
E	= equation of time
h	= solar hour angle (angular distance of sun from meridian of the observer)
I_o	= solar energy reaching unit area of earth in unit time
I_h	= total solar energy reaching unit area of earth in 1 hr
I_t	= solar energy reaching unit area of earth in time interval t
J_o	= solar constant
L	= day length from sunrise to sunset
M	= actual time meridian for the standard time belt in which the station is located
Q_o	= vertical component of daily total solar energy at the surface of the earth in absence of an atmosphere (or at the top of the atmosphere)
r	= radius vector of the orbit of the earth around sun (expressed as a fraction of the mean radius)
S	= solar time
S'_a	= true solar time when the sun is at an elevation, a, in morning
S''_a	= true solar time when the sun is at an elevation, a, in afternoon
t	= time
t_s	= standard time
y	= dependent variable in harmonic equation
z	= angular distance of sun from zenith (z = 90 − a),
λ	= longitude of the station
δ	= declination of the sun
Θ	= $2\pi D/365$ (approximately proportional to the solar longitude)
φ	= latitude of the observer

Table 7
ASTRONOMICAL EQUATIONS FOR SUNRISE AND SUNSET AND CALCULATION OF SOLAR ENERGY

The angular elevation, a, of the sun at any time of the day (hour angle, h) for any day of the year and for any latitude is given by the relationship:

$$\sin a = \sin \phi \sin \delta + \cos \phi \cos \delta \cos h \tag{1}$$

In terms of angular distance from the zenith, the equation is

$$\cos z = \sin \phi \sin \delta + \cos \phi \cos \delta \cos h \tag{2}$$

Rearranging terms in Equation 1

$$h = \cos^{-1} \frac{\sin a - \sin \phi \sin \delta}{\cos \phi \cos \delta} \tag{3}$$

Now h = 0 at solar noon. Let the duration of time between any solar hour angle, h, and solar noon = 1/2 L. Then

$$1/2\ L = \frac{h}{2\pi} \times 24 \text{ (hours)}$$

Because of symmetry, the duration of time from solar angle, h, before solar noon to the same angle after noon is

$$L = \frac{24h}{\pi} = 7.639\ h \tag{4}$$

Therefore

$$L = 7.639 \cos^{-1} \frac{\sin a - \sin \phi \sin \delta}{\cos \phi \cos \delta} \tag{5}$$

Strictly speaking, the above symmetry around solar noon is not exact because of the slight change in the solar declination during the course of the interval involved. This is very small, however, and for our purpose can be ignored.

In order to avoid unrealistic values of day length resulting from calculations when the sun is continuously above or below the horizon (at high latitudes) it is convenient to set

$$L = 24 \text{ when } \frac{\sin a - \sin \phi \sin \delta}{\cos \phi \cos \delta} \geq +1$$

$$L = 0 \text{ when } \frac{\sin a - \sin \phi \sin \delta}{\cos \phi \cos \delta} \geq -1$$

Local standard time is related to true solar time at a specific site by

$$t_s = S + (\lambda - M)\frac{4}{60} - E \tag{6}$$

From Equation 4 the true solar time when the sun is at a given hour angle corresponding to a given elevation in the morning is

$$S'_a = 12 - 1/2\ L$$

$$= 12 - \frac{24}{2\pi} h \tag{7}$$

Table 7 (continued)
ASTRONOMICAL EQUATIONS FOR SUNRISE AND SUNSET AND CALCULATION OF SOLAR ENERGY

Combining Equations 3, 7, and 8 with 6, the time at which the sun is at any solar angle in the morning and afternoon may be calculated for any day of the year and for any location.

The calculation of values for E and δ is discussed below in "Harmonic Analysis". It is assumed that observed sunrise occurs when the upper limb of the sun just becomes visible above the horizon. The semidiameter of the sun is 16' of arc and refraction due to atmosphere is about 34' of arc. Therefore, when the upper limb of the sun just becomes visible, the center of the sun is actually below the horizon (i.e., a = −50' of arc). This same correction applies at sunset. For the duration of light beginning and ending with twilight, no correction is made for semidiameter or atmospheric refraction.

SOLAR ENERGY (Q_o)

The basic formula relating to solar energy received on a horizontal surface on the earth, assuming no influence of the atmosphere, involved three factors: the solar constant, J_o; the radius vector, r, of the orbit of the earth around the sun; and the zenith angle of the sun, z, or its elevation, a, at the site and time in question. These are related by

$$\frac{dI_t}{dt} = \frac{J_o}{r^2} \cos z \quad (8)$$

$$\frac{dI_t}{dt} = \frac{J_o}{r^2} \sin a \quad (9)$$

where a = 90 − z.

Integration gives the total energy for any period t_1 to t_2:

$$I_t = \int_{t_1}^{t_2} \frac{J_o}{r^2} \cos z \, dt \quad (10)$$

For our purpose it is sufficiently accurate to calculate

$$I_o = \frac{J_o}{r^2} \cos z \quad (11)$$

at each hour of the day, working both ways from true solar noon, and find the average for an hour (or fraction of an hour at sunrise or sunset)

$$I_h = \frac{60 J_o}{2 r^2} (\cos z_h + \cos z_{h \pm 1}) \quad (12)$$

These hourly values are added together to obtain the daily total (Q_o).

Symmetry around solar noon is assumed; it is therefore necessary to calculate hourly values for only half a day. The hourly energy will be the same for hours ending at 1300 and 1200 hours, at 1400 and 1100 hours. As the first and last periods of the day (just after sunrise and just prior to sunset) are not complete hours, it is necessary to calculate the times of sunrise and sunset (solar time) in order to determine the fractional hours at the beginning and end of the day, as described earlier (Equation 7). In this case, atmospheric refraction is disregarded and solar elevation is measured from the center of the sun and therefore a = 0 at sunrise and sunset. It is assumed that I_o = 0 at these times.

The solar constant, J_o, is assumed to be 2.00 g cal cm^{-2} min^{-1}. Values of r were calculated for every day of the year by means of a harmonic equation.

The zenith angle of the sun can be calculated for each solar hour by Equation 2 where the hour angle, h, for a given solar time, S, is:

$$h = \frac{2\pi (S - 12.0)}{24} \text{ (radians)}$$

$$= 0.2618 (S - 12.0) \quad (13)$$

Note that at S_{12} (solar noon), $h_{12} = 0$.

Table 7 (continued)
ASTRONOMICAL EQUATIONS FOR SUNRISE AND SUNSET AND CALCULATION OF SOLAR ENERGY

HARMONIC ANALYSIS

The use of Equations 1, 3, 6, and 12 involves tables of daily values for solar-ephemeric data (declination, the equation of time, and the radius vector of the earth). Since *Smithsonian Tables* (see Reference 2, p. 495) show values for every 4th day, a method for systematic interpretation is desirable.

As these values are all harmonic functions of the day of the year, it was decided to determine the coefficients in the harmonic equation

$$Y = a_0 + a_1 \sin \theta + b_1 \cos \theta + a_2 \sin 2\theta + b_2 \cos 2\theta + ... \quad (14)$$

where Y might be any one of the three parameters mentioned above and is an angle corresponding to the day of the year, D, counting January 1 as one. Thus,

$$\theta = \frac{D}{365} 2\pi = 0.01721 \, D \text{ (radians)} \quad (15)$$

Values of Y for declination, equation of time, and radius vector, respectively, were obtained from *Smithsonian Tables* for about every 10th day of the year starting at D = 5 for January 5. Corresponding values of the sine and cosine of θ, 2θ, 3θ, and so on were calculated. The regression coefficients in Equation 14 for each of these three parameters were calculated using partial regression analysis[1] (Table 8).

Table 8
REGRESSION COEFFICIENTS IN THE HARMONIC EQUATION 14 FOR CALCULATING SOLAR EPHEMERIC PARAMETERS FROM DAY OF YEAR

Regression coefficient	Declination	Equation of time	Radius vector
a_0	+0.3964E−00	+0.2733E−02	+0.1000E+01
a_1	+0.3631E+01	−0.7343E+01	−0.9464E−03
a_2	+0.3838E−01	−0.9470E+01	—
a_3	+0.7659E−01	−0.3289E−00	−0.2917E−04
a_4	—	−0.1935E−00	—
b_1	−0.2297E+02	+0.5519E−00	−0.1671E−01
b_2	−0.3885E−00	−0.3020E+01	−0.1489E−03
b_3	−0.1587E−00	−0.7581E−01	—
b_4	−0.1021E−01	−0.1245E−00	−0.3438E−04

Table 9
INPUT CARD FOR PROGRAM 1044-0437
ASTROMETEOROLOGICAL ESTIMATOR INPUT CARDS (COLUMNS)

1—13	Job and program number
15—30	Station identification in alphanumeric
31	Sign of station latitude; N = +; S = −
32—33	Station latitude; whole degrees
35—36	Station latitude; whole minutes
38	Sign of station longitude; W = +; E = −
39—41	Station longitude; whole degrees
43—44	Station longitude; whole minutes
46	Sign of solar elevation; above horizon = +; below horizon = −
47—48	Solar elevation; whole degrees
50—51	Solar elevation; whole minutes
53—55	Meridian of standard time zone; whole degrees
57—59	Day of year for beginning calculations; 1 January = 000
61—63	Day of year for ending calculations; 31 December = 365
65—66	Increment in whole days for calculations
68—69	Beginning date for accumulation; day of month (if zero there will be no output)
70	Blank, slash, or dash
71—72	Beginning date for accumulation; month number
74—75	Ending date for accumulation; day of month
76	Blank, slash, or dash
77—78	Ending date for accumulation; month
80	Code for table output
	1 — list of day lengths only (Table 5)
	2 — table of day lengths only (Table 6)
	3 — both list and table
	4 — tables of times only (Table 7)
	5 — tables of day length and times
	6 — list and all tables
	7 — calculate and list Q_o (Table 4)

Table 10
FORTRAN PROGRAM 1044-04-37 ASTROMETEOROLOGICAL ESTIMATOR

```
      DIMENSION IUS(365),ENERGY(12)
      DIMENSION IS(31,12), JEARS(31,12),NAME(8),WISE(12),MONDAY(1
     11),DCLNTN(365), IR(31,12),IQUAT(365),TITLE(5)
      DIMENSION A1(3,5),B(3,4),SUM(3)
C
C             STATEMENT FUNCTION FOR ARC COSINE
C
      ACOS(X)=ATAN(-1.*(SQRT(1.-X*X)/X))
C
      OPEN(UNIT=1,NAME='ASTRO.DAT',STATUS='OLD')
      OPEN(UNIT=2,NAME='ASTRO.OUT',STATUS='NEW',
     *CARRIAGE CONTROL='LIST')
C              INITIALIZE FOR ENTIRE RUN.
      DATA A1/.3964, .002733, 1.0, 3.631, -7.343, -.0009464, .03838,
     *-9.470, 0., .07659, -.3289, -.00002917, 0.0, -.1935, 0.0/
      DATA B/-22.97, .5519, -.01671, -.3885, -3.020, -.0001489,
     *-.1587, -.07581, 0.0, -.01021, -.1245, -.00003438/
      DO 46 I=1,3
C     WRITE(2,*) I,(A1(I,J),J=1,5)
C     WRITE(2,*) I,(B(I,J),J=1,4)
   46 CONTINUE
      MONDAY(1)=31
      MONDAY(2)=59
      MONDAY(3)=90
      MONDAY(4)=120
      MONDAY(5)=151
      MONDAY(6)=181
      MONDAY(7)=212
      MONDAY(8)=243
```

Table 10 (continued)
FORTRAN PROGRAM 1044-04-37 ASTROMETEOROLOGICAL ESTIMATOR

```
      MONDAY(9)=273
      MONDAY(10)=304
      MONDAY(11)=334
C         FORTRAN PROGRAM 1044-04-32 HARMONIC ESTIMATOR FOR SOLAR EPHEMERIC DATA
      DO 15 I =1,365
      DAY=I
      THETA=DAY/365.*6.2832
      DO 139 J=1,3
  139 SUM(J)=A1(J,1)+A1(J,2)*SIN(THETA)+B(J,1)*COS(THETA)
     *        +A1(J,3)*SIN(2*THETA)+B(J,2)*COS(2*THETA)
     *        +A1(J,4)*SIN(3*THETA)+B(J,3)*COS(3*THETA)
     *        +A1(J,5)*SIN(4*THETA)+B(J,4)*COS(4*THETA)
      IQUAT(I)=SUM(2)*100.0
      IUS(I)=SUM(3)*1000.
      DCLNTN(I)=SUM(1)*.017457
      FLOAT=I
C     WRITE(2,*) I,(SUM(J),J=1,3)
   15 CONTINUE
   10 DO 5 I=1,12
      DO 5 J=1,31
      IS(J,I)=0
      IR(J,I)=0
    5 JEARS(J,I)=0
      SWICH1=0.0
      SWICH2=0.0
      SWICH3=0.0
      SWICH4=0.0
      SUML=0.0
C                     READ RUN-CONTROL CARD (I-CARD-1)
      READ(1,2,END=99)TITLE,NAME,ISTAL,IDECL,ISTL,IDCL,A,
     *IELIVT,IELI,IZON,IFIRST,LAST,JUMP,MIN,MINMON,MAX,MAXMON,IBOSS
C                     WRITE ,RUN-CONTROL CARD FOR IDENTIFICATION (TABLE-3)
      WRITE(2,2)TITLE,NAME,ISTAL,IDECL,ISTL,IDCL,A,
     *IELIVT,IELI,IZON,IFIRST,LAST,JUMP,MIN,MINMON,MAX,MAXMON,IBOSS
    2 FORMAT(4A3,9A2,2I3,I5,I3,1X,A1,I2,I3,3I4,I3,2(I3,1X,I2),I2)
      WRITE(2,2)
C                     INITIALIZING CONSTANTS FOR RUN.
      STALAT=ISTAL+IDECL/60.
      STLONG=ISTL+IDCL/60.
      ZONACT=IZON
      SITN=1.0
      IF(A-0.200)430,420,420
  420 SITN=-1.
  430 ELIVTN=(IELIVT+IELI/60.)*SITN
      IF(MINMON-1)320,320,310
  310 MIN=MONDAY(MINMON-1)+MIN
  320 IF(MAXMON-1)340,340,330
  330 MAX=MONDAY(MAXMON-1)+MAX
C     SELECTION OF COMBINATIONS FOR OUTPUT AND SETTINGS OF SWITCHES.
  340 GO TO (360,370,350,400,390,380,440),IBOSS
  350 SWICH2 =1
  360 SWICH1 =1
      GO TO 410
  370 SWICH2=1
      GO TO 410
  440 SWICH3=1
C                     HEADING FOR TABLE-4.
      WRITE(2,917)
  917 FORMAT(12HSOLAR ENERGY,24X,15HFOR HOUR ENDING /72X,5HDAILY/17X,40H
     1DAY  12  11  10   9   8   7   6   5   4 ,20H  3   2   1   TOTAL/
     218X,40HNO  13  14  15  16  17  18  19  20  21 ,20H22  23  24    E
     3NERGY/)
      GO TO 410
  380 SWICH1=1
  390 SWICH2=1
  400 SWICH4=1
  410 ELEVTN=ELIVTN*.017457
      STALAT=STALAT*.017457
C                     DAILY CALCULATIONS OF DAY LENGTH.
      BIGM=4.0*(STLONG-ZONACT)*.01667
      IFIX=FLOAT*10.+.5
      FL=IFIX
      FL=FL*.1
      DO 520 M=1,12
  520 ENERGY(M)=ENERGY(M)*10.+.5
      TOTAL=TOTAL+.05
```

Table 10 (continued)
FORTRAN PROGRAM 1044-04-37 ASTROMETEOROLOGICAL ESTIMATOR

```
C                     WRITE  HOURLY AND DAILY ENERGY (Q0) (TABLE-4)
      WRITE(2,8)NAME,IDAY,ENERGY,TOTAL
    8 FORMAT (8A2,I3,I4,F8.1)
   29 IFIX=HOURS*10.+.5
      HOURS=IFIX
      HOURS=HOURS/10.
   32 IF(IDAY-MAX)35,35,45
   35 IF(IDAY-MIN)45,45,40
   40 SUML=HOURS+SUML
C                     WRITE DAY LENGTH LIST (TABLE-5)
   45 IF (SWICH1)70,70,50
    3 FORMAT(8A2,3X,I4,3X,F6.2   )
   50 WRITE(2,3)NAME,IDAY,FLOAT
   70 DO 80 I=1,11
      IF(IDAY-MONDAY(I))90,90,80
   80 CONTINUE
      GO TO 100
   90 IF(I-1)100,95,100
   95 IDENT=IDAY
      GO TO 105
  100 IDENT=IDAY-MONDAY(I-1)
  105 JEARS(IDENT,I)=HOURS*10.
C          CALCULATE LOCAL STANDARD TIME OF SUNRISE AND SUNSET.
      IF (SWICH4)110,110,108
  108 IF(HOURS) 110,110,107
  107 IF(HOURS-24.0)111,110,110
  111 EQUAT=IQUAT(IDAY)
      CALA=12.-EQUAT*0.0001667+BIGM
      CALB=0.5*FLOAT
      CALC=CALA-CALB
      IFIX=CALC
      FLOAT=IFIX
      IF(CALC-FLOAT-.991666)113,113,112
  112 IR(IDENT,I)=(FLOAT+1.)*100.0
      GO TO 114
  113 IR(IDENT,I)=(((CALC-FLOAT)*0.600)+FLOAT+.005)*100.
  114 CALC=CALA+CALB
      IFIX=CALC
      FLOAT=IFIX
      IF(CALC-FLOAT-.991666)116,116,115
  115 IS(IDENT,I)=(FLOAT+1.)*100.0
      GO TO 110
  116 IS(IDENT,I)=(((CALC-FLOAT)*0.600)+FLOAT+0.005)*100.
  110 CONTINUE
C                     WRITE DAY LENGTH TABLE (TABLE-6)
      IF (SWICH2)137,137,120
  120 K=5
      WRITE(2,909)
      WRITE(2,901)NAME
      WRITE(2,905)A,IELIVT,IELI
      WRITE(2,902)
      DO 130 I=1,31
      DO 132 J=1,12
      FLOAT=JEARS(I,J)
  132 WISE(J)=FLOAT*.1
      WRITE(2,4)I,WISE
    4 FORMAT(I4,3X,12F6.1)
      IF(I-K)130,135,130
  135 K=K+5
      WRITE(2,7)
  130 CONTINUE
C             WRITE SUNRISE AND SUNSET TABLES (TABLE-7)
  137 IF(SWICH4)250,250,195
  195 K=5
      WRITE(2,909)
      WRITE(2,901)NAME
      WRITE(2,906)A,IELIVT,IELI
      WRITE(2,908)
      WRITE(2,902)
      DO 200 I=1,31
      DO 205 J=1,12
      FLOAT=IR(I,J)
  205 WISE(J)=FLOAT*.01
      WRITE(2,904)I, WISE
      IF(I-K) 200,210,200
```

Table 10 (continued)
FORTRAN PROGRAM 1044-04-37 ASTROMETEOROLOGICAL ESTIMATOR

```
  210 K=K+5
      WRITE(2,7)
  200 CONTINUE
      WRITE(2,909)
      K=5
      WRITE(2,901)NAME
      WRITE(2,908)
      WRITE(2,907)A,IELIVT,IELI
      WRITE(2,902)
      DO 220 I=1,31
      DO 225 J=1,12
      FLOAT=IS(I,J)
  225 WISE(J)=FLOAT*.01
      WRITE(2,904)I, WISE
      IF(I-K) 220,230,220
  230 K=K+5
      WRITE(2,7)
  220 CONTINUE
  250 IF(MIN)150,150,140
  140 WRITE(2,909)
C           WRITE ACCUMULATED HOURS OF DAY LENGTH (TABLE-8)
      WRITE(2,7) NAME,MIN,MAX,JUMP,SUML
    7 FORMAT(8A2,14HTOTAL FROM DAY ,I4,3X,6HTO DAY ,I4,3X,2HBY,I4,2X,8HD
     1AYS IS ,F7.1,//)
  150 CONTINUE
  901 FORMAT(33X,8A2,16X,15HSOLAR ELEVATION /)
  902 FORMAT(40H DATE      JAN    FEB    MAR    APR    MAY  J ,39HUNE    JULY
     1  AUG   SEPT    OCT    NOV    DEC  /)
  904 FORMAT(I4,3X,12F6.2)
  905 FORMAT(32X,18HDAY LENGTH IN HOURS  ,20X,A1,2I3 //)
  906 FORMAT(32X,15HTIME OF SUNRISE  ,23X,A1,2I3 //)
  907 FORMAT(32X,14HTIME OF SUNSET  ,24X,A1,2I3 //)
  908 FORMAT(29X,20HIN HOURS AND MINUTES /  )
  909 FORMAT(79X1H+)
      GO TO 10
   99 CLOSE(UNIT=1,DISP='KEEP')
      CLOSE(UNIT=2,DISP='KEEP')
      STOP
      END
      DO 110 IDAY= IFISRT,LAST,JUMP
      DECLIN= DCLNTN(IDAY)
      CALA=SIN(STALAT)
      CALB=SIN(DECLIN)
      BIGA=CALA*CALB
      CALB=SIN(ELEVTN)
      CALA=BIGA-CALB
      CALB=COS(STALAT)
      CALC=COS(DECLIN)
      BIGB=CALB*CALC
      CALA=CALA/BIGB
      IF(CALA+1.) 21,21,22
   21 HOURS=0.0
      GO TO 27
   22 IF(CALA-1.)24,23,23
   23 HOURS=24.0
      GO TO 27
   24 HOURS=ACOS(CALA)
      HOURS=HOURS*7.639
      IF(HOURS)25,27,27
   25 HOURS=HOURS+24.0
   27 FLOAT=HOURS
      IF(SWICH3)29,29,480
C                  CALCULATE SOLAR TIME OF SUNSET.
  480 TRUESS=0.5*FLOAT+12.0
      DO 485 IUSE=1,12
C                  CALCULATE HOURLY ENERGY.
  485 ENERGY(IUSE)=0.0
      TOTAL=0.0
      NUMBER=1
C          SOLTME= SOLAR TIME AT END OF HOURLY INTERVAL.
      SOLTME=13.0
      RAD=IUS(IDAY)
      RAD=RAD*.001
C      ZT IS SOLAR ZENITH ANGLE AT BEGINNING OF HOURLY INTERVAL.
      COSZT=BIGB*COS(.2618*(SOLTME-13.0))+BIGA
```

Table 10 (continued)
FORTRAN PROGRAM 1044-04-37 ASTROMETEOROLOGICAL ESTIMATOR

```
  490 IF(TRUESS-SOLTME)510,510,500
  500 H=.2618*(SOLTME-12.0)
C                 Z IS SOLAR ZENITH ANGLE AT END OF HOURLY INTERVAL.
      COSZ=BIGA+BIGB*COS(H)
      ENERGY(NUMBER)=(COSZ+COSZT)/(RAD*RAD)*60.
      COSZT=COSZ
      TOTAL=TOTAL+ENERGY(NUMBER)*2.
      NUMBER=NUMBER+1
      SOLTME=SOLTME+1.
      GO TO 490
C                 ENERGY IN FRACTIONAL PERIODS ENDING AFTER AND BEFORE ZT=0.
  510 ENERGY(NUMBER)=60.*COSZT/(RAD*RAD)*(TRUESS-
     1SOLTME+1.)
      TOTAL=TOTAL+ENERGY(NUMBER)*2.
```

Table 11
CLASSIFICATION OF IMPORTANT CROPS FOR PHOTOPERIOD REACTION AND FOR RATES OF PHOTOSYNTHESIS

Crop (scientific name)	Photoperiod reaction	Type of photosynthesis	Rate of photosynthesis[a] (mg CO_2 dm^{-2} min^{-1})	Crop growth rate[b] (g m^{-2} day^{-1})
Alfalfa (*Medicago sativa* L.)	LD	C3		23
Barley (*Hordeum vulgare* L.)	LD	C3		21
Broad bean (*Vicia faba*)	LD	C3		
Buckwheat (*Fagopyrum* sp.)	LD	C3		
Cassava (*Manihot esculenta* Crantz)	SD	C3		
Chickpea (*Cicer arietinum* L.)	LD			
Clovers (*Trifolium* spp.)	LD	C3		
Common bean (*Phaseolus vulgaris* L.)	SD	C3	12—17	
Cool-season grasses				
Wheatgrass (*Agropyron* spp.)	LD	C3		
Tall oat grass (*Arrhenatherum elatius* [L.] Presl.)	LD	C3		
Smooth bromegrass (*Bromus inermis* Leyss)	LD	C3		
Orchard grass (*Dactylis glomerata* L.)	LD	C3	13—24	40
Russian wild rye (*Elymus junceus* Fisch.)	LD	C3		
Fescue (*Festuca* spp.)	LD	C3		43
Annual ryegrass (*Lolium multiflorum* Lam.)	LD	C3		
Perennial ryegrass (*Lolium perenne* L.)	LD	C3		22
Reed canary grass (*Phalaris arundinacea* L.)	LD	C3		
Timothy (*Phleum pratense* L.)	LD	C3		
Kentucky bluegrass (*Poa pratensis* L.)	LD	C3		
Corn (*Zea mays* L.)	SD	C4	46—63	51
Cotton (*Gossypium* spp.)	SD	C3		27
Cowpea (*Vigna unguiculata* L.)	SD	C3		
Crambe (*Crambe abyssinica* Hochst.)	LD	C3		
Field pea (*Pisum arvense* L.)	LD	C3		
Flax (*Linum usitatissimum* L.)	LD	C3		

Table 11 (continued)
CLASSIFICATION OF IMPORTANT CROPS FOR PHOTOPERIOD REACTION AND FOR RATES OF PHOTOSYNTHESIS

Crop (scientific name)	Photoperiod reaction	Type of photosynthesis	Rate of photosynthesis[a] (mg CO_2 dm^{-2} min^{-1})	Crop growth rate[b] (g m^{-2} day^{-1})
Forage legumes				
Cicer milkvetch (*Astragalus cicer* L.)	LD	C3		
Crown vetch (*Coronilla varia* L.)	LD	C3		
Lespedeza (*Lespedeza* spp.)	SD	C3		
Birdsfoot trefoil (*Lotus corniculatus* L.)	LD	C3		
Lupine (*Lupinus luteus* L.)	LD	C3		
Sainfoin (*Onobrychis viciaefolia* Scop.)	LD	C3		
Common vetch (*Vicia sativa* L.)	SD	C3		
Guar (*Cyaniopsis tetragonoloba* [L] Taub)	SD			
Hop (*Humulus lupulus* L.)	LD	C3		
Jute (*Corchorus* spp.)	SD			
Lentil (*Lens culinaris* Medik.)	LD	C3		
Lupine (*Lupinus luteus* L.)	LD	C3		
Sainfoin (*Onobrychis viciaefolia* Scop.)	LD	C3		
Common vetch (*Vicia sativa* L.)	SD	C3		
Guar (*Cyaniopsis tetragonoloba* [L] Taub)	SD			
Hop (*Humulus lupulus* L.)	LD	C3		
Jute (*Corchorus* spp.)	SD			
Lentil (*Lens culinaris* Medik.)	LD	C3		
Oat (*Avena sativa* L.)	LD	C3		
Peanut (*Arachis hypogaea* L.)	SD			
Pearl millet (*Pennisetum americanum* (L.) Leeke)	SD	C4		
Pigeon pea (*Cajanus cajan* (L.) Millsp.)	SD			
Potato (*Solanum tuberosum*)	LD	C3		23
Rapeseed, mustard, kale (*Brassica* spp.)	LD	C3		19
Rice (*Oryza sativa* L.)	SD	C3	29—36	36
Rye (*Secale cereale* L.)	LD	C3		
Safflower (*Carthamus tinctorius* L.)	LD			
Sesame (*Sesamum indicum* L.)	SD			
Sisal and fiber agaves (*Agave* spp.)	SD			
Sorghum (*Sorghum bicolor* [L.] Moench)	SD	C4	55	51
Soybean (*Glycine max* [L.] Merrill)	SD	C3	18	27
Sugar beet (*Beta vulgaris* L.)	LD	C3	24—28	23—32
Sugarcane (*Saccharum* spp.)	SD	C4	42—49	42
Sunflower (*Helianthus anuus* var. *macrocarpa*)	SD	C3	37—44	
Sweet potato (*Ipomoea batatas* [L.] Lam.)	SD			
Tobacco (*Nicotiana tabacum* L.)	SD	C3		
Tomato (*Lycopersicon esculentum* Mill.)	SD	C3	16—21	
Triticale (×*Triticosecale* Wittmach)	LD	C3		

Table 11 (continued)
CLASSIFICATION OF IMPORTANT CROPS FOR PHOTOPERIOD REACTION AND FOR RATES OF PHOTOSYNTHESIS

Crop (scientific name)	Photoperiod reaction	Type of photosynthesis	Rate of photosynthesis[a] (mg CO_2 dm^{-2} min^{-1})	Crop growth rate[b] (g m^{-2} day^{-1})
Warm-season grasses				
Bluestem (*Andropogon* spp.)		C4		
Grama (*Bouteloua* spp.)		C4		
Buffalo grass (*Buchloe dactyloides* (Nutt.) Engelm.)		C4		
Rhodes grass (*Chloris gayana*)		C4		
Lovegrass (*Eragrostis* spp.)		C4		
Guinea-, para-, switchgrass (*Panicum* spp.)		C4		
Dallis-, Bahia-, vaseygrass (*Paspalum* spp.)	SD	C4		
Common reed (*Phragmites communis* Trin)		C4		
Indian grass (*Sorghastrum nutans* [L.] Nash)	SD	C4		
Johnson grass (*Sorghum halepense* [L.] Pers.)	SD	C4		
Sudan grass (*Sorghum sudanense* Stapf.)	SD	C4		51
Prairie cord grass (*Spartina pectinata* Link)		C4		
Wheat (*Triticum aestivum*)	LD	C3	17—31	18
Wild rice (*Zizania aquatica* L.)	LD	C3		

[a] Data from Reference 10.
[b] Data from Reference 11.

REFERENCES

1. **Bickford, E. D. and Dunn, S.,** *Lighting for Plant Growth,* Kent State University Press, Kent, Ohio, 1972.
2. **List, R. J.,** *Smithsonian Meteorological Tables,* 6th rev. ed., Smithsonian Institution, Washington, D.C., 1963.
3. **Major, D. J., Johnson, D. R., Tanner, J. W., and Anderson, I. C.,** Effects of day length and temperature on soybean development, *Crop Sci.,* 15, 174, 1975.
4. **Johnson, H. W., Borthwick, H. A., and Leffel, R. C.,** Effects of photoperiod and time of planting on rates of development of the soybean in various stages of the life cycle, *Bot. Gaz.,* 122, 77, 1960.
5. **Robertson, G. W. and Russelo, D. A.,** Astrometeorological Estimator for Estimating Time when Sun is at Any Elevation, Elapsed Time Between the Same Elevations in the Morning and Afternoon, and Hourly and Daily Values of Solar Energy, Q_o, *Agric. Meteorol. Tech. Bull.,* No. 14, Agrometerology Section, Agriculture Canada, Ottawa, 1968.
6. **Major, D. J.,** Photoperiod response characteristics controlling flowering of nine crop species, *Can. J. Plant Sci.,* 60, 777, 1980.
7. **Biggs, W. W. and Hansen, M. C.,** Radiation measurement, in Li-COR Instrumentation for Biological and Environmental Sciences, Li-COR Inc., Lincoln, 1981, 42.
8. **Evans, L. T.,** The effect of light on plant growth, development and yield, in *Plant Response to Climatic Factors,* Slatyer, R. O., Ed., UNESCO, Paris, 1973.
9. **Fehr, W. R. and Hadley, H. H.,** *Hybridization of Crop Plants,* American Society of Agronomy, Crop Science Society of America, Madison, 1980.
10. **Zelitch, I.,** *Photosynthesis, Photorespiration and Plant Productivity,* Academic Press, New York, 1971.
11. **Cooper, J. P.,** Control of photosynthetic production in terrestrial systems, in *Photosynthesis and Productivity in Different Environments,* Cambridge University Press, Cambridge, 1975, chap. 27.

PLANT TEMPERATURE STRESS

M. N. Christiansen

INTRODUCTION

The economic culture of crop species is limited and delineated by temperature extremes. Temperature optima and limiting extremes for a crop species are related to the geographic evolutionary site or environment of origin. For example, plants of tropical origin are generally very intolerant of chilling (5 to 15°C) but tend to be tolerant to heat, while crop species derived from temperate zones may be tolerant to chilling and freezing.

The mechanisms for survival under temperature stress are quite diverse. Levitt[1] has divided low temperature tolerance into the following classes:

1. Chilling sensitive — Species sensitive to temperatures above 0°C.
2. Tender — Species sensitive to light frost at or near 0°C.
3. Slightly hardy — Species capable of developing some resistance to frost as a consequence of lipid modification; limited to survival above −5°C.
4. Moderately hardy — Species that develop or modify higher cell osmotic levels during hardening; survival limited to −5 to −10°C.
5. Very hardy — Species that develop highly unsaturated membrane lipids and thereby control cell sap osmoticum at low (−10 to −20°C) temperatures. These species also exhibit cryoprotectant systems and resistance to dehydration.
6. Extremely hardy — Species that tolerate −20 to −40°C via the ability to maintain free water in a super-cooled liquid state, by accumulation of special phospholipids and proteins, and by resistance to loss of membrane-mediated water by lipid peroxidation.

From the descriptions of cold tolerance it is evident that lipid constituents of membranes and function of membranes as osmotic barriers and mediators of water movement are extremely important in cold survival.

Heat tolerance is less well defined, but is related in part to resistance to protein denaturation and selective inactivation of enzymes.

PHYSIOLOGY OF TEMPERATURE STRESS

Most of the temperature ideology presented will be concerned with economic crop species. It will consequently deal with temperature stress effects on growth, function, and economic return rather than with survival of the organism. Chilling, freezing, and heat injury will be reviewed with emphasis on developing an understanding of the way plants physically and chemically manifest stress injury and the defense mechanisms that plants develop. The utility of various plant physiological responses as breeding-selection criteria will be considered in a separate section.

Chilling

Most crops of tropical origin are sensitive to temperatures near freezing or in a range of 2 to 15°C. Cotton, citrus, cucurbits, garden beans, peppers, eggplant, and tomato can be included among the chilling-sensitive crops. Chilling damage to germinating cotton can alone account for a $60 to 80 million annual loss in replanting costs. Chilling temperature can injure crop species from the time of seed hydration through germination, growth, flowering, maturation and even in postharvest storage. Types of visible injury include tissue necrosis,[2] premature senescence,[3] growth inhibition,[4] seed death,[5] and interaction with various fungal and bacterial pathogens.[6]

Chilling injury to germinating seed is reported for cotton,[7] soybean,[8] corn,[9] cacao,[10] lima bean,[11] and snap bean.[12] Cotton can serve well as an example; chilling during seed hydration for only brief periods can cause root tip abortion and death.[13] In later stages of germination, cold causes cortical tissue necrosis,[14] root exudation,[15] and long-term growth inhibition[16] that is linearly correlated with length of stress exposure. Solute leakage from radicles increases incidence of *Fusarium* and *Rhizoctonia* sp. infection of seedlings.[6] After seedling emergence, chilling can inhibit root water-uptake ability resulting in water stress and desiccation death of the seedling.[7]

Most of the results from research implicate membranes and membrane permeability as the primary site of chilling impact. Lyons et al.[18] initially proposed that one of the major differences between chilling-sensitive and -resistant species was in the fatty-acid unsaturation of cell membranes. Functionality at low temperatures of mitochondria from chilling-sensitive tomato was much less than mitochondria of cabbage. The chilling tolerance was related to level of linolenic acid in membranes.

In later work with seedling cotton, St. John and Christiansen[19] showed that cotton seedlings could be hardened to withstand 8°C by gradual lowering of temperature. The hardening process was marked by large increases in linolenic acid of membrane polar lipids. If an increase in linolenic acid was blocked by a specific inhibitor of linolenic acid synthesis, the hardening process was blocked and seedlings were cold sensitive. Thus, evidence appears to support the thesis that chilling impacts on membrane form and function.

Early evidence for membrane involvement in chilling response is that provided by Sachs[20] that protoplasmic streaming in *Cucurbita pepo* ceases at 10 to 12°C. Other researchers have ascribed the reaction to thickening or increase in viscosity of protoplasm. Increased carbohydrate and protein synthesis which occurs at low temperatures could also increase cell content viscosity. Other studies show that membranes respond to chilling by leakage of cell contents. Low temperature-induced loss of cell substances — particularly from roots but also from leaves — has been noted for an array of plant species. It is generally agreed that leakage of cell contents indicates damage to membranes and possibly disruption of cell compartmentalization. Activities of cell organelles are neatly "walled off" by membranes, and should these walls be breached by stress, metabolic chaos prevails.

A considerable mass of reports shows that low temperature disrupts most metabolic parameters of sensitive species. Proteins, amino acids, nucleic acids, carbohydrates, organic acids, and lipids are reported to be altered by chilling. Investigations of plant functions show that photosynthesis, respiration, water and ion uptake, and solute transport are altered by chilling. The investigative ideology of all these studies is to determine the master reaction disturbed by cold so that cultural and genetic efforts can be made to improve the condition. Unfortunately, these studies show that most functions in plants are upset by chilling and definition of the primary site of injury is difficult if not impossible.

Currently the most tenable explanation of chilling effect is that the primary site is in membranes. The exact nature of membrane perturbation remains to be elucidated. Evidence in some species points to a rapid physical alteration of plasma or organelle membranes while slow reactions to chilling, particularly in stored fruit, appears to be due to metabolic dysfunction.

Freezing Injury

Injury to plants as a consequence of temperatures below 0° may occur in most crop species. Early theories held that ice formation in cells caused direct mechanical damage. Research has defined several types of injury caused by freezing temperatures.

Early theories held that ice formation within plant tissues caused plant tissue to rupture.[21] The ideology was based on the physical fact that water expands on freezing and in extreme cold, plants were known to split or disintegrate. A number of researchers dispelled this theory; Goppert[22] examined cells from frozen plants and found no ruptured cells, while Nageli[23] proved that cells could stretch much more than the approximate 10% expansion caused by freezing of water. Later studies showed that ice formed outside cells rather than internally and caused actual shrinking of cells and tissue.[24] When frozen tissue is thawed the spaces between cells are usually filled with water and the cells are flaccid due to water loss. If cells are undamaged, the water in the intercellular space is resorbed by the cells upon thawing.

Laboratory studies have shown that ice can be frozen within cells by very rapid reduction of temperature.[25] If cells are frozen slowly, ice forms extracellularly. Most of the evidence indicates that cold resistance is dependent upon an orderly withdrawal of water from the cell to crystal ice development in the extracellular space. Concurrent increase in cell sap osmotic concentration results in freeze-point depression of cell contents and additional cold resistance. Added evidence that water content of cells contributes to cold resistance is the fact that water-stressed plants are more resistant to freezing.[26]

Thus, the ability of a plant to tolerate freezing temperatures is dependent upon the following:

1. The functionality of cell membranes as osmotic barriers and in rapid transfer of water from the cell.
2. Increase in osmotic concentration of the cell sap.
3. Formation of extracellular ice around nucleating particles with no damage to cells.
4. Synthesis of cryoprotectants including carbohydrates and proteins.

Other types of cold injury can occur. By contrast to plant parts frozen in air, injury can occur due to freeze-smothering as a result of ice encasement.[27] A major contribution to damage is exclusion of oxygen thereby blocking respiration.

Winter injury due to desiccation is also of frequent occurrence. It is a consequence of frozen soil conditions coupled with sufficiently high temperatures and wind to cause desiccation of twigs and leaves.

Defense Mechanisms against Freezing

Several mechanisms exist that protect plants from defined levels of freezing temperature. Increase in cell-sap concentration can afford minor protection.[28] However, even very high levels of osmotic pressure (50 bars) only lower the freezing point of cell sap to a −4°C. While freezing protection is minimal, it is definitely protective.

Many species exhibit the ability to supercool to several degrees below the freezing point of their cell contents.[29] Studies using differential thermal analysis have shown that supercooling is a mode of cold resistance in the meristematic tissue of woody species.[30,31]

Sugars and related substances have been intensively studied in relation to freezing. Most of the data show a correlation between freezing resistance and sugar level partially through an osmotic effect, but also as a metabolic effect.[32]

Heat Injury

High-temperature stress occurs with more frequency than one would suspect, particularly in regard to exertion of an inhibiting but not lethal effect. Many economic species are quite sensitive to moderate temperatures. Garden peas, for example, do not tolerate temperatures much above approximately 35°C.[33] Conversely, many species tolerate temperatures in excess of 40°C.[34] Dehydrated tissues such as seed tolerate much higher temperatures (60 to 70°) for extended time periods.[35] Fleshy organs which have greater mass and volume in relation to surface area often suffer heat buildup due to a lesser ability to disperse absorbed heat. Leaves may be cooled below air temperature due to transpiration. Some thick leaves of succulents may show temperatures as much as 20° higher than the ambient air.[36] Temperatures as high as 55°C were recorded on the southern exposure of pine trees.[37] Thus it is evident that dangerously high temperatures can occur naturally in plant tissue.

Injury Symptoms

Sun scald of tree trunks is common in many thin-barked trees. Apples, grapes, cherries, and tomatoes frequently show heat injury. Movement of greenhouse- or shade-grown plants into full sunlight can result in leaf scald from absorbed radiant energy. Dark soils absorb much heat and heat injury can occur to young seedlings at the soil line.[38] Seed germination in hot soils can also be affected. A major problem in wheat culture in China and Russia is the incidence of hot drying winds that occur at heading and pollination of the crop. Heat combined with desiccation reduces grain set and seed size.

These observations show that heat injury (frequently combined with water stress) occurs under natural conditions and has rather serious impact on localized areas of production.

Injury at marginally high temperature is gradual and expressed primarily as a plant-growth-rate reduction. If high temperatures are maintained for long periods, foliar injury symptoms such as chlorosis will become evident. A latent growth effect may persist for several weeks after an episode of heat.

There are four types of heat injury that occur — starvation, toxicity, protein destruction, and biochemical lesion. Starvation occurs when high temperature inhibits photosynthesis more than respiration resulting in a negative energy balance.[39] Toxicity from heat injury is poorly documented, but there is evidence that toxins are produced in apple trees.[40] The toxic products induced by heat included products of anaerobic respiration, ethanol, and acetaldehyde. Petinov and Molotkovsky[41] suggested that heat injury is the consequence of NH_3 production which can be counteracted by respiratory production of organic acids.

Protein denaturation by heat can have effects on both structural and metabolic systems. Many researchers have demonstrated a reduced incorporation of amino acids into protein at high temperature. Proteolysis has also been observed.[42] Denaturation also is a heat-injury process.[43] Thus there is inhibition of protein synthesis and also heat-induced breakdown of protein. Other notable effects include inhibition of chloroplast ribosomes (70S) which results in a marked chlorophyl deficiency.[44]

Visual evidence of heating injury by optical microscope was first reported by Sachs[45] who reported high temperatures of 46 to 47°C caused solidification of protoplasm. The solidification reversed upon cooling. Most of the optical- and electron-microscope studies indicate membranes as the site of injury. Leakage of cell contents is one of the first physical symptoms of heat damage.

Membranes are composed primarily of proteins and lipids; hence the changes that are induced in these classes of compounds should provide the basis for injury. Since proteins are heat denatured, they are the focus of much study. Other than the solidification of protoplasm discussed above, there is also interest in heat inactivation of enzyme systems. Heat stability of enzymes varies over a wide range of temperatures. As a consequence, heat can differentially upset metabolism; e.g., NRAse is inactivated at 36°C, and NADPH-glutamate dehydrogenase at 65°C.[46] Many key enzymes have relatively low thermolabile temperature points. If plants are heat hardened, the thermolability of enzymes may increase. Acid phosphotase thermostability increases as much as 3° with heat hardening although specific activity declined by 50%.[47]

Functionality of membranes at temperature extremes is reported to be directly related to the unsaturation of fatty acids of the membrane lipids. In *Atriplex lentiformis* Pearcy[48] related increased lipid saturation with greater heat thermostability of chloroplasts. Some of the thermophilic bacteria contain unique glycolipids which are thought to have a role in high temperature resistance.[49]

Of the many physiological alterations induced by heat, cell leakage is the primary selective device for heat resistance used by plant breeders.

BREEDING FOR TEMPERATURE TOLERANCE

Most of the breeding effort to select crop plants for the heat and cold resistance has utilized empirical methods under ambient conditions. Although selection for adaptation has resulted in a general improvement in crop performance under temperature stress, there have been no dramatic recent improvements in germplasm. Many plant characteristics (form and function) have been correlated with temperature stress tolerance. Chemical attributes such as lipid unsaturation or high-soluble sugars are correlated with freeze tolerance; or resistance of protein to denaturation or coagulation is correlated with heat resistance. Unfortunately little of the physiological response information has been utilized to develop genetic screening systems for crop improvement of temperature stress.

Some progress has been made in low-temperature selection, primarily in winter hardiness of small grains. One of the problems in winter grain hardiness has centered on survival vs. yield. Considerable improvement can be made in winter survival of wheat but it is often at the expense of yield. Work with wheat (*Triticum aestivum* L.) provided some improvement. Hayes and Garber[50] developed Minhardi, a variety that has served as a germplasm source for hardiness. Most of the very winter-hardy wheats are old varieties such as the 'Albidums' and 'Uljanovka' from Russia.

Barley is a similar situation in that most of the winter hardiness was developed quite early when 'Tennessee Winter' was introduced. According to Wiebe and Reid,[51] 'Kearney' and 'Dicktoo' are the most winter hardy of the barley varieties and exceed 'Tennessee Winter' in survival.

A major breeding effort was made to improve winter hardiness in alfalfa. Early introductions of *Medicago sativa* L. exhibited no winter hardiness, but crosses of *M. sativa* × *M. falcata* L. resulted in development of *M. media* from which varieties with improved cold hardiness were developed. 'Teton' is a recent variety with improved cold tolerance.

Many crop breeders feel that further progress in cold tolerance is not likely. Future progress may be dependent upon introgression of wild germplasm through interspecific crossing.

Progress in cold-tolerance improvement in winter grains is likely to be minimal with currently available germplasm. Coffman[52] suggests that spring-type wheat and oats may be an additional source of winter hardiness to be used by breeders. Wild hexaploid *Avena fatua* L. has been suggested as a source of cold resistance.[53]

Genetic improvement through interspecific crossing is possible in barley particularly if tissue culture techniques are used.[54] Tissue culture and DNA-transfer techniques may provide infusion of genes of value across wide species barriers.

As earlier stated, selection techniques for heat or cold tolerance are not efficient, accurate, or rapid. Little crop cold hardiness has been developed utilizing anything other than empirical field screening. A number of physiological functions are known to be perturbated by heat and cold, and a number of temperature adaptative mechanisms are known. However, no use of physiological knowledge has been made.

Sullivan and Ross[55] have used electrolyte leakage from leaf discs to successfully select heat-stress-resistant corn cultivars. The test is rapid and based on heat-shock effect on leaf-cell membranes. This technique might work well in selection for chilling resistance also, since chilling can also induce tissue leakage.

Controlled freezing tests have been reviewed and tested by Larson and Smith.[56] Marshall[57] indicates that freezing tests are of little value in breeding systems. He reports on observations of Russian research using freezing chambers where testing was only effective in placing wheat germplasm in broad classes and not a useful selection tool.

In screening germplasm for heat or cold tolerance the first requisite is to determine the primary physical site of temperature impact. For example, in wheat the crown is of primary freezing-tolerance importance, because it contains the meristematic tissue for root and leaf regeneration. In chilling injury, water and ion uptake by roots is of primary importance. Age of plant and environmental history are extremely important. Young tissue may be more sensitive than older tissue.

A second consideration in screening germplasm is the biotic environment. Temperature extremes can have considerable impact on the sensitivity of plants to disease, in part because stressed tissue often exudes cellular contents including sugars, amino acids, and other substances favorable to growth of bacteria and fungi.[6]

Other environmental parameters such as soil type, aeration, pH, nutrient and water content, light, and humidity can all alter plant response to temperature. Thus, in cold and heat testing the researcher must strive for absolute environmental control during growth, hardening, temperature screening, and recovery stages of tests.

There is need for a combining of the talents of geneticists, chemists, biophysicists, and plant physiologists to develop rapid, definitive screening techniques for selection of plants tolerant to heat and cold. It is not an impossible task — only difficult.

REFERENCES

1. **Levitt, J.**, *Responses of Plants to Environmental Stresses*, 2nd ed., Vol. 1, *Chilling, Freezing and High Temperature Stresses*, Academic Press, New York, 1980.
2. **Murata, T.**, Physiological and biochemical studies of chilling injury in bananas, *Physiol. Plant.*, 22, 401, 1969.
3. **Sellschop, J. P. F. and Salmon, S. C.**, The influence of chilling above the freezing point on certain crop plants, *J. Agric. Res.*, 37, 315, 1928.
4. **Christiansen, M. N.**, Influence of chilling upon subsequent growth and morphology of cotton, *Crop Sci.*, 4, 584, 1964.
5. **Christiansen, M. N.**, Periods of hypersensitivity to chilling in germinating cotton, *Plant Physiol.*, 42, 431, 1967.
6. **Shao, F. M. and Christiansen, M. N.**, Cotton seedling radicle exudates in relation to susceptibility to *Verticillium* wilt and *Rhizoctonia* root rot, *Phytopathol. Z.*, 105, 351, 1982.
7. **Christiansen, M. N.**, The influence of chilling upon seedling development of cotton, *Plant Physiol.*, 38, 520, 1963.
8. **Obendorf, R. L. and Hobbs, P. R.**, Effect of seed moisture on temperature sensitivity during imbibition of soybean, *Crop Sci.*, 10, 563, 1970.
9. **Cal, J. P. and Obendorf, R. L.**, Imbibitional chilling injury in *Zea mays*, L. altered by initial kernel moisture and maternal parent, *Crop Sci.*, 12, 369, 1972.
10. **Boroughs, H. and Hunter, J. R.**, Effect of temperature on cacao seeds, *Proc. Am. Hortic. Soc.*, 82, 222, 1963.
11. **Pollock, B. M.**, Imbibition temperature sensitivity of lima bean seeds controlled by initial seed moisture, *Plant Physiol.*, 44, 907, 1969.
12. **Pollock, B. M., Ross, E. E., and Manalo, J. R.**, Vigor of garden bean seed and seedlings influenced by initial seed moisture, substrate oxygen and imbibitional temperature, *J. Am. Soc. Hortic. Sci.*, 94, 577, 1969.
13. **Christiansen, M. N.**, Induction and prevention of chilling injury to radicle tips of imbibing cottonseed, *Plant Physiol.*, 43, 743, 1968.
14. **Christiansen, M. N.**, Biochemical and physical responses to chilling by germinating cotton, *Proc. Cotton Production Research Conf.*, National Cotton Council, Memphis, Tenn., 1971.
15. **Christiansen, M. N., Carns, H. R., and Slyter, D. V.**, Stimulation of solute loss from radicles of *Gossypium hirsutum* L. by chilling, anaerobioses and low pH, *Plant Physiol.*, 46, 53, 1970.
16. **Christiansen, M. N. and Thomas, R. O.**, Season-long effects on chilling treatments applied to germinating cottonseed, *Crop Sci.*, 9, 672, 1969.
17. **Christiansen, M. N. and Ashworth, E. N.**, Prevention of chilling injury to seedling cotton with antitranspirants, *Crop Sci.*, 18, 907, 1978.
18. **Lyons, J. M., Wheaton, T. A., and Pratt, H. K.**, Relationship between the physical nature of mitochondrial membranes and chilling sensitivity in plants, *Plant Physiol.*, 39, 262, 1964.

19. **St. John, J. B. and Christiansen, M. N.,** Inhibition of linolenic acid synthesis and modification of chilling resistance in cotton seedlings, *Plant Physiol.,* 57, 257, 1976.
20. **Sachs, J.,** Über die obere Tempertur-grenze der Vegetation, *Flora (Jena),* 47, 5, 1864.
21. **Goetz, A. and Goetz, S. S.,** Vitrification and crystallization of protophyta at low temperatures, *Proc. Am. Phil. Soc.,* 97, 361, 1938.
22. **Goppert, H. R.,** *Über das Gefrieren, Erfrieren der Pflanzen und Schutzmittel degagen Altes and Neves,* Enke, Stuttgart, 1883, 1.
23. **Nägeli, C.,** Über die Wirkung des Frostes auf die Pflanzenzellen, *Sitzungsber. Math. Phys. Kl. Bayer Akad. Wiss. München,* 1861, 264.
24. **Ilijin, W. S.,** The point of death of plants at low temperature, Bull. Ass. Russe Rech Sci., Prague Sect. Sci. Nature Math., 1(6), 135, 1922.
25. **Siminovitch, D., Singh, J., and de la Roche, J. A.,** Freezing behavior of free protoplasts of winter rye, *Cryobiology,* 15, 205, 1978.
26. **Chen, P. M. and Li, P. H.,** Induction of frost hardiness in stem cortical tissues of *Cornus stolonifera* Michx. by water stress. II. Biochemical changes, *Plant Physiol.,* 59, 240, 1977.
27. **Andrews, C. J. and Pomeroy, M. K.,** Mitochondrial activity and ethanol accumulation in ice-encased winter cereal seedlings, *Plant Physiol.,* 59, 1174, 1978.
28. **Levitt, J.,** *Responses of Plants to Environmental Stress,* 2nd ed., Vol. 1, *Chilling, Freezing, and High Temperature Stresses,* Academic Press, New York, 1980, 119.
29. **George, M. F., Burke, M. J., Pellet, H. M., and Johnson, A. G.,** Low temperature exotherms and woody plant distribution, *HortScience,* 9, 519, 1974.
30. **Weiser, C. J.,** Cold resistance and injury in woody plants, *Science,* 169, 1269, 1970.
31. **George, M. F. and Burke, M. J.,** The occurrence of deep super-cooling in cold hardy plants, *Curr. Adv. Plant Sci.,* 22, 349, 1976.
32. **Sakai, A.,** Studies on the frost hardiness of woody plants. I. The causal relation between sugar content and frost-hardiness, *Control Instrum. Low Temp. Sci. Ser.,* 1311, 1, 1962.
33. **Collander, R.,** Beobachtungen über die Quantitativen Beziehungen, zwischen Totungsgeschwindigkeit und Temperatur beim Warmtod Pflanzlicher Zellen, *Commentat. Biol. Soc. Sci. Fenn.,* 1, 1, 1924.
34. **Rouschal, E.,** Zum Warmehavshalt der Macchienpflanzen, *Oesterr. Bot. Z.,* 87, 42, 1938.
35. **Just, L.,** Über die Einwirkung hoher Temperaturen auf die Erhaltung der Keimfahigeit der Samen, *Beitr. Biol. Pflanz.,* 2, 311, 1877.
36. **Askenasy, E.,** Über die Temperatur, Welche Pflanzen in Sonnenlicht Annehmen, *Bot. Z.,* 33, 441, 1875.
37. **Sorauer, P.,** Handbuch der Pflanzen Naturwiss, *Z. Forst. Landwirtsch.,* 12, 169, 1913.
38. **Munch, E.,** Hitzeschaden an Waldpflanzen Naturwiss, *Z. Forst. Landwirtsch.,* 12, 169, 1913.
39. **Lundegardh, H.,** *Klima und Boden,* 4th ed., Fischer, Jena, 1957.
40. **Gur, A., Bravo, B., and Mizrahi, Y.,** Physiological responses of apple trees to supraoptimal root temperature, *Physiol. Plant.,* 27, 130, 1972.
41. **Petinov, N. W. and Molotkovsky, Yu. G.,** Heat stability of plants and ways of increasing it, *Vestn. Acad. Nauk SSSR,* 8, 62, 1962.
42. **Engelbrecht, L. and Mothes, K.,** Kinetin als Faktor der Hitzeresistenz, *Ber. Dtsch. Bot. Ges.,* 73, 246, 1960.
43. **Lepeschkin, W. W.,** Zur Kenntnis des Hitzetodes des Protoplasmas, *Protoplasma,* 23, 349, 1935.
44. **Feierabend, J.,** Inhibition of chloroplast ribosome formation by heat in higher plants, in *Methods in Chloroplast Molecular Biology,* Edeman, M., Hallick, R. B., and Chua, N. H., Eds., Elsevier, Amsterdam, 1982, 671.
45. **Sachs, J.,** Über die obere Temperatur-Grenze der Vegetation, *Flora (Jena),* 47, 5, 1864.
46. **Magalhaes, A. C., Peters, D. B., and Hageman, R. H.,** Influence of temperature on nitrate metabolism and leaf expansion in soybean (*Glycine max.* [L.] Merr.) seedlings, *Plant Physiol.,* 58, 12, 1976.
47. **Feldman, N. L., Lutova, M. I., and Shcherbakova, A. M.,** Resistance of some proteins of *Pisum sativum* L. leaves after heat hardening to elevated temperature proteolysis and shifts in pH, *J. Therm. Biol.,* 1, 47, 1975.
48. **Pearcy, R. W.,** Effect of growth temperature on the fatty acid composition in *Atriplex lentiformis* (Torr.) Wats., *Plant Physiol.,* 61, 484, 1978.
49. **Oshima, M. and Yamakawa, T.,** Chemical structure of a novel glycolipid from an extreme thermophile, *Flavobacterium thermophilum, Biochemistry,* 13, 1140, 1974.
50. **Hayes, H. K. and Garber, R. J.,** *Minn. Agric. Exp. Stn. Bull.,* 182, 1919.
51. **Wiebe, G. A. and Reid, D. A.,** *U.S. Dep. Agric. Tech. Bull.,* No. 176, 1958.
52. **Coffman, F. A.,** Results from uniform winter hardiness nurseries of oats for the five years 1947 - 1951 inclusive, *Agron. J.,* 47, 54, 1955.
53. **Suneson, C. A. and Marshall, H. G.,** Cold resistance in wild oats, *Crop Sci.,* 7, 667, 1967.
54. **Grafius, J. E.,** Breeding for winter hardiness, in *Analysis and Improvement of Plant Cold Hardiness,* Olien, C. R. and Smith, M. N., Eds., CRC Press, Boca Raton, Fla., 1980, 161.

55. **Sullivan, C. Y. and Ross, R. M.,** *Proc. Int. Conf. Stress Physiology in Plants,* Boyce Thompson Institute, Ithaca, N.Y., Mussell, H. and Staples, R., Eds., John Wiley & Sons, New York, 1977, 263.
56. **Larson, K. L. and Smith, D.,** Reliability of various plant constituents and artificial freezing methods in determining winter hardiness of alfalfa, *Crop Sci.,* 4, 413, 1964.
57. **Marshall, H. G.,** Breeding for tolerance to heat and cold, in *Breeding Plants for Less Favorable Environments,* Christiansen, M. N. and Lewis, C. F., Eds., Wiley Interscience, New York, 1982, chap. 3.

SENSITIVITY OF CROP PLANTS TO GASEOUS POLLUTION STRESS

Douglas P. Ormrod, Barbara Young, and Beverley Marie

INTRODUCTION

Crop plants respond to gaseous pollutants in many different ways because of the many factors that affect the response. First, there are several gases that cause injurious effects and each results in a different response pattern. Ozone (O_3) and peroxyacetyl nitrate (PAN) are strong oxidizing agents generated in sunlight from the photochemical reaction of mixtures of nitrogen oxides and hydrocarbons. Sulfur dioxide (SO_2), nitrogen oxides (NO and NO_2), and hydrogen fluoride (HF) are direct products of combustion. Other injurious gases such as chlorine, ammonia, and hydrogen sulfide may be present in localized situations. Also, gases may be present as mixtures at some locations. The concentration of gaseous pollutants will vary spatially and temporally, according to proximity to pollutant sources, wind speed and direction, atmospheric conditions, and other factors.

Second, one must consider crop responses and the diversity of plant sensitivity in response to gaseous pollutants. Visible symptoms and growth responses vary greatly among pollutants. Species, cultivars, and even individual plants of the same cultivar may vary greatly in sensitivity. The sensitivity of a species is based on evaluation of responses of one or more cultivars or strains of that species. In most species, cultivars differ greatly in sensitivity so a number of cultivars or lines must be evaluated to obtain a useful assessment of the species. Also, visible injury symptoms vary widely among species and may not relate particularly well to growth and yield responses.

Finally, environmental factors have a very important role in determining the injury to crop plants from gaseous pollution. Sensitivity is markedly affected by temperature, irradiance, atmospheric water, soil water, air movement, nutrition, and possibly other factors. There are important interactions with other pollutants, such as pesticides or heavy metals, with plant diseases, and with growth regulators.

There are several reference books that provide thorough coverage of the scientific and technical information on crop responses to gaseous pollutants, only three of which are referred to here.[1-3] Other reference books deal with the recognition of gaseous pollution by visible symptoms[4] and with the use of plants as monitors of gaseous pollution.[5] A composite sensitivity chart has been prepared for the U.S. Environmental Protection Agency[6] and is reproduced in part here as Table 1.* This chart should be used with caution because species responses may depend greatly on the cultivars or lines chosen to represent the species in tests and because relative sensitivity can be modified by the use of different environmental conditions in different tests.

The published information on crop responses to gaseous pollutants has been placed in tabular form to facilitate its efficient utilization (Table 2).* The crop species have been grouped in conventional categories within field crops and horticultural crops. For each species, the available references on gaseous pollution are considered in this order: O_3, PAN, SO_2, F, NO_2, others, and mixtures. Then it is indicated whether the study evaluated cultivar differences and whether comparisons were made with other species in the same study. The bases for evaluation of gaseous pollutant injury in each study, whether visible symptoms and/or growth respones and/or yield responses, are next indicated. If environmental factors were varied experimentally, this is indicated in the next column and then it is indicated whether any interactions with biological or chemical factors were evaluated. Finally, supplementary notes are provided to indicate the nature of the study.

* Tables 1 and 2 follow the text.

Table 1
COMPOSITE SENSITIVITY CHART AND SCIENTIFIC NAMES OF PLANTS

Plant	O$_3$	PAN	SO$_2$	F	NO$_2$	NH$_3$	Cl	Mix
Field Crops								
Alfalfa (*Medicago sativa* L.)	S[a]	I[b]	S[c]	T	S		S	
Barley (*Hordeum vulgare* L.)	S	I	S		S			
Bean (*Phaseolus* spp. L.)	S	S	S					
Bush (*P. vulgaris* L.)					I			
Pinto (*P. vulgaris* L.)					S			I
Lima (*P. limensis* Macf.)		T						
Beet (*Beta vulgaris* L.)		I						
Clover (*Trifolium* spp. L.)		S						
Crimson (*T. incarnatum* L.)				I				
Red (*T. pratense* L.)	S		S		S		I	
Corn (*Zea mays* L.)		T	T		I			I
Cotton (*Gossypium* sp. L.)	T	T	S	T			T	
Cowpea (*Vigna sinensis* Savi.)	I							
Grass								
Brome (*Bromus inermis* Leyes.)	S							
Johnson (*Sorghum halepense* Pers.)				I				
Kentucky blue (*Poa pratensis* L.)					T		T	
Orchard (*Dactylis glomerata* L.)	S							
Oats (*Avena sativa* L.)	S	S	S		ST			
Peanut (*Arachis hypogaea* L.)	I							
Rye (*Secale cereale* L.)	S				I			
Safflower (*Carthamus tinctorius* L.)	S							
Sorghum (*Sorgham vulgare* Pers.)	I	T						
Soybean (*Glycine max* Merr.)	S	I		IT				
Sugarcane (*Saccharum officinarum* L.)				T				
Sunflower (*Helianthus* spp. L.)				I	S	S		S
Timothy (*Phleum pratense* L.)	I							
Tobacco (*Nicotiana tabacum* L.)	S	I		T	S		S	
Vetch, spring (*Vicia sativa* L.)					S			
Wheat (*Triticum aestivum* L.)	S	I	S	I	I			
Vegetable Crops								
Asparagus (*Asparagus officinalis* L.)				T	T			
Bean, snap (*Phaseolus vulgarus* L.)				T				
Beet (*Beta* spp. L.)	T		S					
Beet, table (*B. vulgaris* L.)		I						
Broccoli (*Brassica oleracea* var. *italica* L.)	I	T	S					
Cabbage (*B. oleracea* var. *capitata* L.)	I	T	T	T	T			
Carrot (*Daucus carota* L.)	I	I	S	T	S			
Cauliflower (*B. oleracea* var. *botrytis* L.)		T		T				
Celery (*Apium graveolens* L.)		S	T	T	I			
Corn, sweet (*Zea mays* L.)	S			SI				
Cucumber (*Cucumis sativus* L.)	I	T	S	T			I	
Eggplant (*Solanum melongena* var. *esculentum* Nees.)				T				
Endive (*Cichorium endivia* L.)	I	S						

Table 1 (continued)
COMPOSITE SENSITIVITY CHART AND SCIENTIFIC NAMES OF PLANTS

Plant	O_3	PAN	SO_2	F	NO_2	NH_3	Cl	Mix
Vegetable Crops (continued)								
Kohlrabi (*B. oleracea* var. *gongylodes* L.)			S		T			
Leek (*Allium porrum* L.)					S			
Lettuce (*Lactuca sativa* L.)	T	S	S		S			
Muskmelon (*Cucumis melo* L.)	S		T					
Mustard (*Brassica* spp. L.)		S			S		S	
Onion (*Allium cepa* L.)	S	T	T		T			I
Parsley (*Petroselinum crispum* Nym.)	I							
Parsnip (*Pastinaca sativa* L.)			I					
Pea (*Pisum sativum* L.)	I		S	T	S			
Pepper (*Piper nigrum* L.)	I	S					T	
Pepper, bell (*Capsicum frutescens* var. *grossum* Bailey)				T				
Potato (*Solanum tuberosum* L.)	S		T	T	I			
Potato, sweet (*Ipomea batatas* (L.) Lam.)	T		S	S1				
Radish (*Raphanus sativus* L.)	S	T	S				S	
Rhubarb (*Rheum rhaponticum* L.)		T	S		S			
Spinach (*Spinacea oleracea* L.)	S	I	S	T				
Squash (*Cucurbita pepo* L.)		T	S				I	
Swiss chard (*Beta vulgaris* var. *cicla* L.)		S	S					
Tomato (*Lycopersicon esculentum* Mill.)	S	S	S	I	I	T		I
Turnip (*Brassica rapa* L.)	I		S					
Fruit Crops								
Apple (*Malus sylvestris* Mill.)		T	S	I	S			
Apricot (*Prunus armeniaca* L.)	T							
Chinese (*P. armeniaca* L. var. Chinese)	I			S				
Moorpark (*P. armeniaca* L. var. Moorpark)				I				
Royal (*P. armeniaca* L. var. Royal)				S				
Tilton (*P. armeniaca* L. var. Tilton)				I				
Avocado (*Persea americana* Mill.)	T							
Blackberry (*Rubus* spp. L.)							S	
Blueberry (*Vaccinium* spp. L.)				S				
Cherry, sweet (*Prunus avium* L.)				I				
Bing (*P. avium* L. var. Bing)	S							
Lambert (*P. avium* L. var. Lambert)	I							
Citrus (*Citrus* spp. L.)	S		T					
Crabapple (*Malus baccata* Borkh.)	I						S	
Currant (*Ribes* spp. L.)				T				
Grape (*Vitis* spp. L.)	S							
Concord (*V. labruscana* Bailey)	S			I				
European (*V. vinifera* L.)				S				
Grapefruit (*Citrus paradisi* Macf.)				I				
Lemon (*C. limon* Burm.)	T			I				

Table 1 (continued)
COMPOSITE SENSITIVITY CHART AND SCIENTIFIC NAMES OF PLANTS

Plant	O_3	PAN	SO_2	F	NO_2	NH_3	Cl	Mix
Fruit Crops (continued)								
Orange (*C. sinensis* Osbeck)				I	I			
Peach (*Prunus persica* Batsch.)	T							
Fruit				S				
Foliage				I				
Pear (*Pyrus communis* L.)				T	S			
Bartlett (*P. communis* L. var. Bartlett)	T							
Plum, Bradshaw (*Prunus domestica* L. var. Bradshaw)				S				
Raspberry, red (*Rubus idaeus* L.)				T				
Strawberry (*Fragaria* spp. L.)	T	T		I				
Tangerine (*Citrus reticulata* Blanco)				I				
Walnut, English (*Juglans regia* L.)	SI			I				

[a] S = sensitive to pollutant.
[b] I = intermediately sensitivity to pollutant.
[c] T = tolerant to pollutant.

Table 2
PUBLISHED INFORMATION ON CROP RESPONSES TO GASEOUS POLLUTANTS

Species	Pollutant	Cultivar differences	Species differences	Visible injury	Growth responses	Yield responses	Environmental factors	Interactions	Comments	Ref.
Field Crops										
Cereals										
Barley (*Hordeum vulgare* L.)	O_3		×		×				Photosynthetic response	7
	O_3		×		×				Chlorophyll response	8
	O_3	×		×				×	Disease interaction study	9
	O_3		×	×	×	×	×		Moisture effects	10
	SO_2		×	×	×				Foliar response	11
	SO_2		×	×	×				Photosynthetic response	7
	SO_2				×	×	×		Moisture effects	10
	SO_2	×	×		×			×	Heavy-metal interaction study	12
	SO_2				×		×		Moisture effects	13
	SO_2		×		×				Field chamber study	14
	SO_2		×		×	×			SO_2 deposition study	15
	SO_2		×						Foliar response	11
	F		×	×	×				Foliar response	16
	F		×		×				Photosynthetic response	7
	F		×		×				Photosynthetic response	7
	NO_2		×		×				Photosynthetic response	7
	Cl_2		×						Foliar response	11
	Mix		×							
Corn (*Zea mays* L.)	O_3				×				Field study	17
	O_3				×				Pollen response	18
	O_3	×			×				Threshold value study	19
	SO_2		×	×	×	×			Species comparison study	20
	SO_2		×	×		×	×		Nutrition effects	21

Table 2 (continued)
PUBLISHED INFORMATION ON CROP RESPONSES TO GASEOUS POLLUTANTS

Species	Pollutant	Cultivar differences	Species differences	Visible injury	Growth responses	Yield responses	Environmental factors	Interactions	Comments	Ref.
Field Crops (continued)										
Corn (cont.)	SO_2		x				x		Moisture effects	13
	SO_2	x	x		x				Injury mechanism study	22
	SO_2		x		x				Greenhouse study	20
	SO_2		x	x	x				Species comparison study	23
	SO_2		x	x					Relative susceptibility study	24
	F		x	x	x				Species comparison study	24
	F			x	x				F Accumulation study	25
	F								Relative susceptibility study	24
	NO_2		x	x	x				Species comparison study	23
	NO_2		x	x	x				Species comparison study	26
	NH_3		x		x				Continuous stirred tank reactor study	27
	Mix		x	x	x				Species comparison study	23
Oat (*Avena sativa* L.)	O_3	x	x	x					Relative susceptibility study	28
	O_3	x	x	x				x	Disease interaction study	29
	O_3	x		x			x		Air velocity effects	30
	SO_2				x		x		Temperature, relative humidity effects	31
	SO_2				x				Seedling response	32
	SO_2		x	x					Controlled environment study	33
	NO_2		x	x					Controlled environment study	33
	NO_2		x		x				Photosynthetic response	34
	NH_3			x	x				Continuous stirred tank reactor study	27
	Mix		x						Controlled environment study	33

Rice (*Oryza sativa* L.)	O_3				×		Disease comparison study	35
	O_3						Use as pollution indicator	36
	SO_2	×		×			Use as pollution indicator	36
	SO_2	×		×		×	Injury mechanism study	37
	NO_2	×		×			Use as pollution indicator	36
Wheat (*Triticum* sp. L.)	O_3				×		Disease interaction study	38
	O_3				×		Disease interaction study	39
	O_3			×			Relative susceptibility study	40
	SO_2	×		×	×	×	Species comparison study	20
	SO_2	×			×	×	Photosynthetic response	41
	NO_2			×			Photosynthetic response	42
	NH_3			×			Continuous stirred tank reactor study	27
Protein crops Bean (*Phaseolus* spp. L.)	O_3				×		Ethylene response	43
	O_3			×	×	×	Ozone protectant study	44
	O_3			×	×	×	Disease interaction study	45
	O_3	×		×			Ethylene response	46
	O_3	×		×	×		Ozone protectant study	47
	O_3			×	×	×	Disease interaction study	48
	O_3			×	×		Transpiration response	49
	O_3		×	×			Nutrition, temperature effects	50
	O_3		×	×	×	×	Disease interaction study	51
	O_3			×	×		Relative humidity effects	52
	O_3	×		×	×		Species comparison study	53
	O_3		×	×	×	×	Ozone protectant study	54
	O_3		×	×	×		Nutrition effects	55
	O_3			×			Salinity effects	56
	O_3			×	×		Nutrition, temperature effects	57
	O_3			×	×	×	Ozone protectant study	58
	O_3			×	×		Insecticide interaction	59

Table 2 (continued)
PUBLISHED INFORMATION ON CROP RESPONSES TO GASEOUS POLLUTANTS

Species	Pollutant	Cultivar differences	Species differences	Visible injury	Growth responses	Yield responses	Environmental factors	Interactions	Comments	Ref.
Field Crops (continued)										
Bean (cont.) (*Phaseolus* spp. L.)	O_3			x				x	Disease interaction study	60
	O_3			x				x	Ozone protectant study	61
	O_3				x			x	Ozone protectant study	62
	O_3				x			x	Fungi interaction study	63
	O_3			x		x		x	Ozone protectant study	64
	O_3				x				Field study	65
	O_3								Composition study	66
	O_3								Respiration response	67
	O_3		x				x	x	Disease, heat interaction study	68
	O_3		x	x			x		Light, relative humidity effects	69
	O_3			x			x		Temperature, light, relative humidity effects	70
	O_3			x			x		Field and chamber study	71
	O_3			x	x				Photosynthetic response	72
	O_3		x	x					Relative susceptibility study	73
	O_3	x		x	x	x		x	Ozone protectant study	74
	O_3		x	x				x	Disease interaction study	75
	O_3	x		x	x			x	Ozone protectant study	76
	O_3				x				Enzyme activity study	77
	O_3			x					Controlled environment study	78
	O_3			x					Composition study	79
	O_3		x				x	x	Hormone interaction study	80
	O_3	x			x	x			Air velocity effects	30
	O_3	x			x	x			Productivity study	81
	O_3							x	Ozone protectant study	82
	O_3			x	x	x	x		Foliar response	83
	O_3			x	x	x	x		Moisture effects	10
	O_3			x					Temperature effects	84

O₃		Controlled environment study	85
O₃		Composition study	86
O₃		Stomatal response	87
O₃		Controlled environment study	88
O₃		Species comparison study	89
O₃		Foliar response	90
O₃		Controlled environment study	91
O₃		Enzyme activity study	92
O₃	×	Herbicide interaction study	93
O₃	×	Ozone protectant study	94
O₃	×	Disease, ozone protectant interaction study	95
PAN	×	Ozone protectant study	96
PAN		Photosynthetic response	97
PAN		Leaf age effects	98
PAN		Water balance effects	99
PAN	×	Controlled environment study	85
PAN		Composition study	86
PAN		Stomatal response	87
SO₂		Field study	100
SO₂		Foliar response	11
SO₂		Metabolic effects	101
SO₂		Moisture effects	10
SO₂	×	Disease interaction study	102
SO₂		Temperature, relative humidity effects	103
SO₂		Moisture effects	104
SO₂		Metabolic response	105
SO₂		Photosynthesis response	106
SO₂		Moisture effects	13
SO₂		Stomatal response	107
SO₂		Species comparison study	23
SO₂		Transpiration effects	108
SO₂	×	Temperature effects	84
SO₂		Species comparison study	89
SO₂		Foliar response	90
SO₂		Controlled environment study	91

Table 2 (continued)
PUBLISHED INFORMATION ON CROP RESPONSES TO GASEOUS POLLUTANTS

Species	Pollutant	Cultivar differences	Species differences	Visible injury	Growth responses	Yield responses	Environmental factors	Interactions	Comments	Ref.
Field Crops (continued)										
Bean (cont.)	F				×	×			Controlled environment study	109
	F				×				Respiration response	110
	F			×			×		Nutrition effects	111
	F		×	×				×	Disease interaction study	112
	F		×	×					Foliar response	16
	F		×		×				Metabolic response	113
	F		×		×				Field study	100
	F		×		×				Foliar response	11
	NO₂				×				Foliar response	114
	NO₂		×		×				Gas exchange response	115
	NO₂				×		×		Salinity effects	116
	NO₂				×				Short exposure period	117
	NO₂				×		×		Temperature, light, relative humidity effects	118
	NO₂		×	×	×				Species comparison study	23
	NO₂				×				Transpiration response	119
	H₂S		×		×	×			Field study	65
	Cl₂		×	×	×				Field study	100
	HCl		×	×			×		Diurnal response	120
	HCl			×	×				Foliar response	121
	HCl		×		×				Enzyme activity study	77
	HCl		×		×		×		Age effects	122
	Mix				×	×			Field study	100
	Mix		×		×				Field study	65
	Mix						×		Temperature effects	84
	Mix		×	×	×				Species comparison study	23
	Mix				×				Transpiration response	119

Bean (*Vicia* spp.)

Pollutant	Study	Ref.
Mix	Ozone protectant study	96
Mix	Controlled environment study	85
Mix	Temperature effects	84
Mix	Controlled environment study	88
Mix	Foliar response	11
Mix	Species comparison study	89
Mix	Foliar response	90
Mix	Controlled environment study	91
O_3	Photosynthetic response	123
O_3	Photosynthetic response	124
SO_2	Photosynthetic response	124
SO_2	Relative susceptibility study	24
SO_2	Moisture effects	13
SO_2	Stomatal response	125
SO_2	Stomatal response	107
SO_2	Photosynthetic response	126
F	Yield response	127
F	Relative susceptibility study	24
Mix	Photosynthetic response	123
Mix	Photosynthetic response	124

Pea (*Pisum sativum* L.)

Pollutant	Study	Ref.
O_3	Growth response	81
O_3	Heavy-metal interaction study	128
O_3	Species comparison study	129
O_3	Stomatal response	130
SO_2	Species comparison study	129
SO_2	Stomatal response	130
SO_2	Enzyme activity study	131
SO_2	Species comparison study	23
NO_2	Species comparison study	129
NO_2	Enzyme activity study	131
NO_2	Short exposure period	117
Mix	Species comparison study	23
Mix	Species comparison study	23
Mix	Enzyme activity study	131
Mix	Species comparison study	129

Table 2 (continued)
PUBLISHED INFORMATION ON CROP RESPONSES TO GASEOUS POLLUTANTS

Species	Pollutant	Cultivar differences	Species differences	Visible injury	Growth responses	Yield responses	Environmental factors	Interactions	Comments	Ref.
Field Crops (continued)										
Peanut (*Arachis hypogaea* L.)	SO_2		×	×					Species comparison study	24
	F		×	×					Species comparison study	24
Oil crops										
Rapeseed (*Brassica napus* L.)	SO_2		×		×				Species comparison study	15
Safflower (*Carthamus tinctorius* L.)	F				×	×			Enzyme activity study	132
Soybean (*Glycine max* Merr.)	O_3	×		×					Relative susceptibility study	133
	O_3			×	×	×			Field chamber study	134
	O_3	×		×			×		Diurnal response	135
	O_3	×		×					Ozone flux study	136
	O_3		×	×			×		Media, fertilizer effects	137
	O_3			×			×		Nutrition, temperature effects	57
	O_3	×						×	Ozone protectant study	64
	O_3					×			Progeny response	138
	O_3			×				×	Disease interaction study	139
	O_3	×		×	×				Enzyme activity study	140
	O_3	×		×	×				Enzyme activity study	141
	O_3					×			Field chamber study	142
	O_3			×					Histopathological study	143

Pollutant	Study	Page
O₃	Metabolic response	144
O₃	Seed quality study	145
O₃	Controlled environment study	91
O₃	Composition study	146
O₃	Early growth response	147
O₃	Foliar response	90
O₃	Species comparison study	89
O₃	Nematode study	148
O₃	Leaf diffusive resistance study	149
O₃	Nematode study	150
O₃	Growth response	151
SO₂	Controlled environment study	91
SO₂	Field chamber study	174
SO₂	Composition study	146
SO₂	Early growth response	147
SO₂	Nematode study	148
SO₂	Species comparison study	89
SO₂	Growth response	151
SO₂	Nematode study	150
SO₂	Leaf diffusive resistance study	149
SO₂	Light effects	152
SO₂	Acid precipitation effects	153
SO₂	Acid precipitation effects	154
SO₂	Age effects	155
SO₂	Seed quality response	156
SO₂	Stomatal response	157
NO₂	Composition study	146
NO₂	Stomatal response	157
NH₃	Continuous stirred tank reactor study	27
Mix	Controlled environment study	91
Mix	Genetic resistance study	158
Mix	Field chamber study	174
Mix	Early growth response	147
Mix	Pollutant interaction study	159
Mix	Foliar response	90
Mix	Species comparison study	89

Table 2 (continued)
PUBLISHED INFORMATION ON CROP RESPONSES TO GASEOUS POLLUTANTS

Species	Pollutant	Cultivar differences	Species differences	Visible injury	Growth responses	Yield responses	Environmental factors	Interactions	Comments	Ref.
Field Crops (continued)										
Soybean (cont.)	Mix									148
	Mix		×	×	×				Nematode study	151
	Mix			×	×				Growth response	150
	Mix		×	×				×	Nematode study	149
Sunflower (*Helianthus* sp. L.)									Leaf diffusive resistance study	
	O_3				×				Photosynthetic response	160
	O_3			×					Salinity effects	161
	SO_2		×		×		×		Stomatal response	107
	SO_2		×		×		×		Moisture effects	13
	SO_2		×			×	×		Nutrition effects	21
	SO_2				×				Photosynthetic response	160
	NO_2				×				Photosynthetic response	160
	NO_2		×		×				Species comparison study	26
	Mix				×				Photosynthetic response	160
Forage Crops										
Alfalfa (*Medicago sativa* L.)	O_3				×	×	×		Salinity effects	162
	O_3	×			×	×			Controlled environment study	163
	O_3	×		×					Genetic resistance study	164
	O_3	×			×				Chlorophyll response	8
	O_3		×	×					Species comparison study	165
	O_3		×		×				Metabolic response	144

					Response	Page
	O_3				Growth response	166
	O_3		×		Growth response	167
	O_3				Growth response	168
	O_3	×			Quality response	169
	O_3				Growth response	170
	O_3				Photosynthetic response	7
	SO_2		×		Species comparison study	24
	SO_2		×	×	Carbon dioxide interaction study	171
	SO_2				Photosynthetic response	172
	SO_2		×		Photosynthetic response	7
	SO_2		×		Growth response	170
	SO_2		×		Controlled environment study	173
	SO_2		×		Metabolic response	101
	SO_2		×		Photosynthetic response	174
	SO_2		×		Enzyme activity study	175
	SO_2	×	×		Nutrition, moisture, age effects	176
	SO_2		×		Species comparison study	15
	SO_2		×		Growth response	168
	SO_2		×		Quality response	169
	F		×		Species comparison study	24
	F		×		Photosynthetic response	7
	F		×		Species comparison study	177
	F				Growth response	170
	F		×		Burn effects	178
	NO_2			×	Photosynthetic response	34
	NO_2		×		Carbon dioxide interaction study	171
	NO_2		×		Photosynthetic response	172
	NO_2	×			Photosynthetic response	7
	NO_2	×	×		Growth response	170
	H_2S		×		Species comparison study	179
	Cl_2				Photosynthetic response	7
	Mix				Quality response	169
	Mix				Growth response	168
	Mix			×	Carbon dioxide interaction study	171
Clover, alsike (*Trifolium hybridum* L.)	O_3	×			Species comparison study	165

Table 2 (continued)
PUBLISHED INFORMATION ON CROP RESPONSES TO GASEOUS POLLUTANTS

Species	Pollutant	Cultivar differences	Species differences	Visible injury	Growth responses	Yield responses	Environmental factors	Interactions	Comments	Ref.
Forage Crops (continued)										
Clover, crimson (*T. incarnatum* L.)	O_3				x				Greenhouse study	180
Clover, ladino (*T. repens* L.)	O_3		x	x	x				Controlled environment study	181
	O_3		x	x	x	x			Species comparison study	182
	O_3								Species comparison study	165
Clover, red (*T. pratense* L.)	O_3	x		x	x			x	Ozone protectant study	183
	O_3		x	x					Species comparison study	165
Clover, white sweet (*Melilotus alba* Desr.)	O_3		x	x					Species comparison study	165
	SO_2		x	x					Species comparison study	24
	F		x	x					Species comparison study	24
Ryegrass, perennial (*Lolium perenne* L.)	SO_2			x	x				Growth response	184
	SO_2				x	x			Growth response	185
	SO_2				x	x			Air velocity effects	186
	SO_2		x				x		Metabolic response	101
	SO_2				x				SO_2 tolerance study	187
	SO_2				x	x			Growth response	188
	SO_2				x				Growth response	189
	SO_2				x		x		Nutrition effects	190
	SO_2				x				Photoassimilated ^{14}C response	191
	SO_2				x	x	x		Low-exposure concentration study	192

Vegetable Crops

Bean, snap (*Phaseolus vulgaris* var. *humulis*).						
	O_3	×				Foliar response	193
	O_3		×		×	Disease interaction study	194
	O_3		×			Fruit injury in storage	195
	O_3	×		×		Exposures at different growth stages	196
	O_3	×		×		Field chamber study	197
	O_3	×	×			Exposures at different growth stages	73
	O_3		×	×		Ozone protectant study	198
	O_3	×			×	Temperature effects	199
	O_3		× ×	×		Yield response	200
	O_3	×		×		Ozone protectant study	201
	O_3			×	×	Disease interaction study	202
	O_3			×	×	Relative humidity effects	203
	SO_2			×	×	Relative humidity effects	203
	SO_2			×	×	Photosynthetic transpiration response	204
	SO_2	×		×		Field study	205
	F			×	×	Progeny response	206
	F	×		×		Seed production effects	127
	NO_2		× ×	×		Nutrition effects	207
	NH_3	×	×	×		Continuous stirred tank reactor study	27
	Mix	×		×		Foliar response	193
Beet (*Beta vulgaris* L.)						
	O_3			×		Salinity effects	208
Cabbage (*Brassica oleracea* var. *capitata* L.)						
	O_3		×			Composition study	209
	O_3		×	× ×	×	Disease interaction study	210
	SO_2		×	× ×		Nutrition, moisture, age effects	176

Table 2 (continued)
PUBLISHED INFORMATION ON CROP RESPONSES TO GASEOUS POLLUTANTS

Species	Pollutant	Cultivar differences	Species differences	Visible injury	Growth responses	Yield responses	Environmental factors	Interactions	Comments	Ref.
Vegetable Crops (continued)										
Carrot (*Daucus carota* L.)	O_3					×			Root tissue study	211
	O_3		×	×	×				Composition study	209
	F		×	×	×				Burn effects	178
Celery (*Apium graveolens* L. var *dulce*)	O_3	×		×					Field and chamber study	212
	SO_2		×	×					Relative susceptibility study	24
	F		×	×					Relative susceptibility study	24
Corn, sweet (*Zea mays* L.)	O_3	×		×					Susceptibility study	213
	O_3	×		×		×			Low-exposure concentration study	214
	O_3	×		×					Genetic effects	215
	O_3		×		×				Controlled environment study	216
	O_3		×		×	×			Composition study	209
	O_3	×			×	×			Field chamber study	217
	O_3	×			×	×			Water balance effects	218
	PAN	×		×			×	×	Temperature, days to maturity interaction	219
	SO_2		×	×	×	×			Field study	220
	SO_2		×	×	×				Foliar response	11
	SO_2		×	×					Relative susceptibility study	24
	F		×	×	×				Foliar response	11
	F		×						Relative susceptibility study	24
	F			×					F accumulation study	221

	NO$_2$				×	F accumulation study	221
	Mix		×	×		Field study	220
	Mix		×	×		Foliar response	11
	Mix			×		F accumulation study	221
Cress *(Lepidium sativum L.)*							
	O$_3$				×	Heavy-metal interaction study	222
	O$_3$		×	×		Heavy-metal interaction study	223
	O$_3$			×		Heavy-metal interaction study	224
Cucumber *(Cucumis sativus L.)*							
	O$_3$		×	×		Leaf diffusive resistance study	149
	O$_3$		×	×		Foliar response	90
	O$_3$	×	×	×		Ozone protectant study	225
	O$_3$		×	×		Air velocity effects	30
	O$_3$	×	×	×	×	Relative susceptibility study	226
	O$_3$		×	×		Starch hydrolysis response	86
	PAN		×			Starch hydrolysis response	86
	SO$_2$		×	×	×	Leaf diffusive resistance study	149
	SO$_2$		×	×	×	Relative susceptibility study	227
	SO$_2$	×	×			Leaf age effects	228
	SO$_2$		×	×		Controlled environment study	227
	SO$_2$		×	×		Relative susceptibility study	24
	F		×	×		Relative susceptibility study	24
	Mix		×	×		Leaf diffusive resistance study	149
	Mix		×	×		Foliar response	90
Eggplant *(Solanum melongena* L. var. *esculentum)*							
	O$_3$	×	×			Controlled environment study	229
	SO$_2$		×			Relative susceptibility study	24
Lettuce *(Lactuca sativa L.)*							
	O$_3$	×	×			Relative susceptibility study	230
	O$_3$		×	×	×	Relative humidity effects	231

Volume I   243

Table 2 (continued)
PUBLISHED INFORMATION ON CROP RESPONSES TO GASEOUS POLLUTANTS

Vegetable Crops (continued)

Species	Pollutant	Cultivar differences	Species differences	Visible injury	Growth responses	Yield responses	Environmental factors	Interactions	Comments	Ref.
Lettuce (cont.)	O_3		×					×	Heavy-metal interaction	224
	O_3			×					Composition study	79
	O_3		×			×			Composition study	209
	SO_2		×	×				×	Heavy-metal interaction	232
	SO_2		×	×					Relative susceptibility study	24
	F		×	×					Relative susceptibility study	24
	F		×	×					Foliar response	16
	H_2S		×		×	×			Species comparison study	179
Muskmelon (*Cucumis melo* L.)	O_3		×	×				×	Ozone protectant study	198
	O_3		×	×					Ozone protectant study	201
	O_3		×	×					Fruit injury in storage	195
Onion (*Allium cepa* L.)	O_3			×		×		×	Protectant disease interaction	233
	O_3	×	×	×	×				Species comparison study	226
	O_3			×		×		×	Disease interaction	234
	O_3					×			Burn effects	235
	O_3					×			Genetic effects	236
Parsley (*Petroselinum crispum* Nym.)	O_3			×	×				Assimilate partitioning study	237
Parsnip (*Pastinaca sativa* L.)	F		×	×	×				Species comparison study	178

244 CRC Handbook of Plant Science in Agriculture

Pea *(Pisum sativum L.)*	O_3	×		×		Relative susceptibility study	238
	O_3	×		×		Species comparison study	239
	O_3			×		Foliar response	240
	O_3			×		Injury mechanism study	224
	SO_2			×		Foliar response	240
	Mix			×		Foliar response	240
Pepper *(Capsicum frutescens L.)*	O_3		×	×		Assimilate partitioning study	241
	F			×		Species comparison study	178
Potato *(Solanum tuberosum L.)*	O_3	×		×	×	Ozone protectant study	242
	O_3	×		×		Species comparison study	226
	O_3	×		×		Foliar response	243
	O_3	×		×	×	Field and chamber study	244
	O_3	×		×	×	Field and chamber study	245
	O_3	×		×		Days to maturity interaction study	246
	O_3	×		×		Relative susceptibility study	28
	O_3			×	×	Disease interaction	247
	O_3			×		Species comparison study	248
	O_3	×				Composition study	249
	SO_2		×	×		Species comparison study	23
	NO_2		×	×		Species comparison study	23
	Mix		×	×		Species comparison study	23
Potato, sweet *(Ipomoea batatas Lam.)*	SO_2		×	×		Species comparison study	24
	F		×	×		Species comparison study	24
Pumpkin *(Cucurbita pepo L.)*	SO_2		×	×		Species comparison study	24
	F		×	×		Species comparison study	24

246 CRC Handbook of Plant Science in Agriculture

Table 2 (continued)
PUBLISHED INFORMATION ON CROP RESPONSES TO GASEOUS POLLUTANTS

Species	Pollutant	Cultivar differences	Species differences	Visible injury	Growth responses	Yield responses	Environmental factors	Interactions	Comments	Ref.
Vegetable Crops (continued)										
Radish (*Raphanus sativus* L.)	O_3								Continuous stirred tank reactor study	250
	O_3			x	x				Greenhouse study	251
	O_3		x		x				Growth response	168
	O_3				x	x			Low-exposure concentration study	252
	O_3				x		x		Nutrition, temperature effects	253
	O_3			x	x				Root growth effects	252
	O_3			x	x				Growth regulator interaction	255
	O_3	x				x			Relative susceptibility study	230
	O_3				x				Adaptive plant response	256
	O_3	x			x		x		Temperature effects	257
	O_3			x	x				Composition study	258
	O_3		x		x				Metabolic response	144
	O_3		x		x				Membrane permeability response	89
	O_3		x	x					Leaf diffusive resistance study	149
	O_3		x	x					Species comparison study	259
	O_3		x	x					Foliar response	90
	SO_2				x				Greenhouse study	251
	SO_2			x	x				Continuous stirred tank reactor study	250
	SO_2		x		x				Growth response	168
	SO_2		x		x				Membrane permeability response	89
	SO_2		x	x					Foliar response	33
	SO_2		x	x	x				Species comparison study	259
	SO_2		x	x					Leaf diffusive resistance study	149

	SO$_2$	×			×		Heavy-metal interaction study	232
	SO$_2$	×					Stomatal response	107
	SO$_2$	×	×				Heavy-metal interaction study	232
	F	×	×				Foliar response	16
	NO$_2$	×					Short exposure periods	117
	NO$_2$		×	×			Greenhouse study	251
	NO$_2$			×			Continuous stirred tank reactor study	250
	NO$_2$	×	×				Controlled environment study	33
	HCl	×	×		×		Seasonal, diurnal effects	120
	Mix		×	×			Greenhouse study	251
	Mix	×	×				Continuous stirred tank reactor study	250
	Mix		×	×			Growth response	168
	Mix		×	×			Low-exposure concentration study	252
	Mix	×	×	×	×		Membrane permeability response	89
	Mix	×	×	×			Leaf diffusive resistance study	149
	Mix	×	×				Species comparison study	259
	Mix		×				Foliar response	90
	Mix						Controlled environment study	33
Rutabaga (*Brassica napobrassica* Mill.)	SO$_2$	×	×		×		Nutrition, moisture, age effects	176
Spinach (*Spinacia oleracea* L.)	O$_3$	×					Species comparison study	129
	O$_3$		×		×		Composition study	260
	O$_3$	×	×				Controlled environment study	261
	O$_3$	×	×		×	×	Nutrition effects	262
	O$_3$				×	×	Relative humidity effects	231
	O$_3$	×			×		Photosynthetic response	263
	O$_3$		×				Species comparison study	129
	SO$_2$		×	×			Foliar response	264
	SO$_2$	×	×	×			Photosynthetic response	265
	SO$_2$	×					Short exposure periods	117
	NO$_2$	×	×				Species comparison study	129
	Mix	×	×				Species comparison study	129

Table 2 (continued)
PUBLISHED INFORMATION ON CROP RESPONSES TO GASEOUS POLLUTANTS

Species	Pollutant	Cultivar differences	Species differences	Visible injury	Growth responses	Yield responses	Environmental factors	Interactions	Comments	Ref.
Vegetable Crops (continued)										
Squash (*Cucurbita* sp. L.)	F									178
Swiss chard (*Beta vulgaris* var. *circla*)	SO$_2$		x	x	x				Species comparison study	33
	NO$_2$		x	x					Controlled environment study	33
	Mix		x	x					Controlled environment study	33
Tomato (*Lycopersicon esculentum* Mill.)	O$_3$				x		x		Controlled environment study	266
	O$_3$	x						x	Moisture effects	64
	O$_3$	x			x			x	Ozone protectant study	93
	O$_3$	x		x	x				Herbicide interaction study	267
	O$_3$			x	x	x			Pollen response	268
	O$_3$				x	x			Age effects	269
	O$_3$	x			x	x			Diurnal effects	268
	O$_3$			x	x	x			Composition study	270
	O$_3$			x		x			Stem elongation study	271
	O$_3$						x		Field study	272
	O$_3$	x		x	x				Relative susceptibility study	273
	O$_3$				x	x			Nutrition effects	274
	O$_3$				x	x		x	Transpiration response	275
	O$_3$								Ozone protectant study	79
	O$_3$		x		x	x			Composition study	200
	O$_3$								Species comparison study	

								Reference
Tomato (cont.)	O₃				×		Species comparison study	276
	O₃			×	×		Composition study	209
	O₃	×		×			Species comparison study	46
	O₃			×	×		Nutrition effects	277
	O₃	×		×	×	×	Dose, sensitivity interaction study	278
	O₃	×		×	×	×	Virus interaction study	279
	O₃				×		Enzyme activity study	77
	O₃			×	×		Foliar response	90
	SO₂			×	×	×	Nutrition effects	280
	SO₂			×	×		Disease interaction study	281
	SO₂	×		×			Controlled environment study	282
	SO₂			×	×		Species comparison study	24
	F			×	×	×	Nutrition effects	283
	F			×	×		Foliar response	16
	F			×	×	×	Controlled environment study	109
	F			×	×		F accumulation study	25
	F	×		×	×		Species comparison study	178
	F			×	×	×	Nutrition effects	284
	F			×	×	×	Nutrition effects	285
	F			×	×		Nutrition effects	286
	F		×	×	×		Species comparison study	24
	F		×	×	×		Metabolic response	113
	NO₂				×		Photosynthetic response	108
	NO₂			×	×		Composition study	26
	NO₂	×			×		Greenhouse study	287
	NO₂			×	×		Nutrition effects	288
	NO₂	×			×		Controlled environment study	284
	NO₂		×		×		Controlled environment study	289
	NH₃			×	×		Continuous stirred tank reactor study	27
	Mix			×	×		Photosynthetic response	108
	Mix			×	×		Greenhouse study	287
	Mix			×	×		Controlled environment study	282
	Mix			×	×		Foliar response	90
Turnip (*Brassica rapa* L.)	SO₂				×		Nutrition, moisture, age effects	176

Table 2 (continued)
PUBLISHED INFORMATION ON CROP RESPONSES TO GASEOUS POLLUTANTS

Fruit Crops

Species	Pollutant	Cultivar differences	Species differences	Visible injury	Growth responses	Yield responses	Environmental factors	Interactions	Comments	Ref.
Apple (*Malus sylvestris* Mill.)	O_3			×				×	Ozone protectant study	201
	O_3	×		×					Field chamber study	290
	O_3		×	×				×	Ozone protectant study	291
	O_3	×		×					Ozone protectant study	198
	O_3	×		×					Leaf age effects	292
	SO_2			×	×				Container-grown trees	293
	SO_2	×		×					Leaf age effects	292
	SO_2		×	×	×				Container-grown trees	293
	F	×	×	×	×				Species comparison study	24
	F		×	×					Burn effects	178
	Mix	×		×					Species comparison study	24
									Leaf age effects	292
Apricot (*Prunus armeniaca* L.)	SO_2				×				Pollen response	294
	SO_2		×	×					Species comparison study	24
	F		×	×					Species comparison study	24
	F	×		×	×				Pollen response	295
	F			×					Burn effects	178

Species	Pollutant				Description	Ref.
Blueberry (*Vaccinium corymbosum* L.)	O₃		×		Fruit injury in storage	195
	SO₂	×			Moisture, leaf age effects	296
	SO₂		×		Species comparison study	24
	F		×	×	Burn effects	178
	F		×		Species comparison study	24
Cherry, sweet (*Prunus avium* L.)	SO₂		×	×	Pollen response	294
	F	×	×	×	Nutrition effects	297
	F		×		Burn effects	178
	F		×		Pollen response	298
	F		×		Pollen response	299
Grape (*Vitis* sp. L.)	O₃		×		Field study	300
	O₃	×	×		Fruit injury in storage	195
	O₃	×	×		Field chamber study	301
	O₃	×	×	×	Vineyard management effects	302
	O₃	×	×		Ozone protectant study	303
	O₃		×		Relative susceptibility study	304
	O₃		×		Symptom description	305
	O₃		×	×	Ozone protectant study	306
	O₃	×		×	Greenhouse study	307
	PAN	×		×	PAN protectant study	306
	SO₂	×	×		Greenhouse study	307
	SO₂	×	×		Greenhouse study	308
	SO₂		×		Species comparison study	24
	F		×		Species comparison study	24
	F				Growth response	309
	F		×		Field chamber study	310
	F				F accumulation study	311
	F				F accumulation study	312

Table 2 (continued)
PUBLISHED INFORMATION ON CROP RESPONSES TO GASEOUS POLLUTANTS

Fruit Crops (continued)

Species	Pollutant	Cultivar differences	Species differences	Visible injury	Growth responses	Yield responses	Environmental factors	Interactions	Comments	Ref.
Grape (cont.)	F			×	×			×	F protectant study	313
	F	×			×				Growth response	314
	F		×	×	×				Burn effects	178
	F	×	×		×			×	F protectant study	315
	NO$_2$					×		×	NO$_2$ protectant study	306
	H$_2$S	×		×	×				Species comparison study	179
	Mix			×	×				Greenhouse study	307
Grapefruit (Citrus paradisi Macf.)	O$_3$		×	×	×				Species comparison study	316
	F	×	×		×				Controlled environment study	317
	F	×	×		×	×			Field and greenhouse study	318
	F		×		×				Leaf age effects	319
	NO$_2$		×		×				Leaf age effects	319
Lemon (C. limon Burm.)	O$_3$				×				Photosynthetic response	320
	O$_3$				×				Metabolic response	321
	O$_3$		×	×	×				Species comparison study	316
	F	×	×		×	×			Field and greenhouse study	318
	F	×	×		×				Controlled environment study	317
	F		×	×	×				Species comparison study	322
Mandarin (Citrus L. sp.)	SO$_2$		×	×	×				Species comparison study	323
	HF		×	×	×				Species comparison study	323
	Mix		×	×	×				Species comparison study	323

Species	Pollutant					Study	Page
Orange (*C. sinensis* Osbeck)	O_3					Greenhouse study	324
	O_3			×		Ozone protectant study	325
	O_3	×		×		Species comparison study	326
	PAN	×		×		Species comparison study	326
	PAN			×		Greenhouse study	327
	SO_2		×			Leaf age effects	328
	F			×	×	Species comparison study	326
	F	×		×	×	Leaf age effects	319
	F			×		Controlled environment study	323
	F			×	×	Field study	329
	F		×	×		Composition study	330
	F			×	×	Species comparison study	322
	F	×			×	Field and greenhouse study	318
	F	×			×	F protectant study	331
	NO_2	×	×			Leaf age effects	319
	NO_2			×	×	Species comparison study	326
	Mix			×		Controlled environment study	323
Peach (*Prunus persica* Batsch.)	O_3		×			Fruit injury in storage	195
	O_3		×		×	Disease interaction study	332
	SO_2		×	×		Species comparison study	24
	F		×	×		Burn effects	178
	F		×	×		Species comparison study	24
Pecan (*Carya illinoensis* Koch)	SO_2			×		Photosynthetic response	174
Plum (*Prunus domestica* L.)	SO_2	×				Species comparison study	24
	F	×				Species comparison study	24

Table 2 (continued)
PUBLISHED INFORMATION ON CROP RESPONSES TO GASEOUS POLLUTANTS

Species	Pollutant	Cultivar differences	Species differences	Visible injury	Growth responses	Yield responses	Environmental factors	Interactions	Comments	Ref.
Fruit Crops (continued)										
Prune (*Prunus* L. sp.)	SO_2		x	x					Species comparison study	24
	F		x	x					Burn effects	178
	F		x	x	x				Species comparison study	24
Raspberry (*Rubus idaeus* L.)										
Strawberry (*Fragaria* L. spp.)	F		x	x	x				Burn effects	178
	O_3		x	x					Fruit injury in storage	195
	O_3		x						Composition study	209
	O_3			x				x	Disease interaction study	332
	O_3		x	x	x	x			Controlled environment study	333
	SO_2	x	x						Species comparison study	24
	SO_2	x			x	x			Controlled environment study	333
	F			x	x				Fruiting response	334
	F		x	x					Species comparison study	24
	Mix				x	x			Controlled environment study	333
Tangerine (*Citrus reticulata* Blanco)	F		x		x	x			Field and greenhouse study	318

Crop	Pollutant								Study type	Ref.
Cotton (*Gossypium hirsutum* L.)										
	O_3						×		Field chamber study	335
	O_3					×			Sensitivity factors	336
	O_3				×				Ozone protectant study	64
	SO_2				×		×		Field study	100
	SO_2				×				Species comparison study	24
	F				×		×		Field study	100
	F				×				Species comparison study	24
	NH_3				×				Continuous stirred tank reactor study	27
	Cl_2					×	×		Field study	100
	Mix					×	×		Field study	100
Fiber Crops										
Ginseng (*Panax* L. sp.)										
	O_3			×		×			Species comparison study	259
	SO_2			×		×			Species comparison study	259
	Mix			×		×			Species comparison study	259
Drug Crops										
Tobacco (*Nicotiana tabacum* L.)										
	O_3		×		×			×	Disease interaction study	337
	O_3		×		×			×	Ozone protectant study	198
	O_3	×	×		×			×	Ozone protectant study	338
	O_3		×		×			×	Ozone protectant study	339
	O_3		×		×			×	Relative humidity effects	340
	O_3	×	×		×			×	Light relative humidity effects	69
	O_3		×		×			×	Moisture effects	341
	O_3		×		×			×	Previous exposure effects	342
	O_3		×		×			×	Air velocity effects	30
	O_3		×		×			×	Nutrition effects	343

Table 2 (continued)
PUBLISHED INFORMATION ON CROP RESPONSES TO GASEOUS POLLUTANTS

Species	Pollutant	Cultivar differences	Species differences	Visible injury	Growth responses	Yield responses	Environmental factors	Interactions	Comments	Ref.
Drug Crops (continued)										
Tobacco (cont.)	O_3		×	×			×		Temperature, light intensity, humidity effects	70
	O_3	×		×				×	Disease interaction study	344
	O_3		×	×				×	Ozone protectant study	201
	O_3	×		×				×	Ozone protectant study	345
	O_3			×				×	Disease interaction study	75
	O_3			×				×	Herbicide interaction study	346
	O_3				×				Respiration response	347
	O_3		×		×				Metabolic response	144
	O_3		×	×				×	Ozone protectant study	47
	O_3		×		×				Metabolic response	46
	O_3	×			×				Growth study	348
	O_3			×			×		Nutrition, temperature, moisture effects	349
	O_3			×				×	Ozone protectant study	350
	O_3			×			×		Moisture effects	351
	O_3			×			×		Temperature effects	352
	O_3	×		×				×	Disease interaction study	353
	O_3			×		×			Ozone flux study	354
	O_3			×			×		Several environmental effects	355
	O_3	×		×			×		Temperature, light effects	356
	O_3				×				Composition study	357
	O_3	×		×			×		Light effects	358
	O_3	×			×			×	Herbicide interaction study	93
	O_3			×				×	Nutrition effects	359
	O_3	×		×					Leaf age effects	360
	O_3	×		×					Stomatal response	361

O_3			×		Relative susceptibility study	362
O_3	×		×		Field and controlled environment study	363
O_3			×		Field study	364
O_3	×			×	Ozone protectant study	64
O_3			×	×	Ozone protectant study	365
O_3				×	Ozone protectant study	366
O_3	×		×	×	Ozone protectant study	367
O_3	×		×		Field and controlled environment study	368
O_3			×		Leaf age effects	369
O_3		×			Species comparison study	168
O_3		×			Species comparison study	259
O_3	×		×		Controlled environment study	88
O_3	×		×		Relative susceptibility study	370
O_3	×		×		Disease interaction study	371
O_3			×	×	Relative susceptibility study	372
O_3			×		Leaf age effects	369
SO_2		×	×		Species comparison study	168
SO_2		×	×		Species comparison study	259
SO_2		×	×		Controlled environment study	88
SO_2	×	×	×		Disease interaction study	371
SO_2	×	×	×		Relative susceptibility study	372
SO_2		×	×		Species comparison study	23
SO_2		×	×		Field study	100
SO_2		×	×		Species comparison study	24
SO_2		×		×	Moisture effects	13
SO_2		×	×		Species comparison study	107
SO_2		×	×		Species comparison study	373
SO_2		×			Species comparison study	21
SO_2		×	×		Nutrition effects	280
SO_2		×	×		Controlled environment study	229
F		×	×		Field study	100
F		×	×		Species comparison study	24
NO_2		×	×		Species comparison study	23
NO_2		×	×		Species comparison study	117

Table 2 (continued)
PUBLISHED INFORMATION ON CROP RESPONSES TO GASEOUS POLLUTANTS

Species	Pollutant	Cultivar differences	Species differences	Visible injury	Growth responses	Yield responses	Environmental factors	Interactions	Comments	Ref.
Drug Crops (continued)										
Tobacco (cont.)	NH₃		×		×				Continuous stirred tank reactor study	27
	Cl₂		×		×				Field study	100
	Mix		×		×				Field and controlled environment study	368
	Mix	×		×					Leaf age effects	369
	Mix		×	×	×				Species comparison study	168
	Mix		×	×	×				Species comparison study	259
	Mix			×	×				Controlled environment study	88
	Mix	×	×	×					Relative susceptibility study	370
	Mix		×		×				Field study	100
	Mix	×	×					×	Disease interaction study	371
	Mix		×	×	×				Species comparison study	23
	Mix	×	×	×					Relative susceptiblity study	372
Sugar Crops										
Beet, sugar (*Beta vulgaris* L.)	O₃			×	×			×	Nutrition effects	262
	O₃	×		×	×				Controlled environment study	374
	H₂S		×		×	×			Species comparison study	179

Other Starch Crops

Buckwheat (*Fagopyrum esculentum* Moench)	SO_2	×	Species comparison study	24
	F	×	Species comparison study	24

Others

Mustard (*Brassica campestris* L.)	SO_2	× ×	Nutrition, moisture, age effects	176

REFERENCES

1. **Mudd, J. B. and Kozlowski, T. T., Eds.**, *Responses of Plants to Air Pollution*, Academic Press, New York, 1975.
2. **Ormrod, D. P.**, *Pollution in Horticulture*, Elsevier, Amsterdam, 1978.
3. **Unsworth, M. H. and Ormrod, D. P., Eds.**, *Effects of Gaseous Air Pollution in Agriculture and Horticulture*, Butterworths, London, 1982.
4. **Jacobson, J. S. and Hill, A. C., Eds.**, *Recognition of Air Pollution Injury to Vegetation: A Pictorial Atlas*, Air Pollution Control Association, Pittsburgh, 1970.
5. **Manning, W. J. and Feder, W. A.**, *Biomonitoring Air Pollutants with Plants*, Applied Science, London, 1980.
6. **Lacasse, N. L. and Treshow, M., Eds.**, Composite sensitivity chart and scientific names of plants, Appendix C, in Diagnosing Vegetation Injury Caused by Air Pollution, Contract 68-02-1344, U.S. Environmental Protection Agency, Washington, D.C., 1976.
7. **Bennett, J. H. and Hill, A. C.**, Inhibition of apparent photosynthesis by air pollutants, *J. Environ. Qual.*, 2, 526, 1973.
8. **Rabe, R. and Kreeb, K. H.**, Bioindication of air pollution by chlorophyll destruction in plant leaves, *Oikos*, 34, 163, 1980.
9. **Heagle, A. S. and Strickland, A.**, Reaction of *Erysiphe graminis* f. sp. *hordei* to low levels of ozone, *Phytopathology*, 62, 1144, 1972.
10. **Markowski, A. and Grzesiak, S.**, Influence of sulphur dioxide and ozone on vegetation of bean and barley plants under different soil moisture conditions, *Bull. Acad. Pol. Sci. Ser. Sci. Biol.*, 22, 12, 1974.
11. **Mandl, R. H., Weinstein, L. H., and Keveny, M.**, Effects of hydrogen fluoride and sulphur dioxide alone and in combination on several species of plants, *Environ. Pollut.*, 9, 133, 1975.
12. **Toivonen, P. M. A. and Hofstra, G.**, The interaction of copper and sulphur dioxide in plant injury, *Can. J. Plant Sci.*, 59, 475, 1979.
13. **Markowski, A., Grzesiak, S., and Schramel, M.**, Susceptibility of six species of cultivated plants to sulphur dioxide under optimum soil moisture and drought conditions, *Bull. Acad. Pol. Sci. Ser. Sci. Biol.*, 22, 889, 1974.
14. **Buckenham, A. H., Parry, M. A. J., and Whittingham, C. P.**, Effects of aerial pollutants on the growth and yield of spring barley, *Ann. Appl. Biol.*, 100, 179, 1982.
15. **Walker, D. R., Dick, A. C., and Nyborg, M.**, Deposition of sulphur gases from multiple scattered sources in Alberta, *Water Air Soil Pollut.*, 16, 223, 1981.
16. **Granett, A. L.**, Pictorial keys to evaluate foliar injury caused by hydrogen fluoride, *HortScience*, 17, 587, 1982.
17. **Cirelli, B.**, Observations on the effects of some air pollutants on *Zea mays* leaf tissue, *Phytopathol. Z.*, 86, 233, 1976.
18. **Mumford, R. A., Lipke, H., and Laufer, D. A.**, Ozone-induced changes in corn pollen, *Environ. Sci. Technol.*, 6, 427, 1972.
19. **Heagle, A. S., Philbeck, R. B., and Knott, W. M.**, Thresholds for injury growth, and yield loss caused by ozone on field corn hybrids, *Phytopathology*, 69, 21, 1979.
20. **Laurence, J. A.**, Response of maize and wheat to sulfur dioxide. *Plant Dis. Rep.*, 63, 468, 1979.
21. **Faller, N.**, Effects of atmospheric SO_2 on plants, *Sulphur Inst. J.*, 6, 5, 1970.
22. **Klein, H., Jager, H. J., Domes, W., and Wong, C. H.**, Mechanisms contributing to differential sensitivities of plants to SO_2, *Oecologia*, 33, 203, 1978.
23. **Elkiey, T., Ormrod, D. P., and Marie, B. A.**, Growth Responses of Crop Plants in the Vegetative Stage to Sulfur Dioxide and/or Nitrogen Dioxide, Research Report, Department of Horticultural Science, University of Guelph, Ontario, 1981.
24. **Zimmerman, P. W. and Hitchcock, A. E.**, Susceptibility of plants to hydrofluoric acid and sulfur dioxide gas, *Contrib. Boyce Thompson Inst.*, 18, 263, 1956.
25. **Leone, I. A., Brennan, E., and Daines, R. H.**, Atmospheric fluoride: its uptake and distribution in tomato and corn plants, *Plant Physiol.*, 31, 326, 1956.
26. **Matsumaru, T., Yoneyama, T., Totsuka, T., and Shiratori, K.**, Absorption of atmospheric NO_2 by plants and soils. I. Quantitative estimation of absorbed NO_2 in plants by ^{15}N method, *Soil Sci. Plant Nutr.* (Tokyo), 25, 255, 1979.
27. **Rodgers, H. H. and Aneja, V. P.**, Uptake of atmospheric ammonia by selected plant species, *Environ. Exp. Bot.*, 20, 251, 1980.
28. **Brennan, E., Leone, I. A., and Daines, R. H.**, The importance of variety in ozone plant damage, *Plant Dis. Rep.*, 48, 923, 1964.
29. **Heagle, A. S.**, Effect of low level ozone fumigations on crown rust of oats, *Phytopathology*, 60, 252, 1970.

30. **Heagle, A. S., Heck, W. W., and Body, D.,** Ozone injury to plants as influenced by air velocity during exposure, *Phytopathology,* 61, 1209, 1971.
31. **Heck, W. W. and Dunning, J. A.,** Response of oats to sulfur dioxide interactions of growth temperature with exposure temperature or humidity, *J. Air Pollut. Control Assoc.,* 28, 243, 1978.
32. **Marchesani, V. J. and Leone, I. A.,** A bioassay for assessing the effect of sulfur dioxide on oat seedlings, *J. Air Pollut. Control Assoc.,* 30, 163, 1980.
33. **Bennett, J. H., Hill, A. C., Soleimani, A., and Edwards, W. H.,** Acute effects of combination of sulphur dioxide and nitrogen dioxide on plants, *Environ. Pollut.,* 9, 127, 1975.
34. **Hill, A. C. and Bennett, J. H.,** Inhibition of apparent photosynthesis by nitrogen oxides, *Atmos. Environ.,* 4, 341, 1970.
35. **Nakamura, H. and Ota, Y.,** An injury to rice plants caused by photochemical oxidants in Japan, *Jpn. Agric. Res. Q.,* 12, 69, 1978.
36. **Fujinuma, Y. and Aiga, I.,** Selected rice (*Oryza sativa* L.) strains as an indicator plant for air pollution, *Res. Rep. Natl. Inst. Environ. Stud.,* 11, 255, 1980.
37. **Matsuoka, Y.,** Injury of rice plants caused by sulfur dioxide and its mechanism, *Jpn. Agric. Res. Q.,* 12, 183, 1978.
38. **Heagle, A. S. and Key, L. W.,** Effect of *Puccinia graminis* f. sp. *tritici* on ozone injury in wheat, *Phytopathology,* 63, 609, 1973.
39. **Heagle, A. S. and Key, L. W.,** Effect of ozone on the wheat stem rust fungus, *Phytopathology,* 63, 397, 1973.
40. **Shannon, J. G. and Mulchi, C. L.,** Ozone damage to wheat varieties at anthesis, *Crop Sci.,* 14, 335, 1974.
41. **Sij. J. W., Kanemasu, E. T., and Goltz, S. M.,** Some preliminary results of sulfur dioxide effects on photosynthesis and yield in field-grown wheat, *Trans. Kans. Acad. Sci.,* 76, 199, 1974.
42. **Prasad, B. J. and Rao, D. N.,** Influence of nitrogen dioxide (NO_2) on photosynthetic apparatus and net primary productivity of wheat plants, *Acta Bot. Indica,* 7, 16, 1979.
43. **Stan, H. J., Schicker, S., and Kassner, H.,** Stress ethylene evolution of bean plants — a parameter indicating ozone pollution, *Atmos. Environ.,* 15, 391, 1981.
44. **Pellissier, M., Lacasse, N. L., Ercegovich, C. D., and Cole, H., Jr.,** Effects of hydrocarbon wax emulsion sprays in reducing visible ozone injury to *Phaseolus vulgaris* 'Pinto III', *Plant Dis. Rep.,* 56, 6, 1972.
45. **Yarwood, C. E. and Middleton, J. T.,** Smog injury and rust infection, *Plant Physiol.,* 29, 393, 1954.
46. **Craker, L. E.,** Ethylene production from ozone injured plants, *Environ. Pollut.,* 1, 299, 1971.
47. **Miller, P. M., Tomlinson, H., and Taylor, G. S.,** Reducing severity of ozone damage to tobacco and beans by combining benomyl or carboxin with contact nematicides, *Plant Dis. Rep.,* 60, 433, 1976.
48. **Davis, D. D. and Smith, S. H.,** Reduction of ozone sensitivity of Pinto bean by virus-induced local lesions, *Plant Dis. Rep.,* 60, 31, 1976.
49. **Runeckles, V. C. and Rosen, P. M.,** Effects of ambient ozone pretreatment on transpiration and susceptibility to ozone injury, *Can. J. Bot.,* 76, 310, 1976.
50. **Adedipe, N. O., Hofstra, G., and Ormrod, D. P.,** Effects of sulfur nutrition on phytotoxicity and growth responses of bean plants to ozone, *Can. J. Bot.,* 50, 1789, 1972.
51. **Davis, D. D. and Smith, S. H.,** Bean common mosaic virus reduces ozone sensitivity of Pinto bean, *Environ. Pollut.,* 9, 97, 1975.
52. **Otto, H. and Daines, R. H.,** Plant injury by air pollutants: influence of humidity on stomatal apertures and plant response to ozone, *Science,* 163, 1209, 1969.
53. **Larsen, R. I. and Heck, W. W.,** An air quality data analysis system for interrelating effects, standards and needed source reductions. III. Vegetation injury, *J. Air Pollut. Control Assoc.,* 26, 325, 1976.
54. **Pellissier, M., Lacasse, N. L., and Cole H., Jr.,** Effectiveness of benzimidazole, benomyl, and thiabendazole in reducing ozone injury to Pinto beans, *Phytopathology,* 62, 580, 1972.
55. **McIlveen, W. D., Spotts, R. A., and Davis, D. D.,** The influence of soil zinc on nodulation, mycorrhizae, and ozone-sensitivity of Pinto bean, *Phytopathology,* 65, 645, 1975.
56. **Hoffman, G. J., Maas, E. V., and Rawlins, S. L.,** Salinity-ozone interactive effects on yield and water relations of Pinto bean, *J. Environ. Qual.,* 2, 148, 1973.
57. **Dunning, J. A. and Heck, W. W.,** Response of bean and tobacco to ozone: effect of light intensity, temperature, and relative humidity, *J. Air Pollut. Control Assoc.,* 27, 882, 1977.
58. **Kendrick, J. B., Jr., Darley, E. F., and Middleton, J. T.,** Chemotherapy for oxidant and ozone induced plant damage, *Int. J. Air Water Pollut.,* 6, 391, 1962.
59. **Teso, R. R., Oshima, R. J., and Carmean, M. I.,** Ozone-pesticide interactions, *Calif. Agric.,* p. 13, April 1979.
60. **Davis, D. D. and Smith, S. H.,** Reduction of ozone sensitivity of Pinto bean by bean common mosaic virus, *Phytopathology,* 64, 383, 1974.

61. **Pellissier, M., Lacasse, N. L., and Cole, H., Jr.,** Effectiveness of benomyl-folicote treatments in reducing ozone injury to Pinto beans, *J. Air Pollut. Control Assoc.,* 22, 722, 1972.
62. **Manning, W. J., Feder, W. A., and Vardaro, P. M.,** Benomyl in soil and response of Pinto bean plants to repeated exposures to a low level of ozone, *Phytopathology,* 63, 1539, 1973.
63. **Manning, W. J., Feder, W. A., Papia, P. M., and Perkins, I.,** Influence of foliar ozone injury on root development and root surface fungi of Pinto bean plants, *Environ. Pollut.,* 1, 305, 1971.
64. **Rich, S., Ames, R., and Zukel, J. W.,** 1,4-Oxathiin derivatives protect plants against ozone, *Plant Dis. Rep.,* 58, 163, 1974.
65. **Bennett, J. P., Barnes, K., and Shinn, J. H.,** Interactive effects of H_2S and O_3 on yield of snap beans (*Phaseolus vulgaris* L.), *Environ. Exp. Bot.,* 20, 107, 1980.
66. **Spotts, R. A., Lukezic, F. L., and Lacasse, N. L.,** The effect of benzimidazole, cholesterol, and a steroid inhibitor on leaf sterols and ozone resistance of bean, *Phytopathology,* 65, 45, 1975.
67. **Todd, G. W.,** Effect of ozone and ozonated 1-hexene on respiration and photosynthesis of leaves, *Plant Physiol.,* 33, 416, 1958.
68. **Yarwood, C. E.,** Virus infection and heating reduce smog damage, *Plant Dis. Rep.,* 43, 129, 1959.
69. **Dunning, J. A. and Heck, W. W.,** Response of Pinto bean and tobacco to ozone as conditioned by light intensity and/or humidity, *Environ. Sci. Technol.,* 7, 824, 1973.
70. **Dunning, J. A. and Heck, W. W.,** Response of bean and tobacco to ozone: effect of light intensity, temperature and relative humidity, *J. Air Pollut. Control Assoc.,* 27, 882, 1977.
71. **Lewis, E. and Brennan, E.,** A disparity in the ozone response of bean plants grown in a greenhouse, growth chamber, or open-top chamber, *J. Air Pollut. Control Assoc.,* 27, 889, 1977.
72. **Knudson, L. L., Tibbitts, T. W., and Edwards, G. E.,** Measurement of ozone injury by determination of leaf chlorophyll concentration, *Plant Physiol.,* 60, 606, 1977.
73. **Meiners, J. P. and Heggestad, H. E.,** Evaluation of snap bean cultivars for resistance to ambient oxidants in field plots and to ozone in chambers, *Plant Dis. Rep.,* 63, 273, 1979.
74. **Hofstra, G., Littlejohns, D. A., and Wukasch, R. T.,** The efficacy of the anti-oxidant ethylene-diurea (EDU) compared to carboxin and benomyl in reducing yield losses from ozone in navy bean, *Plant Dis. Rep.,* 62, 350, 1978.
75. **Brennan, E.,** On exclusion as the mechanism of ozone resistance in virus infected plants, *Phytopathology,* 65, 1054, 1975.
76. **Carnahan, J. E., Jenner, E. L., and Wat, E. K. W.,** Prevention of ozone injury to plants by a new protectant chemical, *Phytopathology,* 68, 1225, 1978.
77. **Endress, A. G., Suarez, S. J., and Taylor, O. C.,** Peroxidase activity in plant leaves exposed to gaseous HCl or ozone, *Environ. Pollut.,* 22, 47, 1980.
78. **Prasad, K., Weigle, J. L., and Sherwood, C. H.,** Variation in ozone phytotoxicity among *Phaseolus vulgaris* cultivars, *Plant Dis. Rep.,* 54, 1026, 1970.
79. **Freebairn, H. T. and Taylor, O. C.,** Prevention of plant damage from air-borne oxidizing agents, *Proc. Am. Soc. Hortic. Sci.,* 76, 693, 1960.
80. **Runeckles, V. C. and Resh, H. M.,** Effects of cytokinins on responses of bean leaves to chronic ozone treatment, *Atmos. Environ.,* 9, 749, 1975.
81. **Todd, G. W. and Garber, M. J.,** Some effects of air pollutants on the growth and productivity of plants, *Bot. Gaz.,* 120, 75, 1958.
82. **Bennett, J. H., Lee, E. H., and Heggestad, H. E.,** Apparent photosynthesis and leaf stomatal diffusion in EDU treated ozone-sensitive bean plants, in *Proc. 5th Annu. Meet., Plant Growth Regulator Working Group,* Blacksburg, Va., 242, June 25—29, 1978.
83. **Butler, L. K. and Tibbitts, T. W.,** Variation in ozone sensitivity and symptom expression among cultivars of *Phaseolus vulgaris* L., *J. Am. Soc. Hortic. Sci.,* 104, 208, 1979.
84. **Miller, C. A. and Davis, D. D.,** Response of Pinto bean plants exposed to O_3, SO_2 or mixtures at varying temperatures, *HortScience,* 16, 548, 1981.
85. **Kohut, R. J. and Davis, D. D.,** Response of Pinto bean to simultaneous exposure to ozone and PAN, *Phytopathology,* 68, 567, 1978.
86. **Hanson, G. P. and Stewart, W. S.,** Photochemical oxidants: effect on starch hydrolysis in leaves, *Science,* 168, 1223, 1970.
87. **Dugger, W. M., Jr., Taylor, O. C., Cardiff, E., and Thompson, C. R.,** Stomatal action in plants as related to damage from photochemical oxidants, *Plant Physiol.,* 37, 487, 1962.
88. **Jacobson, J. S. and Colavito, L. J.,** The combined effect of sulfur dioxide and ozone on bean and tobacco plants, *Environ. Exp. Bot.,* 16, 277, 1976.
89. **Beckerson, D. W. and Hofstra, G.,** Effects of sulphur dioxide and ozone, singly or in combination, on membrane permeability, *Can. J. Bot.,* 58, 451, 1980.
90. **Hofstra, G. and Beckerson, D. W.,** Foliar responses of five plant species to ozone and a sulphur dioxide/ozone mixture after a sulphur dioxide pre-exposure, *Atmos. Environ.,* 15, 383, 1981.

91. **Hofstra, G. and Ormrod, D. P.**, Ozone and sulphur dioxide interaction in white bean and soybean, *Can. J. Plant Sci.*, 57, 1193, 1977.
92. **Dass, H. C. and Weaver, G. M.**, Enzymatic changes in intact leaves of *Phaseolus vulgaris* following ozone fumigation, *Atmos. Environ.*, 6, 759, 1972.
93. **Carney, A. W., Stephenson, G. R., Ormrod, D. P., and Ashton, G. C.**, Ozone-herbicide interactions in crop plants, *Weed Sci.*, 21, 508, 1973.
94. **Curtis, L. R., Edgington, L. V., and Littlejohns, D. J.**, Oxathiin chemicals for control of bronzing of white beans, *Can. J. Plant Sci.*, 55, 151, 1975.
95. **Temple, P. J. and Bisessar, S.**, Response of white bean to bacterial blight, ozone and antioxidant protection in the field, *Phytopathology*, 69, 101, 1979.
96. **Pell, E. J.**, Influence of benomyl soil treatment on Pinto bean plants exposed to peroxyacetyl nitrate and ozone, *Phytopathology*, 66, 731, 1976.
97. **Dugger, W. M., Jr., Mudd, J. B., and Koukol, J.**, Effects of PAN on certain photosynthetic reactions, *Arch. Environ. Health*, 10, 195, 1965.
98. **Starkey, T. E., Davis, D. D., and Merrill, W.**, Symptomology and susceptibility of ten bean varieties exposed to peroxyacetyl nitrate, *Plant Dis. Rep.*, 60, 480, 1976.
99. **Starkey, T. E., Davis, D. D., Pell, E. J., and Merrill, H.**, Influence of peroxyacetyl nitrate (PAN) on water stress in bean plants, *HortScience*, 16, 547, 1981.
100. **Hindawi, I. J.**, Injury by sulfur dioxide, hydrogen fluoride, and chlorine as observed and reflected on vegetation in the field, *J. Air Pollut. Control Assoc.*, 18, 307, 1968.
101. **Starsed, S. G. and Read, D. J.**, The uptake and metabolism of $^{35}SO_2$ in plants of differing sensitivity to sulphur dioxide, *Environ. Pollut.*, 13, 173, 1977.
102. **Laurence, J. A., Aluisio, A. L., Weinstein, H., and McCune, D. C.**, Effects of sulphur dioxide on southern bean mosaic and maize dwarf mosaic, *Environ. Pollut.*, 24, 185, 1981.
103. **Rist, D. L. and Davis, D. D.**, The influence of exposure temperature and relative humidity on the response of Pinto bean foliage to sulfur dioxide, *Phytopathology*, 69, 231, 1979.
104. **Davids, J. A., Davis, D. D., and Pennypacker, S. P.**, The influence of soil moisture on macroscopic sulfur dioxide injury to Pinto bean foliage, *Phytopathology*, 71, 1208, 1981.
105. **Pierre, M. and Queiroz, O.**, Enzymic and metabolic changes in bean leaves during continuous pollution by subnecrotic levels of SO_2, *Environ. Pollut.*, 25, 41, 1981.
106. **Noyes, R. D.**, The comparative effects of sulfur dioxide on photosynthesis and translocation in bean, *Physiol. Plant Pathol.*, 16, 73, 1980.
107. **Black, V. J. and Unsworth, M. H.**, Stomatal responses to sulphur dioxide and vapour pressure deficit, *J. Exp. Bot.*, 31, 667, 1980.
108. **Capron, T. M. and Mansfield, T. A.**, Inhibition of net photosynthesis in tomato in air polluted with NO and NO_2, *J. Exp. Bot.*, 27, 1161, 1976.
109. **MacLean, D. C., Schneider, R. E., and McCune, D. C.**, Effects of chronic exposure to gaseous fluoride on yield of field-grown bean and tomato plants, *J. Am. Soc. Hortic. Sci.*, 102, 297, 1977.
110. **McNulty, I. B. and Newman, D. W.**, Effects of atmospheric fluoride on the respiration rate of bush bean and gladiolus leaves, *Plant Physiol.*, 32, 121, 1957.
111. **Adams, D. F. and Sulzbach, C. W.**, Nitrogen deficiency and fluoride susceptibility of bean seedlings, *Science*, 133, 1425, 1961.
112. **Treshow, M., Dean, G., and Harner, F. M.**, Stimulation of tobacco mosaic virus-induced lesions on bean by fluoride, *Phytopathology*, 57, 756, 1967.
113. **Weinstein, L. H.**, Effects of atmospheric fluoride on metabolic constituents of tomato and bean leaves, *Contrib. Boyce Thompson Inst.*, 21, 215, 1961.
114. **Taylor, O. C. and Eaton, F. M.**, Suppression of plant growth by nitrogen dioxide, *Plant Physiol.*, 41, 132, 1966.
115. **Srivastava, H. S., Jolliffe, P. A., and Runeckles, V. C.**, Inhibition of gas exchange in bean leaves by NO_2, *Can. J. Bot.*, 53, 466, 1975.
116. **Fuhrer, J. and Erismann, K. H.**, Uptake of NO_2 by plants grown at different salinity levels, *Experentia*, 36, 409, 1980.
117. **Zeevaart, A. J.**, Some effects of fumigating plants for short periods with NO_2, *Environ. Pollut.*, 11, 87, 1976.
118. **Srivastava, H. S., Jolliffe, P. A., and Runeckles, V. C.**, The effects of environmental conditions on the inhibition of leaf gas exchange by NO_2, *Can. J. Bot.*, 53, 475, 1975.
119. **Ashenden, T. W.**, Effects of SO_2 and NO_2 pollution on transpiration in *Phaseolus vulgaris* L., *Environ. Pollut.*, 18, 45, 1979.
120. **Granett, A. L. and Taylor, O. C.**, Diurnal and seasonal changes in sensitivity of plants to short exposures of hydrogen chloride gas, *Agric. Environ.*, 6, 33, 1981.
121. **Granett, A. L. and Taylor, O. C.**, Effect of gaseous hydrogen chloride on seed germination and early development of seedlings, *J. Am. Soc. Hortic. Sci.*, 105, 548, 1980.

122. **Endress, A. G., Oshima, R. J., and Taylor, O. C.**, Age dependent growth and injury responses of Pinto bean leaves to gaseous hydrogen chloride, *J. Environ. Qual.*, 8, 260, 1979.
123. **Black, V. J., Ormrod, D. P., and Unsworth, M. H.**, Effects of low concentrations of ozone, singly, and in combination with sulphur dioxide on net photosynthesis rates of *Vicia faba* L., *J. Exp. Bot.*, 33, 1302, 1982.
124. **Ormrod, D. P., Black, V. J., and Unsworth, M. H.**, Depression of net photosynthesis in *Vicia faba* L. exposed to sulphur dioxide and ozone, *Nature (London)*, 291, 585, 1981.
125. **Black, C. R. and Black, V. J.**, Light and scanning electron microscopy of SO_2-induced injury to leaf surfaces of field bean *(Vicia faba* L.*)*, *Plant Cell Environ.*, 2, 329, 1979.
126. **Black, V. J. and Unsworth, M. H.**, Effects of low concentrations of sulphur dioxide on net photosynthesis and dark respiration of *Vicia faba*, *J. Exp. Bot.*, 30, 1, 1979.
127. **Pack, M.**, Effects of hydrogen fluoride on production and organic reserves of bean seed, *Environ. Sci. Technol.*, 5, 1128, 1971.
128. **Ormrod, D. P.**, Cadmium and nickel effects on growth and ozone sensitivity of pea, *Water Air Soil Pollut.*, 8, 263, 1977.
129. **Fujiwara, T. and Umezawa, T.**, Effects of Mixed Air Pollutants on Vegetation. I. Sulfur Dioxide, Nitrogen Dioxide and Ozone to Injure Peas and Spinach, Rep. 73007, Noden Institute, Central Institute of Electric Power, Akibo, Chiba, Japan, 1973, 12.
130. **Olszyk, D. M. and Tibbitts, T. W.**, Stomatal response and leaf injury of *Pisum sativum* L. with SO_2 and O_3 exposures. II. Influence of moisture stress and time of exposure, *Plant Physiol.*, 67, 545, 1981.
131. **Horsman, D. C. and Wellburn, A. R.**, Synergistic effect of SO_2 and NO_2 polluted air upon enzyme activity in pea seedlings, *Environ. Pollut.*, 8, 123, 1975.
132. **Moeri, P. B.**, Effects of fluoride emissions on enzyme activity in metabolism of agricultural plants, *Fluoride*, 13, 122, 1980.
133. **Howell, R. K. and Thomas, C. A.**, Relative tolerances of twelve safflower cultivars to ozone, *Plant Dis. Rep.*, 56, 195, 1972.
134. **Heagle, A. S., Body, D. E., and Neely, G. E.**, Injury and yield responses of soybean to chronic doses of ozone and sulfur dioxide in the field, *Phytopathology*, 64, 132, 1974.
135. **Tingey, D. T., Reinert, R. A., and Carter, H. B.**, Soybean cultivars: acute foliar response to ozone, *Crop Sci.*, 12, 268, 1972.
136. **Taylor, G. E., Jr., Tingey, D. T., and Ratsch, H. C.**, Ozone flux in *Glycine max* (L.) Merr.: sites of regulation and relationship to leaf injury, *Oecologia*, 53, 179, 1982.
137. **Heagle, A. S.**, Effects of growth media, fertilizer rate, and hour and season of exposure on sensitivity of four soybean cultivars to ozone, *Environ. Pollut.*, 18, 313, 1979.
138. **Howell, R. K., Rose, L. P., Jr., and Leffel, R. C.**, Field-testing soybeans for residual effects of air pollution and seed size on crop yield, *J. Environ. Qual.*, 9, 66, 1980.
139. **Vargo, R. H., Pell, E. J., and Smith, S. H.**, Induced resistance to ozone injury of soybean by tobacco ringspot virus, *Phytopathology*, 68, 715, 1978.
140. **Tingey, D. T., Fites, R. C., and Wickliff, C.**, Activity changes in selected enzymes from soybean leaves following ozone exposure, *Physiol. Plant.*, 33, 316, 1975.
141. **Tingey, D. T., Fites, R. C., and Wickliff, C.**, Differential foliar sensitivity of soybean cultivars to ozone associated with differential enzyme activities, *Physiol. Plant.*, 37, 69, 1976.
142. **Howell, R. K., Koch, E. J., and Rose, L. P., Jr.**, Field assessment of air pollution-induced soybean yield losses, *Agron. J.*, 71, 285, 1979.
143. **Pell, E. J. and Weissberger, W. C.**, Histopathological characterization of ozone injury to soybean foliage, *Phytopathology*, 66, 856, 1976.
144. **Tingey, D. T.**, Ozone induced alterations in plant growth and metabolism, Res. Rep., U.S. Environmental Protection Agency, Corvallis, Ore., 1976, 601.
145. **Howell, R. K. and Rose, L. P., Jr.**, Residual air pollution effects on seed quality, *Plant Dis. Rep.*, 64, 385, 1980.
146. **Keen, N. T. and Taylor, O. C.**, Ozone injury in soybeans, *Plant Physiol.*, 55, 731, 1975.
147. **Tingey, D. T., Reinert, R. A., Wickliff, C., and Heck, W. W.**, Chronic ozone or sulfur dioxide exposures, or both, affect the early vegetative growth of soybean, *Can. J. Plant Sci.*, 53, 875, 1973.
148. **Weber, D. E., Reinert, R. A., and Barker, K. R.**, Ozone and sulfur dioxide effects on reproduction and host-parasite relationships of selected plant-parasite nematodes, *Phytopathology*, 69, 624, 1979.
149. **Beckerson, D. W. and Hofstra, G.**, Response of leaf diffusive resistance of radish, cucumber and soybean to O_3 and SO_2 singly or in combination, *Atmos. Environ.*, 13, 1263, 1979.
150. **Weber, D. E.**, The Effects of Ozone on Plant-Parasitic Nematodes and Certain Plant Micro-Organism Interactions, Research Report, U.S. Environmental Protection Agency, Corvallis, Ore., 1976, 621.
151. **Reinert, R. A. and Weber, D. E.**, Ozone and sulfur dioxide-induced changes in soybean growth, *Phytopathology*, 70, 914, 1980.

152. **Garsed, S. G. and Read, D. J.**, Sulfur dioxide metabolism in soybean, *Glycine max* var. Biloxi. I. The effects of light and dark on the uptake and translocation of $^{35}SO_2$, *New Phytol.*, 78, 111, 1977.
153. **Irving, P. M. and Miller, J. E.**, Productivity of field-grown soybeans exposed to acid rain and sulfur dioxide alone and in combination, *J. Environ. Qual.*, 10, 473, 1981.
154. **Irving, P. M. and Miller, J. E.**, Response of field-grown soybeans to acid precipitation alone and in combination with sulfur dioxide, *Proc. Int. Conf. Ecol. Impact Acid Precip.*, Drablos, D. and Tollen, A., Eds., Norway, SNSF Project, As-NHL, 1980, 170.
155. **Davis, C. R.**, Sulfur dioxide fumigation of soybeans: effect on yield, *J. Air. Pollut. Control Assoc.*, 22, 964, 1972.
156. **Sprugel, D. G., Miller, G. E., Muller, R. N., Smith, H. J., and Xerikos, P. B.**, Sulfur dioxide effects on yield and seed quality in field-grown soybeans, *Phytopathology*, 70, 1129, 1980.
157. **Amundson, R. G. and Weinstein, L. H.**, Joint action of sulfur dioxide and nitrogen dioxide on foliar injury and stomatal behaviour in soybean, *J. Environ. Qual.*, 10, 204, 1981.
158. **Miller, V. L., Howell, R. K., and Caldwell, B. E.**, Relative sensitivity of soybean genotypes to ozone and sulfur dioxide, *J. Environ. Qual.*, 3, 35, 1974.
159. **Heagle, A. S. and Johnston, J. W.**, Variable responses of soybeans to mixtures of ozone and sulfur dioxide, *J. Air Pollut. Control Assoc.*, 29, 729, 1979.
160. **Furukawa, A. and Totsuka, T.**, Effects of NO_2, SO_2 and O_3 alone and in combinations on net photosynthesis in sunflower, *Environ. Control Biol.*, 17, 161, 1979.
161. **Oertli, J. J.**, Effect of salinity on susceptibility of sunflower plants to smog, *Soil Sci.*, 87, 249, 1959.
162. **Hoffman, G. J., Maas, E. V., and Rawlins, S. L.**, Salinity-ozone interactive effects on alfalfa yield and water relations, *J. Environ. Qual.*, 4, 326, 1975.
163. **Thompson, C. R., Kats, G., Pippon, E. L., and Isom, W. H.**, Effect of photochemical air pollution on two varieties of alfalfa, *Environ. Sci. Technol.*, 10, 1237, 1976.
164. **Howell, R. K., Devine, T. E., and Hanson, C. H.**, Resistance of selected alfalfa strains to ozone, *Crop Sci.*, 11, 114, 1971.
165. **Brennan, E., Leone, I. A., and Halisky, P. M.**, Response of forage legumes to ozone fumigations, *Phytopathology*, 59, 1458, 1969.
166. **Oshima, R. J., Poe, M. P., Braegelmann, P. K., Baldwin, D. W., and Van Way, V.**, Ozone dosage-crop loss function for alfalfa: a standardized method for assessing crop losses from air pollutants, *J. Air Pollut. Control Assoc.*, 26, 861, 1976.
167. **Rowell, J.**, The Effect of High vs. Low Levels of Ozone on Plant Growth, Research Report, Department of Biology, University of Victoria, Victoria, B.C., 1975.
168. **Tingey, D. T. and Reinert, R. A.**, The effect of ozone and sulfur dioxide singly and in combination on plant growth, *Environ. Pollut.*, 9, 117, 1975.
169. **Neely, G. E., Tingey, D. T., and Wilhour, R. G.**, Effects of Ozone and Sulfur Dioxide Singly and in Combination on Yield, Quality and N-Fixation of Alfalfa, Research Report, U.S. Environmental Protection Agency, Corvallis, Ore., not dated.
170. **Bennett, J. H. and Hill, A. C.**, Absorption of gaseous air pollutants by a standardized plant canopy, *J. Air Pollut. Control Assoc.*, 23, 203, 1973.
171. **Hou, L., Hill, A. C., and Soleimani, A.**, Influence of CO_2 on the effects of SO_2 and NO_2 on alfalfa, *Environ. Pollut.*, 12, 7, 1977.
172. **White, K. L., Hill, A. C., and Bennett, J. H.**, Synergistic inhibition of apparent photosynthesis rate of alfalfa by combinations of sulfur dioxide and nitrogen dioxide, *Environ. Sci. Technol.*, 8, 574, 1974.
173. **Lockyer, D. R. and Cowling, D. W.**, Growth of lucerne (*Medicago sativa* L.) exposed to sulphur dioxide, *J. Exp. Bot.*, 32, 1333, 1981.
174. **Sisson, W. B., Booth, J. A., and Throneberry, G. O.**, Absorption of SO_2 by pecan (*Carya illinoensis* (Wang) K. Koch) and alfalfa (*Medicago sativa* L.) and its effect on net photosynthesis, *J. Exp. Bot.*, 32, 523, 1981.
175. **Rabe, R. and Kreeb, K. H.**, Effects of SO_2 upon enzyme activity in plant leaves, *Z. Pflanzenphysiol.*, 97, 215, 1980.
176. **Setterstrom, C., Zimmerman, P. W., and Crocker, W.**, Effect of low concentrations of sulphur dioxide on yield of alfalfa and Cruciferae, *Contrib. Boyce Thompson Inst.*, 9, 179, 1938.
177. **Brewer, R. F., Sutherland, H. F., Guillemet, F. B., and Creveling, R. K.**, Some effects of hydrogen fluoride gas on bearing navel orange trees, *Proc. Am. Soc. Hortic. Sci.*, 76, 208, 1960.
178. **Adams, D. F., Hendrix, J. W., and Applegate, H. G.**, Relationship among exposure periods, foliar burn, and fluorine content of plants exposed to hydrogen fluoride, *Agric. Food Chem.*, 5, 109, 1957.
179. **Thompson, C. R. and Kats, G.**, Effects of continuous H_2S fumigation on crop and forest plants, *Environ. Sci. Technol.*, 12, 550, 1978.
180. **Bennett, J. P. and Runeckles, V. C.**, Effects of low levels of ozone on growth of crimson clover and annual ryegrass, *Crop Sci.*, 17, 443, 1977.

181. **Letchworth, M. B. and Blum, U.**, Effects of acute ozone exposure on growth, nodulation and nitrogen content of ladino clover, *Environ. Pollut.*, 14, 303, 1977.
182. **Kochhar, M., Blum, U., and Reinert, R. A.**, Effects of O_3 and (or) fescue on ladino clover: interactions, *Can. J. Bot.*, 58, 241, 1980.
183. **Lee, E. H., Bennett, J. H., and Heggestad, H. E.**, Retardation of senescence in red clover leaf discs by a new antiozonant, N-[2-(2-oxo-1-imidazolidinyl)ethyl]-N'-phenylurea, *Plant Physiol.*, 67, 347, 1981.
184. **Bleasdale, J. K. A.**, Atmospheric pollution and plant growth, *Nature (London)*, 169, 376, 1952.
185. **Cowling, D. W. and Koziol, M. J.**, Growth of ryegrass (*Lolium perenne* L.) exposed to SO_2. I. Effects on photosynthesis and respiration, *J. Exp. Bot.*, 29, 1029, 1978.
186. **Ashenden, T. W. and Mansfield, T. A.**, Influence of wind/speed on the sensitivity of ryegrass to SO_2, *J. Exp. Bot.*, 28, 729, 1977.
187. **Horsman, D. C., Roberts, T. M., and Bradshaw, A. D.**, Evolution of sulphur dioxide tolerance in perennial ryegrass, *Nature (London)*, 276, 493, 1978.
188. **Cowling, D. W., Jones, L. H. P., and Lockyer, D. R.**, Increased yield through correction of sulphur deficiency in ryegrass exposed to sulphur dioxide, *Nature (London)*, 243, 479, 1973.
189. **Bell, J. N. B. and Clough, W. S.**, Depression of yield in ryegrass exposed to sulphur dioxide, *Nature (London)*, 241, 47, 1973.
190. **Ayazloo, M., Bell, J. N. B., and Garsed, S. G.**, Modification of chronic sulphur dioxide injury to *Lolium perenne* L. by different sulphur and nitrogen nutrient treatments, *Environ. Pollut.*, 22, 295, 1980.
191. **Koziul, M. J. and Cowling, D. W.**, Growth of ryegrass (*Lolium perenna* L.) exposed to SO_2. II. Changes in the distribution of photoassimilated ^{14}C, *J. Exp. Bot.*, 29, 1431, 1978.
192. **Cowling, D. W. and Lockyer, D. R.**, Growth of perennial ryegrass (*Lolium perenne* L.) exposed to a low concentration of sulphur dioxide, *J. Exp. Bot.*, 27, 411, 1976.
193. **Beckerson, D. W., Hofstra, G., and Wukasch, R.**, The relative sensitivities of 33 bean cultivars to ozone and sulfur dioxide singly or in combination in controlled exposures and to oxidants in the field, *Plant Dis. Rep.*, 63, 478, 1979.
194. **Resh, H. M. and Runeckles, V. C.**, Effects of ozone on bean rust *Uromyces phaseoli*, *Can. J. Bot.*, 51, 725, 1973.
195. **Spalding, D. H.**, Effects of Ozone Atmospheres on Spoilage of Fruits and Vegetables, Marketing Res. Rep. No. 801, United States Department of Agriculture, Washington, D.C., 1968.
196. **Blum, U. and Heck, W. W.**, Effects of acute ozone exposures on snap bean at various stages of its life cycle, *Environ. Exp. Bot.*, 20, 73, 1980.
197. **Heggestad, H. E., Heagle, A. S., Bennett, J. H., and Koch, E. J.**, The effects of photochemical oxidants on yield of snap beans, *Atmos. Environ.*, 14, 317, 1980.
198. **Gilbert, M. D., Maylin, G. A., Elfving, D. C., Edgerton, L. J., Gutermann, W. H., and Lisk, D. J.**, The use of diphenylamine to protect plants against ozone injury, *HortScience*, 10, 228, 1975.
199. **Heck, W. W. and Dunning, J. A.**, Response of snap bean to ozone: effects of growth temperature, in Influence of Environmental Stress on Plant Response to Air Pollutants, paper presented at the 71st Annu. Meet. Air Pollution Control Assoc., Houston, June 25 to 30, 1978.
200. **MacLean, D. C. and Schneider, R. E.**, Photochemical oxidants in Yonkers, New York: effects on yield of bean and tomato, *Environ. Qual.*, 5, 75, 1976.
201. **Gilbert, M. D., Elfving, D. C., and Lisk, D. J.**, Protection of plants against ozone injury using the antiozonant *N*-(1,3-dimethylbutyl)-*N'*phenyl-*p*-phenylenediamine, *Bull. Environ. Contam. Toxicol.*, 18, 783, 1977.
202. **Brennan, E. and Rhoads, A.**, Response of field grown bean cultivars to atmospheric oxidant in New Jersey, *Plant Dis. Rep.*, 60, 941, 1976.
203. **McLaughlin, S. B. and Taylor, G. E.**, Relative humidity: important modifier of pollutant uptake by plants, *Science*, 211, 167, 1981.
204. **McLaughlin, S. B., Shriner, D. S., McConathy, R. K., and Mann, L. K.**, The effects of SO_2 dosage kinetics and exposure frequency on photosynthesis and transpiration of kidney beans (*Phaseolus vulgaris* L.), *Environ. Exp. Bot.*, 19, 179, 1979.
205. **Heggestad, H. and Bennett, J. H.**, Photochemical oxidants potentiate yield losses in snap beans attributable to sulfur dioxide, *Science*, 213, 1008, 1981.
206. **Pack, M. R.**, Effects of hydrogen fluoride on bean reproduction, *J. Air Pollut. Control Assoc.*, 21, 9133, 1971.
207. **Srivastava, H. S., Jolliffe, P. A., and Runeckles, V. C.**, The influence of nitrogen supply during growth on the inhibition of gas exchange and visible damage to leaves by NO_2, *Environ. Pollut.*, 9, 35, 1975.
208. **Ogata, G. and Maas, E. V.**, Interactive effects of salinity and ozone on growth and yield of garden beet, *J. Environ. Qual.*, 2, 518, 1973.
209. **Pippen, E. L., Potter, A. L., Randall, V. G., Ng, K. C., Reuter, F. W., III, and Morgan, A. I., Jr.**, Effect of ozone fumigation on crop composition, *J. Food Sci.*, 40, 672, 1975.

210. **Manning, W. J., Feder, W. A., Papia, P. M., and Perkins, I.**, Effect of low levels of ozone on growth and susceptibility of cabbage plants to *Fusarium oxysporum* f. sp. *conglutinans*, *Plant Dis. Rep.*, 55, 47, 1971.
211. **Bennett, J. P. and Oshima, R. J.**, Carrot injury and yield response to ozone, *J. Am. Soc. Hortic. Sci.*, 101, 638, 1976.
212. **Proctor, J. T. A. and Ormrod, D. P.**, Response of celery to ozone, *HortScience*, 12, 321, 1977.
213. **Cameron, J. W. and Taylor, O. C.**, Injury to sweet corn inbreds and hybrids by air pollutants in the field and by ozone treatments in the greenhouse, *J. Environ. Qual.*, 2, 387, 1973.
214. **Heagle, A. S., Body, D. E., and Pounds, E. K.**, Effects of ozone on yield of sweet corn, *Phytopathology*, 62, 683, 1972.
215. **Cameron, J. W.**, Inheritance in sweet corn for resistance to acute ozone injury, *J. Am. Soc. Hortic. Sci.*, 100, 577, 1975.
216. **Oshima, R. J.**, Effect of ozone on a commercial sweet corn variety, *Plant Dis. Rep.*, 57, 719, 1973.
217. **Thompson, C. R., Kats, G., and Cameron, J. W.**, Effects of ambient photochemical air pollutants on growth, yield and ear characteristics of two sweet corn hybrids, *J. Environ. Qual.*, 5, 410, 1976.
218. **Harris, M. J. and Heath, R. L.**, Ozone sensitivity in sweet corn (*Zea mays* L.) plants: a possible relationship to water balance, *Plant Physiol.*, 68, 885, 1981.
219. **Cameron, J. W., Johnson, H., Jr., Taylor, O. C., and Otto, H. W.**, Differential susceptibility of sweet corn hybrids to field injury by air pollution, *HortScience*, 5, 217, 1970.
220. **Mandl, R. H., Weinstein, L. H., Dean, M., and Wheeler, M.**, The response of sweet corn to HF and SO_2 under field conditions, *Environ. Exp. Bot.*, 20, 359, 1980.
221. **Amundson, R. G., Weinstein, L. H., Van Leuken, P., and Colavito, L. J.**, Joint action of HF and NO_2 on growth, fluorine accumulation, and leaf resistance in Marcross sweet corn, *Environ. Exp. Bot.*, 22, 49, 1982.
222. **Czuba, M. and Ormrod, D. P.**, Irreversible water loss — a significant factor in cadmium-enhanced ozone phytotoxicity, paper presented at the Annu. Meet. Canadian Soc. Plant Physiologists, University of Guelph, Guelph, Ontario, June 1976.
223. **Czuba, M. and Ormrod, D. P.**, Cadmium concentrations in cress shoots in relation to cadmium-enhanced ozone phytotoxicity, *Environ. Pollut.*, 25, 67, 1981.
224. **Czuba, M. and Ormrod, D. P.**, Effects of cadmium and zinc on ozone-induced phytotoxicity in cress and lettuce, *Can. J. Bot.*, 52, 645, 1974.
225. **Siegel, S. M.**, Protection of plants against airborne oxidants: cucumber seedlings at extreme ozone levels, *Plant Physiol.*, 37, 261, 1962.
226. **Ormrod, D. P., Adedipe, N. O., and Hofstra, G.**, Responses of cucumber, onion and potato cultivars to ozone, *Can. J. Plant Sci.*, 51, 283, 1971.
227. **Mejstrik, V.**, The influence of low SO_2 concentrations on growth reduction of *Nicotiana tabacum* L. cv. Samsun, and *Cucumis sativus* L. cv. Unikat, *Environ. Pollut.*, 21, 73, 1980.
228. **Bressan, R. A., Wilson, L. G., and Filner, P.**, Mechanisms of resistance to sulfur dioxide in the Cucurbitaceae, *Plant Physiol.*, 61, 761, 1978.
229. **Rajput, C. B. S. and Ormrod, D. P.**, Response of eggplant cultivars to ozone, *HortScience*, 11, 462, 1976.
230. **Reinert, R. A., Tingey, D. T., and Carter, H. B.**, Ozone induced foliar injury in lettuce and radish cultivars, *J. Am. Soc. Hortic. Sci.*, 97, 711, 1972.
231. **Bennett, J. P.**, The Interaction of Low Levels of Ozone and Relative Humidity on Leafy Vegetables, Research Division, California Air Resources Board, Sacramento, March 1979.
232. **Krause, G. H. M. and Kaiser, H.**, Plant response to heavy metals and sulphur dioxide, *Environ. Pollut.*, 12, 63, 1977.
233. **Wukasch, R. T. and Hofstra, G.**, Ozone and *Botrytis* spp. interaction in onion leaf dieback: field studies, *J. Am. Soc. Hortic. Sci.*, 102, 543, 1977.
234. **Wukasch, R. T. and Hofstra, G.**, Ozone and botrytis interactions in onion-leaf dieback: open top chamber studies, *Phytopathology*, 67, 1080, 1977.
235. **Engle, R. L., Gabelman, W. H., and Romanowski, R. R., Jr.**, Tipburn, an ozone incited response in onion, *Allium cepa* L., *Proc. Am. Soc. Hortic. Sci.*, 86, 468, 1973.
236. **Engle, R. L. and Gabelman, W. H.**, Inheritance and mechanism for resistance to ozone damage in onion, *Allium cepa* L., *Proc. Am. Soc. Hortic. Sci.*, 87, 423, 1966.
237. **Oshima, R. J., Bennett, J. P., and Braegelmann, P. K.**, Effect of ozone on growth and assimilate partitioning in parsley, *J. Am. Soc. Hortic. Sci.*, 103, 348, 1978.
238. **Dijak, M. and Ormrod, D. P.**, Some physiological and anatomical characteristics associated with differential ozone sensitivity among pea cultivars, *Environ. Exp. Bot.*, 22, 395, 1982.
239. **Ormrod, D. P.**, Sensitivity of pea cultivars to ozone, *Plant Dis. Rep.*, 60, 423, 1976.
240. **Olszyk, D. M. and Tibbitts, T. W.**, Evaluation of injury to expanded and expanding leaves of peas exposed to sulfur dioxide and ozone, *J. Am. Soc. Hortic. Sci.*, 107, 266, 1982.

241. **Bennett, J. P., Oshima, R. J., and Lippert, L. F.,** Effects of ozone on injury and dry matter partitioning in pepper plants, *Environ. Exp. Bot.,* 19, 33, 1979.
242. **Clarke, B., Henninger, M., and Brennan, E.,** The effects of two anti-oxidants on foliar injury and tuber production in 'Norchip' potato plants exposed to ambient oxidants, *Plant Dis. Rep.,* 62, 715, 1978.
243. **Hooker, W. J., Yang, T. C., and Potter, H. S.,** Air pollution injury of potato in Michigan, *Am. Potato J.,* 50, 151, 1973.
244. **Heggestad, H. E.,** Photochemical air pollution injury to potatoes in the Atlantic coastal states, *Am. Potato J.,* 50, 315, 1973.
245. **Leone, I. A. and Green, D.,** A field evaluation of air pollution effects on petunia and potato cultivars in New Jersey, *Plant Dis. Rep.,* 58, 683, 1974.
246. **Mosley, A. R., Rowe, R. C., and Weidensaul, T. C.,** Relationship of foliar ozone injury to maturity classification and yield of potatoes, *Am. Potato J.,* 55, 147, 1978.
247. **Manning, W. J., Feder, W. A., Perkins, I., and Glickman, M.,** Ozone injury and infection of potato leaves by *Botrytis cinerea, Plant Dis. Rep.,* 53, 691, 1969.
248. **Tingey, D. T., Standley, C., and Field, R. W.,** Stress ethylene evolution: a measure of ozone effects on plants, *Atmos. Environ.,* 10, 969, 1976.
249. **Speroni, J. J., Pell, E. J., and Weissberger, W. C.,** Glycoalkaloid levels in potato tubers and leaves after intermittent plant exposure to ozone, *Am. Potato J.,* 58, 407, 1981.
250. **Reinert, R. A. and Gray, T. N.,** The response of radish to nitrogen dioxide, sulfur dioxide, and ozone, alone and in combination, *J. Environ. Qual.,* 10, 240, 1981.
251. **Reinert, R. A., Shriner, D. S., and Rawlings, J. O.,** Responses of radish to all combinations of three concentrations of nitrogen dioxide, sulfur dioxide, and ozone, *J. Environ. Qual.,* 11, 52, 1982.
252. **Tingey, D. T., Heck, W. W., and Reinert, R. A.,** Effect of low concentrations of ozone and sulfur dioxide on foliage, growth, and yield of radish, *J. Am. Soc. Hortic. Sci.,* 96, 369, 1971.
253. **Ormrod, D. P., Adedipe, N. O., and Hofstra, G.,** Ozone effects on growth of radish plants as influenced by nitrogen and phosphorus nutrition and by temperature, *Plant Soil,* 39, 437, 1973.
254. **Tingey, D. T., Dunning, J. A., and Jividen, G. M.,** Radish root growth reduced by acute ozone exposures, *Proc. 3rd Int. Clean Air Congr.,* VDI-Verlag, Dusseldorf, 1973, A154.
255. **Adedipe, N. O. and Ormrod, D. P.,** Hormonal regulation of ozone phytotoxicity in *Raphanus sativus, Z. Pflanzenphysiol.,* 68, 254, 1972.
256. **Walmsley, L., Ashmore, M. R., and Bell, J. N. B.,** Adaption of radish *Raphanus sativus* L. in response to continuous exposure to ozone, *Environ. Pollut.,* 23, 165, 1980.
257. **Adedipe, N. O. and Ormrod, D. P.,** Ozone induced growth suppression in radish plants in relation to pre- and post-fumigation temperatures, *Z. Pflanzenphysiol.,* 71, 281, 1974.
258. **Athanassious, R.,** Ozone effects on radish (*Raphanus sativus* L. cv. Cherry Belle): foliar sensitivity as related to metabolite levels and cell architecture, *Z. Pflanzenphysiol.,* 97, 183, 1980.
259. **Proctor, J. T. A. and Ormrod, D. P.,** Sensitivity of ginseng to ozone and sulfur dioxide, *HortScience,* 16, 647, 1981.
260. **Mudd, J. B., McManus, T. T., Ongun, A., and McCullogh, T. E.,** Inhibition of glycolipid biosynthesis in chloroplasts by ozone and sulfhydryl reagents, *Plant Physiol.,* 48, 355, 1971.
261. **Manning, W. J., Feder, W. A., and Perkins, I.,** Sensitivity of spinach cultivars to ozone, *Plant Dis. Rep.,* 56, 832, 1972.
262. **Brewer, R. F., Guillemet, F. B., and Creveling, R. K.,** Influence of N-P-K fertilization on incidence and severity of oxidant injury to mangels and spinach, *Soil Sci.,* 92, 298, 1961.
263. **Coulson, C. and Heath, R. L.,** Inhibition of the photosynthetic capacity of isolated chloroplasts by ozone, *Plant Physiol.,* 53, 32, 1974.
264. **Silvius, J. E., Baer, C. H., Dodrill, S., and Patrick, H.** Photoreduction of sulfur dioxide by spinach leaves and isolated spinach chloroplasts, *Plant Physiol.,* 57, 799, 1976.
265. **Silvius, J. E., Ingle, M., and Baer, C. H.,** Sulfur dioxide inhibition of photosynthesis in isolated spinach chloroplasts, *Plant Physiol.,* 56, 434, 1975.
266. **Knatamien, H., Adedipe, N. O., and Ormrod, D. P.,** Soil-plant-water aspects of ozone phytotoxicity in tomato plants, *Plant Soil,* 38, 531, 1973.
267. **Gentile, A. G., Feder, W. A., Young, R. E., and Santner, Z.,** Susceptibility of *Lycopersicon* spp. to ozone injury, *J. Am. Soc. Hortic. Sci.,* 94, 92, 1971.
268. **Henderson, W. R. and Reinert, R. A.,** Yield response of four fresh market tomato cultivars after acute ozone exposure in the seedling stage, *J. Am. Soc. Hortic. Sci.,* 104, 754, 1979.
269. **Reinert, R. A., Tingey, D. T., and Carter, H. B.,** Sensitivity of tomato cultivars to ozone, *J. Am. Soc. Hortic. Sci.,* 97, 149, 1972.
270. **Neil, L. J., Ormrod, D. P., and Hofstra, G.,** Ozone stimulation of tomato stem elongation, *HortScience,* 8, 488, 1973.
271. **Oshima, R. J., Braegelmann, P. K., Baldwin, D. W., Van Way, V., and Taylor, O. C.,** Reduction of tomato fruit size and yield by ozone, *J. Am. Soc. Hortic. Sci.,* 102, 289, 1977.

272. **Oshima, R. J., Braegelmann, P. K., Baldwin, D. W., Van Way, V., and Taylor, O. C.,** Responses of five cultivars of fresh market tomato to ozone: a contrast of cultivar screening with foliar injury and yield, *J. Am. Soc. Hortic. Sci.,* 102, 286, 1977.
273. **Leone, I. A. and Brennan, E.,** Ozone toxicity in tomato as modified by phosphorus nutrition, *Phytopathology,* 60, 1521, 1970.
274. **Koritz, H. G. and Went, F. W.,** The physiological action of smog on plants. I. Initial growth and transpiration studies, *Plant Physiol.,* 28, 50, 1953.
275. **Legassicke, B. C. and Ormrod, D. P.,** Suppression of ozone-injury on tomatoes by ethylene diurea in controlled environments and in the field, *HortScience,* 16, 183, 1981.
276. **Oshima, R. J., Taylor, O. C., Braegelmann, P. K., and Baldwin, D. W.,** Effect of ozone on the yield and plant biomass of a commercial variety of tomato, *J. Environ. Qual.,* 4, 463, 1975.
277. **Leone, I. A.,** Response of potassium-deficient tomato plants to atmospheric ozone, *Phytopathology,* 66, 734, 1966.
278. **Reinert, R. A. and Henderson, W. R.,** Foliar injury and growth of tomato cultivars as influenced by ozone dose and plant age, *J. Am. Soc. Hortic. Sci.,* 105, 322, 1980.
279. **Ormrod, D. P. and Kemp, W. G.,** Ozone response of tomato plants infected with cucumber mosaic virus and/or tobacco mosaic virus, *Can. J. Plant Sci.,* 59, 1077, 1979.
280. **Leone, I. A. and Brennan, E.,** Modification of sulfur dioxide injury to tobacco and tomato by varying nitrogen and sulfur nutrition, *J. Air Pollut. Control Assoc.,* 22, 544, 1972.
281. **Weinstein, L. H., McCune, D. C., Aluiso, A. L., and Van Leuken, P.,** The effect of sulfur dioxide on the incidence and severity of bean rust and early blight of tomato, *Environ. Pollut.,* 9, 145, 1975.
282. **Marie, B. A. and Ormrod, D. P.,** Tomato plant growth with continuous exposure to sulphur dioxide and nitrogen dioxide, Research Report Department of Horticultural Science, University of Guelph, Ontario, Canada, *Environ. Pollut.,* 33, 257, 1984.
283. **MacLean, D. C., Schneider, R. E., and McCune, D. C.,** Fluoride susceptibility of tomato plants as affected by magnesium nutrition, *J. Am. Soc. Hortic. Sci.,* 101, 347, 1976.
284. **MacLean, D. C., Roark, O. F., Folkerts, G., and Schneider, R. E.,** Influence of mineral nutrition on the sensitivity of tomato plants to hydrogen fluoride, *Environ. Sci. Technol.,* 3, 1201, 1969.
285. **Pack, M. R.,** Response of tomato fruiting to hydrogen fluoride as influenced by calcium nutrition, *J. Air Pollut. Control Assoc.,* 16, 541, 1966.
286. **Brennan, E., Leone, I. A., and Daines, R. H.,** Fluorine toxicity in tomato as modified by alterations in the nitrogen, calcium, and phosphorous nutrition of the plant, *Plant Physiol.,* 25, 736, 1950.
287. **Capron, T. M. and Mansfield, T. A.,** Inhibition of the growth in tomato by air polluted with nitrogen oxides, *J. Exp. Bot.,* 28, 112, 1977.
288. **Anderson, L. S. and Mansfield, T. A.,** The effects of nitric oxide pollution on the growth of tomato, *Environ. Pollut.,* 20, 113, 1979.
289. **Spierings, F. H. F. G.,** Influence of fumigations with NO_2 on growth and yield of tomato plants, *Neth. J. Plant Pathol.,* 77, 194, 1971.
290. **Miller, P. M. and Rich, S.,** Ozone damage on apples, *Plant Dis. Rep.,* 52, 730, 1968.
291. **Elfving, D. C., Gilbert, M. D., Edgerton, L. J., Wilde, M. H., and Lisk, D. J.,** Antioxidant and antitranspirant protection of apple foliage against ozone injury, *Bull. Environ. Contam. Toxicol.,* 15, 336, 1976.
292. **Shertz, R. D., Kender, W. J., and Musselman, R. C.,** Foliar response and growth of apple trees following exposure to ozone and sulfur dioxide, *J. Am. Soc. Hortic. Sci.,* 105, 594, 1980.
293. **Kender, W. J. and Spierings, F. H. F. G.,** Effects of sulfur dioxide, ozone, and their interactions on 'Golden Delicious' apple trees, *Neth. J. Plant Pathol.,* 81, 149, 1975.
294. **Facteau, T. J. and Rowe, K. E,.** Response of sweet cherry and apricot pollen tube growth to high levels of sulfur dioxide, *J. Am. Soc. Hortic. Sci.,* 106, 77, 1981.
295. **Facteau, T. J. and Rowe, K. E.,** Effects of hydrogen fluoride qand hydrogen chloride on pollen tube growth and NaF on pollen germination in 'Tilton' apricot, *J. Am. Soc. Hortic. Sci.,* 102, 95, 1977.
296. **Brennan, E., Leone, I. A., and Daines, R. H.,** Toxicity of sulfur dioxide to highbush blueberry, *Plant Dis. Rep.,* 54, 704, 1970.
297. **Facteau, T. J., Wang, S. Y., and Rowe, K. E.,** Response of sweet cherry leaf tissue to hydrogen fluoride fumigation at different nitrogen levels, *J. Am. Soc. Hortic. Sci.,* 103, 115, 1978.
298. **Facteau, T. J., Wang, S. Y., and Rowe, K. E.,** The effects of hydrogen fluoride on pollen germination and pollen tube growth in *Prunus avium* L. cv 'Royal Ann', *J. Am. Soc. Hortic. Sci.,* 98, 234, 1973.
299. **Adams, D. F., Applegate, H. G., and Hendrix, J. W.,** Relationship among exposure periods, foliar burn, and fluorine content of plants exposed to hydrogen fluoride, *Agric. Food Chem.,* 5, 108, 1957.
300. **Thompson, C. R., Hensel, E., and Kats, G.,** Effects of photochemical air pollutants on Zinfandel grapes, *HortScience,* 4, 222, 1969.

301. **Musselman, R. C., Kender, W. J., and Crowe, D. E.,** Determining air pollutant effects on the growth and productivity of 'Concord' grapevines using open top chambers, *J. Am. Soc. Hortic. Sci.*, 103, 645, 1978.
302. **Kender, W. J. and Shaulis, N. J.,** Vineyard management practices influencing oxidant injury in 'Concord' grapevines, *J. Am. Soc. Hortic. Sci.*, 101, 129, 1976.
303. **Kender, W. J., Taschenberg, E. F., and Shaulis, N. J.,** Benomyl protection of grapevines from air pollution injury, *HortScience*, 8, 396, 1973.
304. **Kender, W. J. and Carpenter, S. G.,** Susceptibility of grape cultivars and selections to oxidant injury, *Fruit Var. J.*, 28, 59, 1974.
305. **Richards, B. L., Middleton, J. T., and Hewitt, W. B.,** Air pollution with relation to agronomic crops. V. Oxidant stipple of grape, *Agron. J.*, 50, 559, 1958.
306. **Thompson, C. R. and Kats, G.,** Antioxidants reduce grape yield reduction, from photochemical smog, *Calif. Agric.*, 24, 12, 1970.
307. **Shertz, R. D., Kender, W. J., and Musselman, R. C.,** Effects of ozone and sulfur dioxide on grapevines, *Sci. Hortic.*, 13, 37, 1980.
308. **Fujiwara, T.,** Sensitivity of grapevines to injury by atmospheric sulfur dioxide, *J. Jpn. Soc. Hortic. Sci.*, 39, 319, 1970.
309. **Greenhalgh, W. J. and Brown, G. S.,** Effect of Airborne Fluoride on Grape Vines. Research Report, Department of Agronomy and Horticultural Science, University of Sydney, Australia, 1981.
310. **Greenhalgh, W. J. and Brown, G. S.,** Towards equity for grapevines in the energy-rich Hunter Valley: studies on the tolerances of *Vitis* cultivars to airborne fluorides, paper presented at the 51st ANZAAS Congr., Brisbane, Australia, May 11, 1981.
311. **Greenhalgh, W. J., Brown, G. S., and Horning, D. S., Jr.,** Accumulation of Airborne Fluoride in the Leaf Lamina of the Grape Cultivar Semillon, Research Report Department of Agronomy and Horticultural Science, University of Sydney, Australia, 1981.
312. **Brewer, R. F., McCulloch, R. C., and Sutherland, F. H.,** Fluorine accumulation in foliage and fruit of wine grapes growing in the vicinity of heavy industry, *Proc. Am. Soc. Hortic. Sci.*, 70, 183, 1957.
313. **Greenhalgh, W. J. and Brown, G. S.,** The effect of three crop protectants on the appearance of injury in the grape cultivar Semillon following fumigation with 0.2 µg/m^{-3} hydrogen fluoride, paper presented at the 51st ANZAAS Congr., Brisbane, Australia, May 11, 1981.
314. **Greenhalgh, W. J. and Brown, G. S.,** Effect of Airborne Fluorides on Growth and Health of Grapevines (*Vitis* sp.), paper presented at 13th Int. Botanical Congr., University of Sydney, Sydney, Australia, August 1981.
315. **Greenhalgh, W. J., and Brown, G. S.,** The effect of airborne fluorides on growth and physiology of grapes, paper presented at 21st Int. Horticultural Congr., Hamburg, Germany, 1982.
316. **Taylor, O. C.,** Air pollution with relation to agronomic crops. IV. Plant growth suppressed by exposure to airborne oxidants (smog), *Agron. J.*, 50, 556, 1958.
317. **Brewer, R. F., Crevelling, R. K., Guillemet, F. B., and Sutherland, F. H.,** The effects of hydrogen fluoride gas on seven citrus varieties, *Proc. Am. Soc. Hortic. Sci.*, 75, 236, 1959.
318. **Brewer, R. F.,** The effects of fluoride air pollution on citrus growth and fruit production, in *Proc. 1st Int. Citrus Symp. Riverside*, 2, 729, 1969.
319. **MacLean, D. C., McCune, D. C., Weinstein, L. H., Mandl, R. H., and Woodruff, G. N.,** Effects of acute hydrogen fluoride and nitrogen dioxide exposures on citrus and ornamental plants of central Florida, *Environ. Sci. Technol.*, 2, 444, 1968.
320. **Taylor, O. C., Cardiff, E. A., and Mersereau, J. D.,** Apparent photosynthesis as a measure of air pollution damage, *J. Air Pollut. Control Assoc.*, 15, 171, 1965.
321. **Dugger, W. M., Jr. and Palmer, R. L.,** Carbohydrate metabolism in leaves of rough lemon as influenced by ozone, in *Proc. 1st Int. Citrus Symp. Riverside*, 2, 711, 1969.
322. **Brewer, R. F., Sutherland, F. H., and Guillemet, F. B.,** Effects of various fluoride sources on citrus growth and fruit production, *Environ. Sci. Technol.*, 3, 378, 1969.
323. **Matsushima, J. and Brewer, R. F.,** Influence of sulfur dioxide and hydrogen fluoride as a mix or reciprocal exposure on citrus growth and development, *J. Air Pollut. Control Assoc.*, 22, 710, 1972.
324. **Thompson, C. R., Kats, G., and Hensel, E.,** Effects of ambient levels of ozone on navel oranges, *Environ. Sci. Technol.*, 6, 1014, 1972.
325. **Thompson, C. R. and Taylor, O. C.,** Reduction of fruit drop by navel oranges with antioxidant dusts and girdling, *HortScience*, 2, 103, 1967.
326. **Thompson, C. R. and Taylor, O. C.,** Effects of air pollutants on the growth, leaf drop, fruit drop, and yield of citrus trees, *Environ. Sci. Technol.*, 3, 934, 1969.
327. **Thompson, C. R. and Kats, G.,** Effects of ambient concentrations of peroxyacetyl nitrate on navel orange trees, *Environ. Sci. Technol.*, 9, 35, 1975.
328. **Darley, E. F., Middleton, J. T., and Kendrick, J. B., Jr.,** Sulfur dioxide injury on citrus, *Calif. Agric.*, 10, 9, 1956.

329. **Brewer, R. F., Sutherland, H. F., Guillemet, F. B., and Creveling, R. K.,** Some effects of hydrogen fluoride gas on bearing navel orange trees, *Proc. Am. Soc. Hortic. Sci.,* 76, 208, 1960.
330. **Kandy, J. C., Bingham, F. T., McCulloch, R. C., Liebig, G. F., and Vanselow, A. P.,** Contamination of citrus foliage by fluorine from air pollution in major California citrus areas, *Proc. Am. Soc. Hortic. Sci.,* 65, 121, 1955.
331. **Brewer, R. F., Sutherland, F. H., and Guillemet, F. B.,** Application of calcium sprays for the protection of citrus from atmospheric fluorides, *J. Am. Soc. Hortic. Sci.,* 94, 302, 1969.
332. **Spalding D. H.,** Appearance and Decay of Strawberries, Peaches, and Lettuce Treated with Ozone, Marketing Res. Rep. No. 756, United States Department of Agriculture, Washington, D.C., Oct. 1966.
333. **Rajput, C. B. S., Ormrod, D. P., and Evans, W. D.,** The resistance of strawberry to ozone and sulfur dioxide, *Plant Dis. Rep.,* 61, 222, 1977.
334. **Pack, M. R.,** Response of strawberry fruiting to hydrogen fluoride fumigation, *J. Air Pollut. Control Assoc.,* 22, 714, 1972.
335. **Brewer, R. F. and Ferry, G.,** Effects of air pollution on cotton in the San Joaquin Valley, *Calif. Agric.,* 28(6), 6, 1974.
336. **Ting, I. P. and Dugger, W. M., Jr.,** Factors affecting ozone sensitivity and susceptibility of cotton plants, *J. Air Pollut. Control Assoc.,* 18, 810, 1968.
337. **Reinert, R. A. and Gooding, G. V., Jr.,** Effect of ozone and tobacco streak virus alone and in combination on *Nicotiana tabacum, Phytopathology,* 68, 15, 1978.
338. **Reinert, R. A. and Spurr, H. W., Jr.,** Differential effect of fungicides on ozone injury and brown spot disease of tobacco, *J. Environ. Qual.,* 1, 450, 1972.
339. **Kisaki, T., Koiwai, A., Kitano, H., and Fukuda, W.,** Piperonyl butoxide: a protectant against ozone injury of plants, *Can. J. Bot.,* 75, 226, 1975.
340. **Otto, H. W. and Daines, R. H.,** Plant injury by air pollutants: influence of humidity on stomatal apertures and plant response to ozone, *Agron. Notes,* May-June, 1969.
341. **Rich, S. and Turner, N. C.,** Importance of moisture on stomatal behaviour of plants subjected to ozone, *J. Air Pollut. Control Assoc.,* 22, 718, 1972.
342. **Heagle, A. S. and Heck, W. W.,** Predisposition of tobacco to oxidant air pollution injury by previous exposure to oxidants, *Environ. Pollut.,* 7, 247, 1974.
343. **Menser, H. A. and Hodges, G. H.,** Nitrogen nutrition and susceptibility of tobacco leaves to ozone, *Tob. Sci.,* 11, 151, 1967.
344. **Bisessar, S. and Temple, P. J.,** Reduced ozone injury on virus-infected tobacco in the field, *Plant Dis. Rep.,* 61, 961, 1977.
345. **Walker, E. K.,** Chemical control of weather fleck in flue-cured tobacco, *Plant Dis. Rep.,* 45, 583, 1961.
346. **Reilly, J. J., and Moore, L. D.,** Influence of selected herbicides on ozone injury in tobacco, *(Nicotiana tabacum), Weed Sci.,* 30, 260, 1982.
347. **Lee, T. T.,** Inhibition of oxidative phosphorylation and respiration by ozone in tobacco mitochondria, *Plant Physiol.,* 42, 691, 1967.
348. **Rhoads, A. and Brennan, E.,** Ozone response of tobacco cultivars unrelated to root system characteristics, *Phytopathology,* 65, 1239, 1975.
349. **MacDowell, F. D. H.,** Predisposition of tobacco to ozone damage, *Can. J. Plant Sci.,* 45, 1, 1965.
350. **Walker, E. K.,** Evaluation of foliar sprays for control of weather fleck on flue-cured tobacco, *Can. J. Plant Sci.,* 47, 99, 1967.
351. **Walker, E. K. and Vickery, L. S.,** Influence of sprinkler irrigation on the incidence of weather fleck on flue-cured tobacco in Ontario, *Can. J. Plant Sci.,* 41, 281, 1961.
352. **Cantwell, A. M.,** Effect of temperature on response of plants to ozone as conducted in a specially designed plant fumigation chamber, *Plant Dis. Rep.,* 52, 957, 1968.
353. **Moyer, J. W. and Smith, S. H.,** Oxidant injury reduction on tobacco induced by tobacco etch virus infection, *Environ. Pollut.,* 9, 103, 1975.
354. **Leuning, R., Unsworth, M. H., Neumann, H. N., and King, K. M.,** Ozone fluxes to tobacco and soil under field conditions, *Atmos. Environ.,* 13, 1155, 1979.
355. **Macdowell, F. D. H., Vickery, L. S., Runeckles, V. C., and Patrick, Z. A.,** Ozone damage to tobacco in Canada, *Can. Plant Dis. Surv.,* 43, 131, 1963.
356. **Menser, H. A., Heggestad, H. E., Street, O. E., and Jeffrey, R. N.,** Response of plants to air pollutants. I. Effects of ozone on tobacco plants preconditioned by light and temperature, *Plant Physiol.,* 38, 605, 1963.
357. **Menser, H. A. and Chaplin, J. F.,** Air pollution: effects on the phenol and alkaloid content of cured tobacco leaves, *Tob. Sci.,* 13, 169, 1969.
358. **Menser, H. A.,** Response of plants to air pollutants. III. A relation between ascorbic acid levels and ozone susceptibility of light-preconditioned tobacco plants, *Plant Physiol.,* 39, 564, 1964.
359. **Leone, I. A., Brennan, E., and Daines, R. H.,** Effect of nitrogen nutrition on the response of tobacco to ozone in the atmosphere, *J. Air Pollut. Control Assoc.,* 16, 191, 1966.

360. **Heggestad, H. E., Burleson, F. R., Middleton, J. T., and Darley, E. F.,** Leaf injury on tobacco varieties resulting from ozone, ozonated hexene-1 and ambient air of metropolitan areas, *Int. J. Air Water Poll.*, 8, 1, 1964.
361. **Turner, N. C., Rich, S., and Tomlinson, H.,** Stomatal conductance, fleck injury, and growth of tobacco cultivars varying in ozone tolerance, *Phytopathology*, 62, 63, 1972.
362. **Menser, H. A. and Hodges, G. H.,** Oxidant injury to shade tobacco cultivars developed in Connecticut for weather fleck resistance, *Agron. J.*, 64, 189, 1972.
363. **Menser, H. A., and Hodges, G. H.,** Tolerance to ozone of flue-cured tobacco cultivars in field and fumigation chamber tests, *Tob. Sci.*, 13, 176, 1969.
364. **Menser, H. A., Jr.,** Effects of air pollution on tobacco cultivars grown in several states, *Tob. Sci.*, 13, 99, 1969.
365. **Lee, T. T.,** Chemical regulation of ozone susceptibility in *Nicotiana tabacum*, *Can. J. Bot.*, 44, 487, 1967.
366. **Taylor, G. C. and Rich, S.,** Ozone injury to tobacco in the field influenced by soil treatments with benomyl and carboxin, *Phytopathology*, 64, 814, 1974.
367. **Koiwai, A., Kiteno, H., Fukuda, M., and Kisake, T.,** Methylenedioxyphenyl and its related compounds as protectants against ozone injury to plants, *Agric. Biol. Chem.*, 38, 301, 1974.
368. **Menser, H. A. and Hodges, G. H.,** Effects of air pollutants on burley tobacco cultivars, *Agron. J.*, 62, 265, 1970.
369. **Hodges, G. H., Menser, H. A., Jr., and Ogden, W. B.,** Susceptibility of Wisconsin Havana tobacco cultivars to air pollutants, *Agron. J.*, 63, 107, 1971.
370. **Menser, H. A., Hodges, G. H., and McKee, C. G.,** Effects of air pollution on Maryland (type 32) tobacco, *J. Environ. Qual.*, 2, 253, 1973.
371. **Grosso, J. J., Menser, H. A., Hodges, G. H., and McKenney, H. H.,** Effects of air pollutants on *Nicotiana* cultivars and species used for virus studies, *Phytopathology*, 61, 945, 1971.
372. **Menser, H. A. and Heggestad, H. E.,** Ozone and sulfur dioxide synergism: injury to tobacco plants, *Science*, 22, 424, 1966.
373. **Mejstrik, V.,** The influence of low SO_2 concentrations on growth reduction of *Nicotiana tabacum* L. cv. Samsun and *Cucumis sativus* L. cv. Unikat, *Environ. Pollut.*, 21, 73, 1980.
374. **Menser, H. A.,** Response of sugar beet cultivars to ozone, *J. Am. Soc. Sugar Beet Technol.*, 18, 81, 1974.

Index

INDEX

A

Abaca, 11
Abyssinian center, 16, 17, 18
Abyssinian mustard, 6
Acacia, 11
Acetylene reduction technique, 176
Ackee, 103
Action spectra, 196—197
Adzuki bean, 4, 17
Aegilops species hybrids, 26—28
African oil palm, 123—124
African rice, 19
Agave species, 11, 47
Agropyron species, see also Wheatgrass
 botany, 153
 introgressive hybridization, 27, 28, 33
Akee, propagation, 103
Akia, chromosome number, 7
Aleurites species, see also Tung, 8, 48
Alfalfa, 10, 17, 31
 botany, 152, 156
 gaseous pollution stress, 226, 238—239
 half-sib progeny selection, 78
 nitrogen fixation, 171, 173
 photoperiod and photosynthesis, 199, 214
 propagation, 63, 95
 temperature tolerance, breeding for, 221
All-heal, 12
Allium
 botany, 144
 cepa, see Onion
 chromosome number, 6
 introgressive hybridization, 35
 porrum, see Leek
 propagation, 100
 sativum, see Garlic
Allspice, 111, 147
Almond, 17, 45
 botany, 138, 140
 chromosome number, 5, 7
 propagation, 103
Alopecurus species hybrids, 33
Altai wild rye, 9
Althaea officinalis, 12
Alysicarpus, 172
Amaranths (*Amaranthus*), 4, 17, 25
American cranberry bush, 5
American hazel, 7
American pawpaw, 40
Ammonia, see Gaseous pollution stress
Amoracia lapathifolia, 6
Amorpha canescens, 172
Anabaena azollae, 174
Ananas species hybrids, 40
Andean potato, 7
Andropogon species hybrids, 33
Andropogon gerardii, 9, 33

Angelica, 12
Angiosperm families, nitrogen fixing, 173
Angleton bluestem, 9
Angola grass, 152
Anise, 12, 17, 147
Annual cycles, 198
Annual ryegrass, see Ryegrass, annual
Anthemis nobilis, 12
Anthyllis, nitrogen fixation, 172
Apium species hybrids, see also Celery, 37, 39
Apomixis, 62, 93
Apple, 5, 17, 41, 104
 botany, 136, 138
 gaseous pollution stress, 227, 250
Apricot, 5, 17, 104
 botany, 136, 138
 gaseous pollution stress, 227, 250
Arabidopsis thaliana, 6
Arachis species, see also Peanut
 introgressive hybridization, 30
 nitrogen fixation, 172, 173, 174
Aramina, 11
Ardisia crispa, nitrogen fixation, 173
Arrowroot, 12, 102, 125
Artemisa dracunculus, 12, 150
Artichoke, 17, 39, 98
Asexual propagation, 91
Asian cotton, 17
Asimina, see also Pawpaw, American
 species hybrids, 40
Asparagus, 37, 63
 botany, 143
 chromosome number, 4, 6
 gaseous pollution stress, 226
 propagation, 98—99
Astragalus, nitrogen fixation, 172
Astrometeorological estimator, 210—214
Astronomical equations, 206—209
Astronomical twilight, 198
Atropa, 45, 46
Attoto yam, 6
Avena, see Oats
Avocado, 5, 17, 42
 botany, 136, 138
 gaseous pollution stress, 227
 propagation, 104
Azolla, nitrogen fixation, 174

B

Backcrossing, 69—71
Bahiagrass
 chromosome number, 9
 photoperiod and photosynthesis, 215
 propagation, 95
Ball clover, 10
Bamboo *Bambusa vulgaris*, 11, 17

Banana, 5, 17, 41, 104
   botany, 139—140
Bard vetch, 10
Barley, 4, 17, 20, 25
   botany, 116, 118
   breeding, 72, 221
   gaseous pollution stress, 226
   introgression, 25—28
   photoperiod and photosynthesis, 214
   propagation, 63, 94
Barnyard millet, 17
Barrel medic, 10
Basil, 49, 147
Bean, see also *Phaseolus*, 4, 17, 30
   botany, 120, 144
   breeding, 72
   cold sensitivity, 217, 218
   gaseous pollution stress, 226, 231—235, 241
   nitrogen fixation, 173
   propagation, 63, 99
Bean, adzuki, 4, 17
Bean, broad, see also *Vigna* species, 4, 17, 173
   botany, 120, 144
   photoperiod and photosynthesis, 214
Bean, castor, see Castor bean
Bean, common, see also *Phaseolus*, 17, 20, 30
   gaseous pollution stress, 226
   photoperiod and photosynthesis, 214
   propagation methods, 63
Bean, lima, 4, 17
   cold sensitivity, 218
   propagation, 63, 99
Bean, mung, 19, 173
Beet (*Beta vulgaris*), 17, 35
   botany, 128—129, 144
   chromosome number, 6, 8
   gaseous pollution stress, 226, 241, 248, 258
   photoperiod and photosynthesis, 215
   propagation, 99, 112, 102
Belladonna, 45, 46
Bent grass, 95
Bermuda grass, 9, 95
Berseem, 17
Beta vulgaris, see Beet
Betel nut, 110
Betel palm, 17
Betel pepper, 110
Beverage crops, botany, 130—133
Big bluestem, 9, 33
Birdsfoot trefoil, 10, 31, 95
   botany, 154, 156
   photoperiod and photosynthesis, 215
Bitter gourd, 17
Blackberry, 43, 104, 227
Black gram, 4, 9
Black mustard, see Mustard, black
Black pepper, see Pepper, black
Black raspberry, see Raspberry, black
Black walnut, see Walnut, black
Blueberry, 5, 44
   botany, 137, 139

   gaseous pollution stress, 227, 251
   propagation, 104
Blueberry, Canada, 5
Blueberry, highbush, 5, 44
Blueberry, rabbiteye, 5
Blue bunch wheatgrass, 9
Blue grama, see Grama, blue
Bluegrass, Kentucky, 9
   botany, 153—154
   photoperiod and photosynthesis, 214
   propagation, 95
Blue-green algae, 169
Blue light, 196
Blue lupin, 10, 31
Bluestem, 9, 33, 215
Boehmeria nivea, see Ramie
Botany, see also Light and photoperiod
   beverage and drug crops
      cacao, 130—131
      cola, 131
      hop, 131—132
      pyrethrum, 133
      quinine, 132
      tea, 132
      tobacco, 132—133
   buckwheat, 157
   cereal crops, 116—119
   fiber crops, 133—135
      cotton, 133—134
      hemp, 134
      jute, 134
      kapok, 134
      linen, 134—135
      sisal, 135
   forage crops, perennial, 150—156
      development, 153—154, 155—156
      grasses, 150—152
      legumes, 152, 154—155
   fruit crops, 135—140
      banana, 139—140
      development of, 136—139
      pineapple, 140
   legumes, grain, 119—121
   life cycles, 115—116
   nut crops, tree, 140—143
      almond, 140
      Brazil nut, 141
      cashew, 141
      chestnut, 141—142
      filberts, 142
      pecans, 142
      walnuts, 142—143
   oil-seed crops, 121—125
      coconut, 123
      oil palm, 123—124
      olive, 124
      tung, 124—125
   pulse crops, 119—121
   reproduction, principles of, 62
   root crops, starch-producing, 125—127
      arrowroot, 125

cassava, 125—126
potatoes, 126
sweet potato, 126—127
taro, 127
yam, 127
rubber, 155—157
spice crops, 145, 147—150
sugar crops, 127—130
sorghum, 130
sugarbeet, 128—129
sugarcane, 127—128
sugar maple, 129—130
vegetable crops, 143—145, 146
asparagus, 143
rhubarb, 145
transplanted crops, production cycle, 146
Bothriochloa species hybrids, see also Bluestem, 33
Bouteloua, see Grama
Bowstring hemp, 97
Brachiaria deflexa, 19
Brassica
botany, 144, 145, 146
chromosome number, 6
gaseous pollution stress, 226, 227
introgressive hybridization, 23, 36, 37—38
photoperiod and photosynthesis, 215
propagation, 99, 100
Brazilian-Paraguayan center, 16, 17, 18
Brazil nut, 17, 141
Breadfruit, 5, 104—105
Breeding, see also Hybridization, introgressive; Propagation
germplasm, 83—85
population improvement
ear-to-row selection, 77
full-sib, 79, 82
genetic gain, 82—83
imbred progeny, 79—80
interpopulation recurrent selection, 80—83
intrapopulation recurrent selection, 76—80
mass selection, 76
progeny selection, 65—75
backcrossing, 69—71
bulk breeding, 71—73
pedigree, 65—69
single-seed descent, 73—75
reproduction, principles of, 62—63
selection methods, 64—65
for temperature tolerance, 221—222
Broad bean, see Bean, broad
Broccoli, 6, 99, 146, 226
Brome/bromegrass, 9, 33, 214
botany, 150, 153
propagation, 95
Bromegrass, smooth, see Smooth bromegrass
Bromus, see Brome/bromegrass
Brown mustard, 6, 38
Brussels sprouts, 6, 99, 146
Buckwheat, 4, 17, 25, 214
botany, 157
gaseous pollution stress, 259

propagation, 94
Budding, 91, 93, see also Propagation, vegetative
Buffalo gourd, 6, 40
Buffalo grass, 153, 215
Buffel grass, 9, 33
Bulb, 91, 93
Bulk-population breeding, 72—73
Butternut, 108—109
Byzantine oats, 17

C

Cabbage
botany, 146
centers of origin, 17
chromosome number, 6
gaseous pollution stress, 226, 241
introgressive hybridization, 37
propagation, 99
Cabbage, celery, 38
Cabbage, Chinese, 6, 17
Cacao, 12
botany, 130—131
centers of origin, 17, 20
cold sensitivity, 218
propagation, 110
Caesalpinioideae, 169
Cajanus species
introgressive hybridization, 30
nitrogen fixation, 172, 173
Calabazilla, 6
Calapogonium, 173
California burr-clover, 10
Calvin cycle, 199
Camellia sinensis, 12
Camphor, 12
Canadian wild rye, 153
Canary grass, reed, see Reed canary grass
Canavalia, 172, 173
Candlenut, 8
Cane bluestem, 9
Canna, 6, 17, 102
Cannabis, see also Hemp, 11, 47
Canola oil, 8
Cantaloupe, 6, 17, 40
propagation, 100
Capillipedium species hybrids, 33
Capsicum, 6, 146, 217
hybrids, 40
pollution stress, 226, 245
propagation, 100
Capsicum peppers, see *Capsicum*; Paprika
Caragana arborescens, 172
Caraway, 12, 17
botany, 147—148
Carbon dioxide fixation, 199—200
Cardamom, 17, 148
propagation, 110
Carib grass, 95
Carob, 17, 109

Carpet grass, 95, 153
Carrot, 6, 17, 35
  botany, 144
  gaseous pollution stress, 226, 242
  propagation, 102
Carthamus species hybrids, 48
Carum carvi, see Caraway
Carya species, see also Pecan, 7, 45
Cashew, 7, 17
  botany, 141
  propagation, 109
Cassava/manioc, 7, 17, 35
  botany, 125—126
  photoperiod and photosynthesis, 214
  propagation, 102
Castanea species hybrids, see also Chestnut, 45
Castor bean, 17, 121
  chromosome number, 8, 12
  propagation, 97
Casuarina, 172
Cauliflower, 6, 226
  botany, 146
  propagation, 99
Cayenne pepper, 6, 12
Ceiba, see Kapok
Celeriac, 102
Celery, 6, 17, 37
  botany, 146
  gaseous pollution stress, 226, 242
  propagation, 99
Celery cabbage, 38
Cenchrus, 9, 33
Centers of origin, 15—20
Central American center of origin, 16—18
Central Asiatic center of origin, 16—18
Centrosema, 31, 172—173
Cercocarpus, 172
Cereal crops, see also crops by name, 4
  botany, 116—119
  carbon fixation pathway, 200
  gaseous pollution stress, 229—231
  propagation methods, 63, 94
C-4 plants
  carbon fixation in, 199—200
  nitrogen fixation, 168
Chamomile, 12
Chard, Swiss, 99
Charlock, 38
Chat, 17
Chayote, 7, 17
Cherimoya, 17
  propagation, 105
Cherry, 5, 42
  botany, 136, 138
  centers of origin, 17
  gaseous pollution stress, 227
  propagation, 105
Cherry, sour, 42
Cherry, sweet, 5, 42, 251
Cherry plum, 5
Chestnut, 7, 17, 45

  botany, 141—142
Chestnut, Chinese, 109
Chestnut, Oriental, 17
Chickpea, 4, 17, 30
  botany, 120
  nitrogen fixation, 172
  photoperiod and photosynthesis, 214
  propagation, 94, 99
Chicory, see also Endive, 6
Chile center of origin, 16—18
Chili pepper, 8, 17
Chilling, see Temperature stress
China grass, 98
China wood oil, 8
Chinese apple, 5
Chinese cabbage, 6, 17
Chinese center of origin, 16—18
Chinese chestnut, 109
Chinese date, 106
Chinese gooseberry, 106
Chinese hazelnut, 17
Chinese kale, 6
Chinese mustard, 6, 38
Chinese onion, 17
Chinese pear, 5, 17
Chinese sugarcane, 17
Chinese tea, 12
Chinese walnut, 17
Chinese yam, 6, 17
Chive, 6, 99
Chlorine, see Gaseous pollution stress
Chloris, 9
Chloroplasts, 200
Cho-cho, 99
Choyote, 99
Chromosome number
  cereals, 4
  drug crops, 12
  dyes, rubbers, hops, 11
  fiber crops, 11
  forage grasses, 9—10
  forage legumes, 10
  fruit crops, 5
  nut crops, 7
  oil crops, 8
  protein crops, 4
  spices and flavorings, 12
  sugar crops, 8
  vegetable and tuber crops, 6—7
Chrysanthemum cinerariaefolium, 12
Chrysanthemum coccineum, 12
Cicer, see Chickpea
Cichorum, see also Endive, 39
Cinchona, 12, 17, 111, 132
Cinchonidine, 12
Cinnamon, 12
  botany, 148
  propagation, 110
Cinnamonum camphora, see Cinnamon
Citron, 5
Citronella grass, 9

Citrullus, 40, 101
Citrus species, 17, 40
    botany, 136, 138
    cold sensitivity, 217
    gaseous pollution stress, 227, 227—228, 252—253, 254
    propagation, 105
Climate, see Botany
Clone, defined, 91
Clove
    botany, 145, 147
    propagation, 110
Clover, alsike, 10
    botany, 156
    gaseous pollution stress, 239
Clover, ball, 10
Clover, crimson, 10, 17
    botany, 156
    gaseous pollution stress, 240
Clover, Egyptian, 10, 11
Clover, red, 10, 32
    botany, 154—155, 156
    gaseous pollution stress, 240
Clover crops, 17, 214
    botany, 154—155, 156
    chromosome number, 10, 11
    gaseous pollution stress, 226, 239—240
    nitrogen fixation, 171, 173
    propagation, 95
Clover, sweet, 10
    botany, 154—155, 156
    gaseous pollution stress, 240
    nitrogen fixation, 171, 173
Clover, white, 10, 32, 155—156
Clover, white sweet, 10, 32, 156, 240
Cobnut, 7
Coca, 17
Cocksfoot, 9
Cocoa, see Cacao
Coconut, 17
    botany, 123
    chromosome number, 7, 8
    propagation, 97, 109
Cocoyam, 103
Cocoyam, see Taro
Coffea, see Coffee
Coffee, 12
    centers of origin, 17, 20
    introgressive hybridization, 45, 46
    propagation, 110
Cola, 12, 131
Cole, see Kale
Collards, 6
    botany, 146
    propagation, 99
Colocasia, see Taro
Common bean, see Bean
Common lespedeza, see also Lespedeza, 10
Common reed, 215
Common vetch, 10, 32, 215
Common wheat, see Wheat

Congo jute, 11
Controlled-environment chambers, 198
Conversion factors, light measurement, 195, 201
Corchorus species hybrids, see also Jute, 47
Coriander, 12, 17, 148
Coriandrum sativum, see Coriander
Coriaria, 172, 173
Corn
    botany, 118, 144
    breeding, 72, 78
    carbon fixation pathway, 200
    centers of origin, 17
    chromosome number, 4, 7, 8
    cold sensitivity, 218
    gaseous pollution stress, 226, 229—230, 242—243
    introgressive hybridization, 28
    photoperiod and photosynthesis, 214
    propagation, 63, 94, 97, 101
Coronilla, 172
Corylus species, see also Filbert, 7, 45
Cotton/cottonseed
    botany, 133—134
    centers of origin, 17
    chromosome number, 8, 11
    cold sensitivity, 217, 218
    gaseous pollution stress, 226, 255
    introgressive hybridization, 47
    photoperiod and photosynthesis, 214
    propagation, 63, 98
Cowpea, 4, 17, 30
    botany, 120
    gaseous pollution stress, 226
    nitrogen fixation, 173
    photoperiod and photosynthesis, 214
Crabapple, 105, 227
Crambe species hybrids, 48
Crambe, 214
Cranberry
    botany, 137, 138
    propagation, 105
Creeping red fescue, 153
Crepis species hybrids, 48
Cress, 243
Cress, common wall, 6
Crested wheatgrass, 9, 153
Crimson clover, see Clover, crimson
Critical photoperiod, 198—199
Crop classification, 214—216
Cross-fertilization, see also Hybridization, introgressive, 62, 78, 80
Crossing, 63
Crotalaria, 11, 172
Crown vetch, 10, 214
Cucumber, 6, 17, 40
    botany, 144
    gaseous pollution stress, 226, 243
    propagation, 99
Cucumis, see Canteloupe; Cucumber
Cucurbits, 6, 40
    botany, 144

cold sensitivity, 217, 218
propagation, 100
Cumin, 17
Currant, 5
    botany, 137, 139
    gaseous pollution stress, 227
    introgressive hybridization, 42—43
    propagation, 105
Currant, black, 5, 42
Currant, red, 5, 43
Cushaw, 40
Custard apple, 105
Cutting, see also Propagation, vegetative, 91, 93
Cyamopsis, 172—173
Cyanobacterium, 174
Cymbopogon, 8
Cynodon, 9, 95
Cynara species hybrids, see also Artichoke, 36

D

Dactylis 9, 33
Dalea alopecuroides, 172
Dallisgrass, 9, 34
    botany, 154
    photoperiod and photosynthesis, 215
    propagation, 95
Dandelion, 48
Dasheen, see Taro
Date, 5
    botany, 136, 139
    propagation, 105
Date, Chinese, 106
Daucus species hybrids, 35
Day length, in hours, 202—206
Development, see Botany
Dichanthium species, see also Bluestem
    chromosome numbers, 9
    hybrids, 33
Digitalis, 33
Digitaria, 9, 96, 152
Digitaria exilis, 19
Dill, 148
Dioecious, defined, 62
Dioscoreaceae, nitrogen fixation, 173
Dioscorea species hybrids, 35
Discaria, 172
Diurnal cycles, 198
DNA-transfer, 221
Dormancy, see Botany
Dorycnium, 172
Drug crops
    botany, 130—133
    chromosome numbers, 12
    gaseous pollution stress, 255—258
    introgressive hybridization, 45
Dryas, 172
Dry-matter production, 199
Dry napier grass, 9
Durian, 17

Durum wheat, see Wheat
Dye crops, 11

E

E-amino caproic acid, 23
Ear-to-row selection, 77—78
Echinochlea, 96
Edible canna, 6, 17
Eggplant, 7, 17, 43
    botany, 146
    cold sensitivity, 217
    gaseous pollution stress, 226, 243
    propagation, 63, 99
Egyptian clover, 10
Egyptian cotton, 11
Einkorn wheat, 4
Elaeis, see Oil palm
Eleagnus, 172
Elymus, see Wild rye
Elephant grass, 152
Eleusine, 96
Elymus, see also Wild rye species hybrids, 27, 33
Emmer, see also Wheat, 4
    centers of origin, 17, 20
Endive, 6, 39
    botany, 144
    gaseous pollution stress, 226
    propagation, 99
English rhubarb, 7
English walnut, 7, 17
Ensete, 17
Eragrostis curvula, 9, 153
Eruca sativa, 6
Escarole, 99
Ethiopian durum wheat, 17
European hazel, 7
European yellow lupin, 10

F

Faba bean, see Bean, broad
Fagopyrum species hybrids, see also Buckwheat, 25
Fennel, 17
Fenugreek, 10, 12, 17
Fertilization (reproductive), 62—63
Fescue, 9, 34
    botany, 153
    photoperiod and photosynthesis, 214
    propagation, 95
Fescue, meadow, 9, 153
Fescue, red, 9
Festuca species hybrids, see also Fescue, 34
Fiber crops, 11, 47
    botany, 133—135
    gaseous pollution stress, 255
    propagation, 97—98
Ficus, see Fig
Field bean, see Bean

Field crops, gaseous pollution stress, 226
Field pea, see Pea
Fig, 5, 17
    botany, 136, 138, 139
    introgressive hybridization, 40—41
    propagation, 105
Filbert, 7, 45
    botany, 142
    propagation, 109
Fineleaf clover, 10
Finger millet, see also Millet, 17
Flax, 17, 48
    botany, 134—135
    chromosome number, 8, 11
    oil-seed crop, 121—122
    photoperiod and photosynthesis, 214
    propagation, 63, 97, 98
Flowering, see Botany
Fluorescent lamps, 197, 201—202
Fonio, 19
Footlambert, 201
Forage crops, see also Grasses, forage; Legumes, forage, 95—96
    botany, 150—156
    gaseous pollution stress, 238—240
    propagation, 95—96
Fox grape, 5
Fragaria, 41
Frankia, 172
Fruit crops, 5
    gaseous pollution stress, 227—228, 250—254
    introgressive hybridization, 40—44
    propagation, 103—108
Full-sib progeny selection, 79, 81—82

G

Gamba grass, 153
Garden cress, 7
Garlic, 6, 17, 102
Gaseous pollution stress
    composite sensitivity, 226—228
    responses to, 229
        cereals, 229—231
        drug crops, 255—258
        fiber crops, 255
        forage crops, 238—240
        fruit crops, 250—254
        mustard, 259
        oil crops, 236—238
        protein crops, 231—236
        starch crops, 259
        sugar crops, 258
        vegetable crops, 241—249
Genetic gain, 82—83
Genetics, see also Breeding; Hybridization, introgressive; Propagation
Geography, see Botany
Germplasm, 83—85
Giant radish, 17

Giber agaves, 215
Ginger, 17, 111, 148
Ginkgo, 17
Ginseng, 12, 17, 111, 255
Globe artichoke, 6
Glycine species, 8, 30
    nitrogen fixation, 171, 173
Glycine max, see Soybean
Glycyrrhiza glabra, 12
Gooseberry, 5, 43
    botany, 137
    propagation, 106
Gooseberry, Chinese, 106, 136
Gossypium, see also Cotton/cottonseed, 11, 47
Gourd, bitter, 17
Gourd, buffalo, 6, 40
Graftage, 91, 93
Grafting, see also Propagation, vegetative, 91, 93
Grain amaranth, see Amaranth
Grain crops, see also specific crop names, 25—29, 72, 94
Grain legumes, see Legumes, grain
Grain sorghums, see Sorghum
Grama
    black, 4, 9
    blue, 9, 153
    chromosome number, 40
    green, 4, 30, 173
    introgressive hybridization, 30
    photoperiod and photosynthesis, 215
Granadilla, 107
Grape, 5, 17, 44
    botany, 137—139
    gaseous pollution stress, 227, 251—252
    propagation, 106
Grapefruit
    botany, 136, 138
    gaseous pollution stress, 227—228, 252
    propagation, 105
Grasses, see also Grasses, forage; individual grasses by name
    botany, 150—152
    gaseous pollution stress, 226
    photoperiod and photosynthesis, 214, 215—216
Grasses, forage, 95—96
    botany, 150—152
    chromosome number, 9—10
    introgressive hybridization, 33—34
    propagation, 63, 95—96
Grasspea, 4, 17
Great millet, 10
Green gram, 4, 30, 40, 173
Green onion, 6
Gro-Lux lamps, 197, 201—202
Ground cherry, 7, 17
Groundnut, see Peanut
Growth habit, see Botany
Growth regulators, 91
Guar, 17, 20, 173, 215
Guava, 5, 17, 106
Guineagrass, 9, 95, 215

Guinea millet, 19
Gum arabic, 11

H

Haber-Bosch process, 165
Hairy litchi, 108
Hairy vetch, 10
Half-sib progeny selection, 78—79
Hardiness, see Temperature stress
Harmonic analysis, 209
Haynaldia species hybrids, 27, 28
Hazelnut American, 7, 17
Hazelnut, Chinese, 17
Heat stress, see Temperature stress
Heilborn, 5
Helianthus species hybrids, 35, 48
Hemp, 17, 47
    botany, 134
    chromosome number, 8, 11
    propagation, 98
Henequen, 11, 17
Hevea, see Rubber
Hibiscus cannabinus, 11, 47, 55, 98
Hibiscus esculentus, 12
Hibiscus species hybrids, 41, 47
Hickory, 7, 109
Highbush blueberry, see Blueberry
High-energy reaction, 196
Hop, 11, 215
    botany, 131—132
    propagation, 63, 111
Hordeum, see also Barley, 25—28
Horseradish, 6, 102, 149
Hot pepper, see Pepper
Hungarian vetch, 10
Hyacinth bean, 4
Hybridization, introgressive, 23—24
    drug crops, 46
    fiber crops, 47
    forage grasses, 33—34
    forage legumes, 31—32
    fruit and vegetable crops, 40—44
    grain crops, 25—29
    latex, rubber, oil, and wax crops, 48
    nut crops, 45
    root crops, 35—36
    seed legumes, 30
    spice and perfume crops, 49
    stem and leaf crops, 37—39
Hydrogen fluoride, see Gaseous pollution stress
Hydrogen sulfide, see Gaseous pollution stress

I

Ilex paraguariensis, 12, 17
Illumination, 195, 201
Immunosuppressants, 23
Inbred progeny selection, 79—80

Incandescent lamps, 196, 197, 201—202
Indian center of origin, 16—18
Indian corn, see also Corn, 17
Indian grass, 10, 215
Indian sugarcane, 17
Indigo, 11
Indo-Malayan center of origin, 16—18
Intermediate wheatgrass, 9, 153
Interspecific crossing, see also Hybridization, introgressive, 221
Introgressive hybridization, see Hybridization, introgressive
Ipomoea species hybrids, see also Potato, sweet, 35
Irradiance, 195
Isotope methods, nitrogen fixation measurement, 176—184
Italian millet, 17
Italian ryegrass, 9

J

Jack bean, 6, 17
Jack fruit, 17, 106
Jambos, 17
Japanese peach, 5
Japanese persimmon, 5
Jerusalem artichoke, 35
Jicama, 17
Job's tears, 17
Johnson grass, 10, 216
Jojoba, 97
Joule, 195, 201
Juglans species hybrids, see also Walnut, 45
Jujube, 17, 106
Jute, 11, 17, 47
    botany, 134
    photoperiod and photosynthesis, 215
    propagation, 98
Jute, Congo, 11

K

Kale
    botany, 144, 145, 146
    chromosome number, 6
    photoperiod and photosynthesis, 215
    propagation, 99
Kale, Chinese, 6
Kans grass, 8
Kaoliang, 17
Kapok, 11, 98, 134
Karashina, 6
Kenaf, 11, 47, 98
Kentucky bluegrass, see Bluegrass, Kentucky
Khuskus, 111
Kikuyu grass, 9
Kilolux, 201
King ranch bluestem, 9
Kiwifruit, 106, 136

Kleberg, bluestem, 9
Klebsiella, 173
Kodo millet, 17
Kohlrabi, 6, 227
    botany, 144
    propagation, 99
Kola nut, 12, 111
Korean lespedeza, 10
Kudzu, 10, 95

L

Lac tree, 17
Latuca, see also Lettuce, 39
Ladino clover, 240
Lambert, 201
Large water grass, 9
Latex crops, 48
Lathyrus, 171
Latitude, and daylength, 202—206
Lavender, 12
Layering, 91, 93
Leek, 6, 17, 227
    botany, 144, 146
    propagation, 100
Legumes
    nitrogen fixation, 169—174
    propagation methods, 63
Legumes, forage, 10
    botany, 152, 154—155
    introgressive hybridization, 31—32
    photoperiod and photosynthesis, 214—215
Legumes, grain and seed 30, 119—121
Leguminosae, 169
Lemon, see also Citrus, 5
    botany, 136
    gaseous pollution stress, 227, 252
    propagation, 105
Lens, see also Lentil; Nitrogen fixation, 30
Lentil, 4, 17, 30
    botany, 120
    nitrogen fixation, 171
    photoperiod and photosynthesis, 215
Lepidium sativum, 7
Lespedeza, 10, 31
    botany, 156
    nitrogen fixation, 172, 173
    photoperiod and photosynthesis, 214
    propagation, 95
Lettuce, 7, 17, 39
    botany, 144, 146
    gaseous pollution stress, 227, 243—244
    propagation, 100
Leucaena, 172
Licorice root, 12
Life cycle, see Botany
Light and photoperiod, see also Botany
    action spectrum, 196—197
    annual and diurnal cycles, 197—198
    astrometeorological estimator program, 210—214
    conversion factors, 195, 201—202
    critical photoperiod, 198—199
    crop classification, 214—216
    daylength tables, 202—206
    harmonic equation, 14
    measurement units, 195—196
    processes controlled by, 199—200
    solar energy calculations, 207—209
Light sources
    conversion factors, 201—202
    irradiance of, 195, 196
Lima bean, see Bean, lima
Lime, see also Citrus species, 136
Linen, see Flax
Linseed, see also Flax, 121
Linum, see also Flax, 48
Litchi, 5, 7, 17, 106, 108
Litchi, hairy, 108
Lolium species, see also Ryegrass, 9, 34
Loofah, 7
Loquat, 17, 106
Lotononis, 172
Lotus species, see also Birdsfoot trefoil
    introgressive hybridization, 31
    nitrogen fixation, 172, 173
Lovegrass, 215
Lumen (unit), 201
Lupine, see also *Lupinus*, 10, 17, 31
    photoperiod and photosynthesis, 215
    propagation, 95
Lupine, blue, 10, 31
Lupinus
    introgressive hybridization, 31
    nitrogen fixation, 171—173
Lux, 210
Lycopersicon, see also Tomato, 41

M

Macadamia, 45, 109
Macadamia integrifolia, 7
Mace, 111
Macroptilium, 172
Madia, 8
Maheleb cherry, 5
Maize, see Corn
Malabar gourd, 6
Malus, see also Apple, 41
Mamey, 106
Manch, 8
Manchurian soya, 4
Mandarin, see also *Citrus* species, 5, 136, 252
Mango, 5, 17
    botany, 136
    propagation, 106—107
Mangosteen, 5, 17
Manihot, see also Cassava/manioc
    introgression, 35
Manila hemp, 11, 17
Manioc, see Cassava/manioc

Maple, 8
   botany, 129—130
   propagation, 111
Maracuja, 41
Maranta arundinacea, see Arrowroot
Marshmallow, 12
Mass selection, 76
Mat bean, 17
Mate, 12, 17
Meadow brome, 9
Meadow fescue, 9, 153
Meadow foxtail, 9, 33
Medicago species, see also Alfalfa, 10, 31
   nitrogen fixation, 171, 173
Medicago lupulina, nitrogen fixation, 173
Mediterranean center of origin, 16—18
Melilotus species, see also Clover, 10, 31, 173
Melon, see also Cucurbits, 6
Mendelian genetics, 61
Mentha, see also Mint, 12
Mentha species hybrids, 49
Mexican rubber guayule, 11
Micropropagation, 92, 93
Milkvetch, 214
Millet, see also Sorghum, 4, 17
   botany, 118
   propagation, 96
Millet, foxtail, 118
Millet, Guinea, 19
Millet, pearl, 4, 17, 118
   introgressive hybridization, 25
   photoperiod and photosynthesis, 215
Millet, proso, 17, 118
Mimosoideae, 169
Minimum optimal photoperiod, 198
Mint, 12, 49
   botany, 149
   propagation, 111
Miscanthus sinensis, 39
Monantha (bitter) vetch, 10
Mother of thyme, 12
Mulberry, 5, 107
Mung bean, 17, 173
Musa, see also Banana; Plantain
   species hybrids, 41, 47
   textiles, 11
Muscadine grape, 44
Muskmelon
   botany, 144
   gaseous pollution stress, 227, 244
   propagation, 100
Mustard, 6, 17, 38
   botany, 149
   gaseous pollution stress, 227, 259
   photoperiod and photosynthesis, 215
   propagation, 100
Mustard, black, 6, 38
Mustard, brown, 6, 38
Mustard, Chinese, 6, 38
Mustard, white, 6, 17
Mutations, 63

Mu-tree, 97
Mutton grass, 9
Myrica, 172
Myristica fragrans, 12
Myrsinaceae, 173

N

Naked oats, 17
Napiergrass, 96
Narrowleaf vetch, 10
Nasturtium, 7
Near Eastern center of origin, 16—18
Nectarine, 5, 107
Nicotiana, see Tobacco
Nigella, 17
Nitrogenase, 165
Nitrogen fixation
   chemistry, 165—166
   estimation of, 174—188
      acetylene reduction technique, 176
      increment in N yield and plant growth, 174—175
      isotope methods, 176—184
      nitrogen balance method, 175—176
      sources and resources, 187—188
   nonsymbiotic, 166—169
   symbiotic, 169—174
      legumes, 169—174
      nonlegumes, 170, 172—174
Nitrogen oxides, see Gaseous pollution stress
Noog (Guizotia), 17
Nut crops, 7, 45
   propagation, 108—109
Nutmeg, 12, 17
   botany, 147
   propagation, 111

O

Oak, 45
Oats, 4, 25
   botany, 116, 118
   breeding, 72, 221
   centers of origin, 17, 20
   gaseous pollution stress, 226, 230
   photoperiod and photosynthesis, 199, 215
   propagation, 63, 94
   temperature tolerance, 221
Ocimum, 49
Oil crops, 8, 48
   botany, 121—125
   gaseous pollution stress, 236—238
   propagation, 97
Oil palm, 8, 20, 48
   botany, 123—124
   propagation, 97
Oil rape, see also Rape/rapeseed, 6
Okra, 12, 17, 41

botany, 144
propagation, 100
Olive, 8, 17
    botany, 124, 138
    propagation, 107, 111
Onion, 6, 17, 35
    botany, 144, 146
    gaseous pollution stress, 227, 244
    propagation, 63, 102
Onion, Chinese, 17
Onobrychis viciaefolia, see Sainfoin
Opium poppy, see also Poppy, 12, 17, 46
Orange, see also *Citrus*, 5
    botany, 136
    gaseous pollution stress, 228, 252
    propagation, 105
Orchard grass, 33
    botany, 151—153
    photoperiod and photosynthesis, 214
    propagation, 96
Oregano, 49
Oriental chestnut, 17
Oriental cotton, 11
Oriental persimmon, 17
Ornithopus sativus, 10
Oryza glaberrima, 19
Oryza, see Rice
    species hybrids, 25
Oxidizing agents, see Gaseous pollution stress
Oxytropis sericea, 172
Ozone, see Gaseous pollution stress

P

Pak-choi, 6
Palm, oil, see Oil palm
Palm pyrethrum, 12
Panax ginseng, see Ginseng
Pangola grass, 9, 96, 152
Panicum, see also Millet, 9
Papaver, see Opium poppy; Poppy
Papaya, 5, 17, 40
    botany, 136
    propagation, 107
Papilionoideae, 169
Paprika, 12
Paragrass, 9, 96, 215
Paraguay tea, 12
Parsley, 7, 17, 39
    botany, 144
    gaseous pollution stress, 227, 244
    propagation, 100
Parsnip, 7, 17
    botany, 144
    gaseous pollution stress, 227, 244
    propagation, 102
Parthenium argentatum, 11
Paspalum species, see also Dallisgrass
    chromosome numbers, 9
    hybrids, 34

nitrogen fixation, 169
Passion flower, see *Passiflora*
Passion fruit, see *Passiflora*
Passiflora
    centers of origin, 17
    chromosome number, 5
    propagation, 107
    species hybrids, 41—42
Pastinaca sativa, see Parsnip
Pavetta zimmermanniana, 173
Pawpaw, American, 40, 107
Pea, 7, 17, 30
    botany, 120, 144
    gaseous pollution stress, 227, 235, 245
    heat sensitivity, 219
    nitrogen fixation, 171, 173
    photoperiod and photosynthesis, 214
    propagation, 63, 100
Pea, pigeon, see Pigeon pea
Peach, 5, 17, 42
    botany, 136, 138
    gaseous pollution stress, 228, 252
    propagation, 107
Peach, Japanese, 5
Peanut (groundnut), 17, 30
    botany, 120
    chromosome numbers, 4, 7, 8
    gaseous pollution stress, 226, 236
    nitrogen fixation, 173, 174
    photoperiod and photosynthesis, 215
    propagation, 97
Pear, 5, 17, 42
    botany, 136, 138
    gaseous pollution stress, 228
    propagation, 107
Pear, Chinese, 5, 17
Pearl millet, see Millet, pearl
Pecan, 7, 45, 252
    botany, 142
    propagation, 109
Pedigree selection, 65—69
Pennisetum, see also Millet, 9, 25
    propagation, 96
PEP-carboxylase, 200
Pepino, 7, 17
Pepper, see *Capsicum*; *Piper*
Pepper, bell, 227
Pepper, black, 12, 17, 110
Pepper, cayenne, 6, 12
Pepper, chili, 8, 17
Peppermint, see also Mint, 12
Perennial ryegrass, see Ryegrass, perennial
Perfume crops, 49
Peroxyacetyl nitrate, see Gaseous pollution stress
Persea species hybrids, see also Avocado, 42
Persian clover, 10
Persian lime, 105
Persimmon, 5, 17, 107
Petalostemum, 172
Petroselinum hybrids, see also Parsley, 39
Phalaris, see Reed canary grass

Phaseolus species, see also Bean
   botany, 144
   gaseous pollution stress, 226, 231—235
   introgressive hybridization, 30
   nitrogen fixation, 171—173
Phenology, see Botany
Phenotypic value, 64
Phleum, see Timothy
Photon flux density, 195
Photoperiod, see also Botany; Light and photoperiod
Photoperiod cycles, 198
Photosynthesis, see also Botany; Light and photoperiod
Phototropism, 196
Physalis peruviana, 7
Phytochrome, 196
Pigeon pea, 17, 30
   botany, 120
   chromosome numbers, 4, 6
   photoperiod and photosynthesis, 215
   propagation, 100
Pimento, 111, 147, 149
Pimpinella anisum, see Anise
Pineapple, 5, 17, 40
   botany, 140
   propagation, 107
Piper, 12, 145, 226
Piper betle, see Betel
Pistachio, 7, 17
   propagation, 109
Pisum species, see also Pea, 30
   nitrogen fixation, 171, 173
Plantain, 17, 107
Plantation crops, 20
Ploidy levels, 62
Plum, 42
   botany, 136, 138
   gaseous pollution stress, 228, 252
   propagation, 107
Poa, see Bluegrass, Kentucky
Podocarpus, 172
Pollution stress, see Gaseous pollution stress
Polyploidy, 62, 63
Pomegranate, 5, 17, 108
Poncirus species hybrids, 40
Poppy, see also Opium poppy
   botany, 149
   chromosome number, 8, 12
Population improvement, 75—83
Potato, 7, 17
   botany, 126, 144
   gaseous pollution stress, 227, 245
   introgressive hybridization, 23, 36
   photoperiod and photosynthesis, 215
   propagation, 63, 102—103
Potato, sweet, 7, 17, 35
   botany, 126—127, 144
   gaseous pollution stress, 227, 245
   photoperiod and photosynthesis, 215
   propagation, 103
Potentilla species hybrids, 41

Prairie cord grass, 216
Progeny performance, 65—75
Propagation, see also Botany
   seed, 91, 93—102
      cereals and grains, 94
      fiber crops, 97—98
      forages, 95—96
      oil crops, 97
      vegetables, 98—101
   vegetative, 91—113
      fruit crops, 103—108
      nut crops, 108—109
      roots and tubers, 102—103
      specialty crops, 110—113
Proso millet, 17, 118
Protein crops
   chromosome numbers, 4
   gaseous pollution stress, 231—236
Prune, 5, 107, 254
Prunus species, see also fruits by name
   botany, 136, 138
   gaseous pollution stress, 228, 253—254
   introgressive hybridization, 42, 45
   propagation, 105, 107
Pseudananas species hybrids, 40
Psophocarpus, 172
Psychotria punctata, 173
Pubescent wheatgrass, 9
Pueraria thunbergiana, 10, 95
Pulse crops, see also Legumes, grain
   botany, 119—121
   propagation, 94
Pummelo, 5, 105
Pumpkin, 6, 40
   gaseous pollution stress, 245
   propagation, 100
Purple granadilla, 5
Purple vetch, 10
Purshia, 172
Pyrethrum, 12
   botany, 133
Pyrus species hybrids, 42

Q

Quackgrass, 33
Queensland nut, 7
Quercus species hybrids, 45
Quillow coffee, 12
Quince, 5, 17, 108
Quinine, 12, 17
   botany, 132
   propagation, 111
Quinoa, 17

R

Rabbiteye blueberry, 139
Radiant energy, 195, 201

Radish, 7, 17, 36
   botany, 144
   gaseous pollution stress, 227, 246—247
   propagation, 103
Rambutan, 17, 108
Ramie, 11, 17, 98
Rape/rapeseed, 17, 37
   chromosome number, 6, 8
   gaseous pollution stress, 236
   oil crop, 121—122
   photoperiod and photosynthesis, 215
   propagation, 97
Raphanus species hybrids, see also Radish, 36, 38
Raspberry, red and black, 5, 43
   botany, 137
   gaseous pollution stress, 228, 254
   propagation, 108
Receptor pigments, 196
Reciprocal full-sib selection, 81—82
Reciprocal recurrent selection, 80—81
Recurrent selection
   interpopulation, 80—83
   intrapopulation, 76—80
Red chili, 12
Red clover, see Clover, red
Red fescue, 9
Red mulberry, 9
Red raspberry, see Raspberry
Redtop, 96, 153
Reed, common, 215
Reed canary grass, 9
   botany, 154
   photoperiod and photosynthesis, 214
   propagation methods, 63, 95
Reproduction, see also Botany; Propagation, 62—63
Rhea, 11, 98
Rheum, see Rhubarb
Rhizobium species, 169—172
Rhizomes, 91, 93
Rhodes grass, 9, 153
Rhodesian timothy, 9
Rhubarb, 7, 227
   botany, 145
   propagation, 100
Rhubarb, English, 7
Ribes species, see also Currant; Gooseberry
   introgressive hybridization, 42—43
   propagation, 106
Rice, 4, 17, 35
   botany, 118, 119
   carbon fixation pathway, 200
   gaseous pollution stress, 231
   photoperiod, 198, 199, 215
   propagation, 94
Rice, African, 19
Rice bean, 17
Robinia pseudoacacia, 172
Rocket salad, 6
Root crops
   botany, 125—127
   introgressive hybridization, 35—36

Roots, propagation, 102—103
Rosa species hybrids, 49
Rose clover, 10
Rose oil, 49
Roughpea, 96
Rubber, 11, 20, 48
   botany, 155—157
   propagation, 112
Rubiaceae, 173
Rubus species, see also Blackberry; Raspberry
   botany, 137
   introgressive hybridization, 43
RuDP carboxylase, 200
Runners, 91, 93
Russian wild rye, 9, 153, 214
Rutabaga, 6, 38, 247
   botany, 144
   propagation, 103
Rye, 4, 17, 26
   botany, 116, 118
   gaseous pollution stress, 226
   half-sib progeny selection, 78
   photoperiod and photosynthesis, 215
   propagation, 63, 94
   Russian wild, see Russian wild rye
   wild, 9, 33, 153
Rye, Altai wild, 9
Rye buckwheat, 4
Ryegrass, 34
   botany, 154
   gaseous pollution stress, 240
   photoperiod and photosynthesis, 214
   propagation, 63, 96
Ryegrass, annual, 34, 63, 214
Ryegrass, perennial, 9, 34, 154, 214, 240

S

Saccharum species hybrids, see also Sugarcane, 26, 39
Safflower, 8, 17, 48
   gaseous pollution stress, 226, 236
   oil-seed crop, 121—122
   photoperiod and photosynthesis, 215
   propagation, 97
Sage, 12, 17, 149—150
Sainfoin, 10, 17
   botany, 156
   photoperiod and photosynthesis, 215
   propagation, 96
Saint Augustine grass, 96
Salsify, 7, 36, 103
Sandalwood, 11
Sann (Sunn) hemp, 98
Sanseviera species hybrids, 47
Santalum album, 11
Sapodilla, 112
Sapote, 17
Sarson, see also *Brassica*, 17
Saskatoon serviceberry, 108

Scorzonera hispanica, see Salsify
Screw pine, 112
Sea kale, 6
Secale species hybrids, 26—27, 28
Sechium edule, see Chayote
Seed legumes, see Legumes, seed
Seed propagation, see Propagation, seed
Selection, see Breeding
Self-fertilization, 62
Self-sterility, 155
Sericea lespedeza, 156
Serradella, 10
Sesame, 8, 17, 48
    as oil-seed crop, 121—122
    photoperiod and photosynthesis, 215
    propagation, 97
Setaria, see also Timothy, 4, 9
Setaria grass, 96, 153, 154
Shagbark hickory, 7
Shallot, 6
Sheeps fescue, 9
Shepherdia, 172
Side oats grama, 9
Silk cotton tree, 11, 17
Silky bluestem, 9
Silver bluestem, 9
Sinapis alba, see Mustard, white, 12
Single-seed descent, 73—75
Sinskaya, E. N., 17
Sisal, 11, 17, 47, 215
    botany, 135
    propagation, 98
Slender wheatgrass, 9, 153
Smooth bromegrass, 9, 214
    botany, 150—151, 153
    propagation methods, 63, 95
Snake gourd, 7
Snap bean, see Bean
Solanum, see also Eggplant; Potato
    introgression, 36, 41, 43
Solar energy calculations, 207—209
Sorghastrum nutans, 10
Sorghum, see also Millet, 226
    botany, 118, 130
    carbon fixation pathway, 200
    centers of origin, 17, 20
    chromosome numbers, 8, 10
    introgressive hybridization, 26, 28, 39
    photoperiod, 199, 215
    propagation, 63, 94
Sorghum halepense, see Johnson grass
Sorghum sudanense, see Sudan grass
Sorghum vulgare, see Millet
Sorrel, 112
Sour cherry, 42
South American center of origin, 16—18
South American lupin, 10
South Mexican center of origin, 16—18
Soybean, 17, 30, 121
    breeding, 72
    carbon fixation pathway, 200

    chromosome numbers, 4, 8
    cold sensitivity, 218
    gaseous pollution stress, 226, 236—238
    nitrogen fixation, 171, 173, 174
    photoperiod and photosynthesis, 215
    propagation, 63, 97
Spearmint, 12
Specialty crops, 110—113
Spectral distribution of energy, 196
Spice crops, see also individual crops, 12, 49
    botany, 145, 147—150
    propagation, 110—113
Spinach, 7, 17
    botany, 144
    gaseous pollution stress, 227, 247
    propagation, 101
Spotted burr-clover, 10
Squash, 17, 40
    botany, 144
    gaseous pollution stress, 227, 248
    propagation, 100
Star apple, 108
Starch, 200
Starch crops, see Tubers; specific crops
Stem lettuce, 17
Stolon, 91, 93
Strawberry, 5, 41
    botany, 137
    gaseous pollution stress, 228, 254
    propagation, 108
Stressors, see Gaseous pollution stress
Strophostyles helvola, 172
Stylosanthes, 173
Subterranean clover, 10, 156
Sudan grass, 10, 216
Sugar apple, 108
Sugarbeet
    botany, 128—129
    chromosome number, 6, 8
    gaseous pollution stress, 258
    photoperiod and photosynthesis, 215
    propagation, 63, 112
Sugarberry, 5
Sugarcane, 8, 17, 39
    botany, 127—128
    carbon fixation pathway, 200
    gaseous pollution stress, 226, 258
    photoperiod and photosynthesis, 215
    propagation, 112
Sugarcane, Chinese, 17
Sugar crops, see also specific crops, 8, 17
    botany, 127—130
    gaseous pollution stress, 258
Sugar maple
    botany, 129—130
    chromosome number, 8
Sugar palm, 17
Sulfur dioxide, see Gaseous pollution stress
Summer savory, 150
Summer squash, 6
Sunflower, 8, 48

gaseous pollution stress, 226, 238
half-sib progeny selection, 78
as oil-seed crop, 121—123
photoperiod and photosynthesis, 215
propagation, 97
Sunn hemp, 10, 11
Sun scald, 220
Swedes, see Rutabaga
Sweet cherry, 5, 42, 251
Sweet clover, see Clover, sweet
Sweet pepper, see also Pepper, 40
Sweet potato, see Potato, sweet
Sweetsop, 5, 108
Sweet sorghum, see Sorghum
Swiss chard
gaseous pollution stress, 227, 248
propagation, 99
Switchgrass, 9
botany, 154
photoperiod and photosynthesis, 215

T

Tall fescue, 9
botany, 151—153
propagation methods, 63
Tall oat grass, 9, 214
Tall wheatgrass, 9, 63
Tamarind, 5
Tangerine, see also *Citrus*, 5
gaseous pollution stress, 228, 254
propagation, 105
Tapioca, see also Cassava/manioc, 102
Taraxacum species hybrids, 48
Taro, 6, 17
botany, 127
propagation, 103
Tarragon, 12, 150
Tar weed, 8
Tea, Chinese, 12, 17
botany, 132
propagation, 112
Teff, 17
Temperature stress
breeding for tolerance, 221—222
chilling, 217—218
freezing, 218—219
heat, 219—220
physiology, 217—220
Tepary bean, 17
Thyme, 12, 112, 150
Thymus, see Thyme
Timothy
botany, 154
chromosome number, 4, 9
gaseous pollution stress, 226
photoperiod and photosynthesis, 214
propagation, 96
Tissue culture, 92, 93, 221
Tobacco, 12, 17
botany, 132—133
gaseous pollution stress, 226, 255—258
introgressive hybridization, 23, 45, 46
photoperiod and photosynthesis, 215
propagation, 63, 112—113
Tomato, 7, 17
botany, 146
cold sensitivity, 217, 218
gaseous pollution stress, 227, 248—249
introgressive hybridization, 23, 41
photoperiod and photosynthesis, 215
propagation, 63, 101
Tomato, cherry, 7
Tragopogon species hybrids, see also Salsify, 36
Transdomestication, 20
Tree alfalfa, 10
Tree cotton, 11, 17
Tree tomato, 17
Trefoil, birdsfoot, see Birdsfoot trefoil
Trichosanthes anguina, 7
Trifolium, see also Clovers, 10, 32
gaseous pollution stress, 226
nitrogen fixation, 171, 173
Trigonella species
chromosome number, 10, 12
nitrogen fixation, 171, 173
Tripsacum species hybrids, 28
Triticale, 4, 28
botany, 116, 118
photoperiod and photosynthesis, 215
propagation, 94
Triticum, see also Wheat
species hybrids, 26—28, 33
Tropaeolum tuberosum, 7
Tubers, see also specific crops, 6—7
propagation, 91, 93, 102—103
Tung, 8, 48
botany, 124—125
propagation, 97
Turmeric, 150
Turnip, 6, 17, 37
botany, 145
gaseous pollution stress, 227, 249
propagation, 102, 103
Turnip rape, 6
Turnip root, 102

U

Udo, 17
Upland cotton, 11, 17
Upright brome, 9
Urd, 17
Urena lobata, 11

V

Vaccinium, see also Blueberry; Cranberry
introgression, 44

Valerian, 12
Vanilla, 12, 49, 113, 147
Vaseygrass, 215
Vavilov, Nikolai Ivanovich, 15
Vegetable crops, 6—7
   botany, 143—146
   gaseous pollution stress, 226—227, 241—249
   introgressive hybridization, 40—44
   propagation, 63, 98—101
Vegetative propagation, see Propagation, vegetative
Velvet bean, 17
Vetch, 10, 17, 32
   gaseous pollution stress, 226, 235
   nitrogen fixation, 171, 173, 175
   photoperiod and photosynthesis, 214, 215
   propagation, 96
Vetch, common, 10, 17, 32, 215
Vetch, crown, 10, 214
Vetch, hairy, 10
Vetiver, 111
Vicia, see Vetch
Vicia villosa, 173
Vigna species, 4, 30
   nitrogen fixation, 172, 173
Vitis species, see also Grape, 44
   botany, 137—139
   propagation, 106

W

Walnut, 7, 45
   botany, 142—143
   gaseous pollution stress, 228
   propagation, 109
Walnut, black, 7, 45
Walnut, Chinese, 17
Walnut, English, 7, 17
Water chestnut, 17
Water cress, 7, 101
Watermelon, 5, 40, 101, 145
Water stress, 200
Wax crops, 48
Weeping lovegrass, 9, 153
Western wheatgrass, 9, 153
Wheat, 4, 17
   botany, 116, 118
   breeding, 72, 221
   gaseous pollution stress, 226, 231
   introgressive hybridization, 26—28
   photoperiod and photosynthesis, 216
   propagation, 63, 94

   propagation methods, 63
   temperature tolerance, 221
Wheatgrass, 9, 33
   botany, 153
   photoperiod and photosynthesis, 214
   propagation, 96
Wheatgrass, blue bunch, 9
Wheatgrass, crested, 9, 153
White clover, see Clover, white
White lupin, 10, 31
White mustard, 12
White sweet clover, see Clover, white sweet
Wikstroemia canescens, 7
Wild black cherry, 5
Wild flax, 11
Wild rice, 28, 216
Wild rye, 9, 33, 153
Wild rye, Altai, 9
Wild rye, Russian, 9, 153, 214
Wild soya, 4, 8
Wild strawberry, 5
Winter pumpkin, 6
Wolf bean, 10
Wood meadow grass, 9
Woolly pod vetch, 10

X

Xanthosoma, 17

Y

Yam, 7, 16, 35
   botany, 127
   propagation, 103
Yam, Attoto, 6
Yam, Chinese, 6, 17
Yellow lucerne, 10
Yellow lupine, 31
Yellow sweet clover, see also Clover, sweet, 10, 31, 156
Yucatan sisal, 11

Z

Zea species hybrids, see also Corn, 28
Zhukovsky, P. M., 17, 18
Zizania rice, 17
Zizania species hybrids, 28

DATE DUE